Frontiers in Stem Cell and Regenerative Medicine Research

(*Volume 9*)

Edited by

Atta-ur-Rahman, *FRS*

Kings College, University of Cambridge, Cambridge, UK

&

Shazia Anjum

Department of Chemistry, Cholistan Institute of Desert Studies, The Islamia University of Bahawalpur, Bahawalpur, Pakistan

Frontiers in Stem Cell and Regenerative Medicine Research

Volume # 9

Editors: Prof. Atta-ur-Rahman, *FRS* and Dr. Shazia Anjum

ISSN (Online): 2352-7633

ISSN (Print): 2467-9593

ISBN (Online): 978-1-68108-762-7

ISBN (Print): 978-1-68108-763-4

need for a court order if at any point you breach any terms of this License Agreement. In no event will any delay or failure by Bentham Science Publishers in enforcing your compliance with this License Agreement constitute a waiver of any of its rights.

3. You acknowledge that you have read this License Agreement, and agree to be bound by its terms and conditions. To the extent that any other terms and conditions presented on any website of Bentham Science Publishers conflict with, or are inconsistent with, the terms and conditions set out in this License Agreement, you acknowledge that the terms and conditions set out in this License Agreement shall prevail.

Bentham Science Publishers Pte. Ltd.
80 Robinson Road #02-00
Singapore 068898
Singapore
Email: subscriptions@benthamscience.net

BENTHAM
SCIENCE

CONTENTS

PREFACE

The ninth volume of 'Frontiers in Stem Cell and Regenerative Medicine Research' presents important recent developments in this fast growing field.

Hasan and Wu review the current literature on the utility of exogenous and endogenous neural stem cells in spinal cord injury. Miguel *et al.* focus on the various sources of somatic cells for human induced pluripotent stem cells (iPSC) generation, their characterization, and the progress on directed differentiation toward several cell types.

Cell signaling and redox reactions are involved in the regulation of neural-lineage cells for reprograming of adipose-derived stem cells. Abrahamse *et al.* discuss the role of reactive oxygen species (ROS) and ROS mediated cellular signaling for differentiation of stem cells.

The human skin represents the largest and most accessible part of the body that is exposed to infections, physical wounds, diseases *etc*. Titorencu *et al.* describe commercially available skin substitutes that are in demand and also described the use of stem cells for skin regeneration. Recently, the role of microRNA in cardiac conduction diseases, arrhythmogenesis and their therapeutic potential has caught the attention of researchers. Sankaranarayanan *et al.* have reviewed the role of miRNAs in cardiac rhythmic disorders and their potential in diagnosis and treatment. Finally, Rivera *et al.* discuss the use of stem cell therapy in the treatment of malaria.

We owe our special thanks to all the contributors for their valuable contribution in bringing together the ninth volume of this book series. We also thank the editorial staff of Bentham Science Publishers for their help and support.

Atta-ur-Rahman, *FRS*
Kings College,
University of Cambridge,
Cambridge,
UK

Shazia Anjum
Department of Chemistry,
The Islamia University of Bahawalpur,
Pakistan

List of Contributors

Hunaid Hasan	Department of Neurology – Clinical Neurophysiology, NYU Langone Medical Center 301 E. 17th Street, Suite 1534, New York, 10003, USA
Ping Wu	Department of Neuroscience and Cell Biology, University of Texas Medical Branch 4.212B Research Building 17 Galveston, Texas, 77555, USA
Maria P De Miguel	Cell Engineering Laboratory, La Paz University Hospital Health Research Institute, IdiPAZ, Madrid, Spain
Adrian Moratilla	Cell Engineering Laboratory, La Paz University Hospital Health Research Institute, IdiPAZ, Madrid, Spain
Alba Cabrera-Fisac	Cell Engineering Laboratory, La Paz University Hospital Health Research Institute, IdiPAZ, Madrid, Spain
P Gonzalez-Molina	Cell Engineering Laboratory, La Paz University Hospital Health Research Institute, IdiPAZ, Madrid, Spain
Sajan George	Laser Research Centre, Faculty of Health Science, University of Johannesburg P.O. Box 17011, Doornfontein, South Africa
Anine Crous	Laser Research Centre, Faculty of Health Science, University of Johannesburg P.O. Box 17011, Doornfontein, South Africa
Heidi Abrahamse	Laser Research Centre, Faculty of Health Science, University of Johannesburg P.O. Box 17011, Doornfontein, South Africa
Ana-Maria Rosca	Institute of Cellular Biology and Pathology, "Nicolae Simionescu", Bucharest, Romania
Raluca Tutuianu	Institute of Cellular Biology and Pathology, "Nicolae Simionescu", Bucharest, Romania
Irina Titorencu	Institute of Cellular Biology and Pathology, "Nicolae Simionescu", Bucharest, Romania
Priyadarshan Kathirvelu	Ion Channel Biology Laboratory, AU-KBC Research Center, Madras Institute of Technologycampus of Anna University, Chennai, India
Bhuvaneshwari Sampath	Ion Channel Biology Laboratory, AU-KBC Research Center, Madras Institute of Technologycampus of Anna University, Chennai, India
Kavitha Sankaranarayanan	Ion Channel Biology Laboratory, AU-KBC Research Center, Madras Institute of Technologycampus of Anna University, Chennai, India
Norma Rivera	Departamento de Microbiología y Parasitología, Facultad de Medicina, UNAM, CDMX 04510, Mexico
Marcela Rojas-Lemus	Departamento de Biología Celular y Tisular, Facultad de Medicina, UNAM, CDMX 04510, Mexico
Nancy G. Flores-Jiménez	Departamento de Microbiología y Parasitología, Facultad de Medicina, UNAM, CDMX 04510, Mexico
Teresa I. Fortoul	Departamento de Biología Celular y Tisular, Facultad de Medicina, UNAM, CDMX 04510, Mexico

CHAPTER 1

Insights into Exogenous, Endogenous and Combination Therapies of Neural Stem Cells in Spinal Cord Injury

Hunaid Hasan[1,*] and **Ping Wu**[2]

[1] *Department of Neurology – Clinical Neurophysiology NYU Langone Medical Center 301 E. 17th Street, Suite 1534 New York, 10003, USA*

[2] *Department of Neuroscience and Cell Biology University of Texas Medical Branch 4.212B Research Building 17 Galveston, Texas, 77555, USA*

Abstract: Advancements in the understanding of spinal cord injury (SCI) and repair have taken great leaps in the past decade. Although understanding has evolved significantly there continues to be major limitations in the clinical interventions available for patients with spinal cord injuries. Exogenous stem cells (ExNSC) have progressed to human clinical studies, but have limitations due to ethical issues, technique and long term outcomes. However, the use of ExSCs through a stimulatory effect on growth factors, cytokine production and neurotrophic factors post injury may be beneficial. Bone marrow derived stem cells, mesenchymal stem cells, embryonic stem cells, umbilical cord stem cells, adipose derived stem cells, NSCs, Schwann cells grafts and olfactory ensheathing cells have been various types of exogenous cell types and techniques used in SCIs. The role of endogenous stem cells (eNSCs) in SCI has been promising, but still requires better lineage analyses to fully understand the responses of NSCs after SCI. It has been demonstrated that there exists a bidirectional interaction within the neuro vascular system forming the neuro vascular niche. Purinergic receptor activation was found to alter the intrinsic properties of the ependymal stem/progenitor cells enhancing regulation of proliferation, differentiation and lineage specification after a SCI. Therapies have been described using nervous tissue in combination with various synthetic bridges to overcome the structural barriers of regeneration through bypassing the injured area. More recently, newer techniques such as electrical stimulation have been described to stimulate mature neuronal differentiation. Various groups have emphasized that the glial scar is counter productive. Anderson *et al.* have shown the beneficial effects of a chronic glial scar in neural tissue preservation after SCI. Moreover, they have demonstrated higher levels of chondroitin sulfate proteoglycans in injured spinal cords independent of the glial scar. In sum, we have reviewed the previous and current literature on NSC and SCI to address the neurobiological utility of NSCs in spinal cord injury.

* **Corresponding author Hunaid Hasan:** Department of Neurology – Clinical Neurophysiology NYU Langone Medical Center 301 E. 17th Street, Suite 1534 New York, 10003, USA; Tel: +17348588947; E-mails: hunaidhasan10@gmail.com, hunaidhasan@hotmail.com

Atta-ur-Rahman and Shazia Anjum (Eds.)

Keywords: Clinical Neurology, Cytokine, Electrical Stimulation, Electroacupuncture, Endogenous Neural Stem Cell, Exogenous Neural Stem Cell, Genetic Fate Mapping, Glial Scar, Growth Factor, Induced Pluripotent Stem Cell, Neural Stem Cell, Neurovascular Niche, Oligodendrocyte Ensheathing Cells, Progenitor, Purinergic Receptors, Schwann Cell Transplantation, Spinal Cord Injury, Stem Cell, Transplantation, Treatment, Transcription Factor.

INTRODUCTION

Advancements in the understanding of spinal cord injury (SCI) and repair have taken great leaps in the past decade. The greatest challenge in SCI research has been the translation of what is known at the basic science level, and applying it to the clinical setting. Although, the understanding of the molecular basis of SCI has become more cogent, the therapeutic options continue to be limited. Although neural stem cells (NSCs) have been studied for over a decade, they continue to be at a translational research stage. Initially, exogenous stem cells (ExSCs) were an exciting breakthrough. However due to various reasons discussed below their clinical application as independent interventions has started to diminish. Recently, endogenous stem cells (eNSCs) have become more of an interest in SCI. Nature's endogenous process of healing spinal cord injuries was once thought to be counterproductive. However, as discussed below there has been evidence to suggest on the contrary. In sum, the function of eNSCs has been encouraging, and could help intervene in altering this natural process of healing. Further, its translation into clinical practice could improve the quality of life for numerous patients affected by SCIs.

SPINAL CORD INJURY

Epidemiology

Epidemiologically, SCI has been and continues to be a huge burden to society. Some reports find the incidence to be as low as 8 and as high as 246 cases per million depending on political, geographical, sociological and economic differences [1 - 4]. The mean age has increased from 28.3 years in the 1970s to 37.1 years surveyed in 2005-2008. Most concerning has been the epidemiological age group affected. As the majority affected have been in their peak productive years of their lives [5]. Also, more than 50% of SCIs affect the cervical region of the spinal cord, which is more likely to cause total body paralysis [6 - 8].

Mechanism of SCIs were either traumatic or nontraumatic [3]. Traumatic causes in young adults (age 15-29 years) were motor vehicle accidents (50%), violence (12%) and sport-related injuries (10%) [7, 9, 10]. Traumatic causes in the elderly (age >65 years) were mainly accidental falls [8 - 12]. Non-traumatic causes in both age groups were congenital and inflammatory spinal cord disorders, tumor compression, vascular ischemia vertebral spondylosis [13].

Pathophysiology

Primary Phase

SCI has been categorized into primary and secondary phases [14, 15]. The primary phase occurs due to compression and contusive injury resulting in shearing, laceration and stretching of the spinal cord. This exerts forces causing disruption of axonal function [16, 17]. It was suggested that at this stage full transection of the spinal cord was rare. There was sparing of demyelinated axons found at the subpial rim [17 - 21]. Animal models suggest that a minimum of 10% preservation of original axons could lead to significant neurological recovery [22, 23].

Secondary Phase

The pathophysiology of the secondary phase injury was directly proportional to the injuries occurring in the primary injury phase [17]. This phase was marked by a negative environment for neuronal regeneration [17]. The pathophysiology was marked by various processes such as ischemia, excitotoxicity, vascular dysfunction, oxidative stress and inflammation resulting in cell death [18, 24, 25]. Secondary phase injury was deleterious to surviving surrounding neurons, which further impaired functional recovery [14, 26]. The secondary phase had sub-phases which were: immediate (<2 hours), acute (2 hours – 2 days), subacute (3 days to 2 weeks), intermediate (2-3 weeks – 6 months) and chronic (>6 months) [17, 18, 27, 28].

Immediate phase consisted of upregulation of tumor necrosis factor (TNF)-alpha, interleukin (IL)-beta and elevation of extracellular glutamate [29 - 32]. This resulted in necrosis of neurons secondary to ischemia, lipid peroxidation, hemorrhage, reactive oxygen species (ROS) production, edema and cell membrane disruption. This lead to early loss in function and neurogenic shock [16, 17, 23, 33, 34].

A patient with SCI has shown to be initially clinically intervened in the acute phase. This phase was further divided into early acute and subacute stages [17]. Genes that change after injury have been classified as either “early” or “late” genes. They were responsible for activating cascades at various stages [35]. After an SCI event there was an upregulation of pro-apoptotic, cell cycle and oxidative stress mediating genes. Moreover, there was downregulation of antiapoptotic genes, genes involved in neural excitation, neurotransmission, electrochemical gradient and preservation of the neuronal cytoskeleton. Both necrosis and apoptosis were processes of cell death after SCI [36 - 38]. Both neuronal and oligodendrocyte cell death were mediated through activation of the Fas receptors and p75 receptor signaling pathways [39 - 44]. Fas-mediated apoptosis caused demyelination and Wallerian degeneration [36, 39, 40, 45 - 48]. Whereas, Fas receptors in the spinal cord were present on oligodendrocytes, activated microglia and lymphocytes [40, 41, 49]. Therefore, blocking of Fas related mechanisms could have clinical implications [38, 39, 50].

Acute Stage

Early acute stage was marked by ischemia due to disruption of the microvasculature, hypotension and elevated interstitial pressures. This lead to decreased perfusion of the spinal cord after injury [33, 51 - 53] (*Refer to* Fig. (**1**)). The mechanism of secondary injury was likely due to direct mechanical disruption and the effect of inflammatory mediators leading to increased permeability of the blood-brain barrier and blood-spinal cord barrier [18, 54]. As this leakiness allowed an immune cascade to occur. This permeability window could potentially be a therapeutic window for intervention [17]. Since this process had both a neuroprotective and detrimental neurobiological effect it was described as a "double-edged sword". Initially, surrounding microglia proliferated and attracted peripheral leukocytes, which upregulated chemokine production [31, 55, 56]. Neutrophil migration helped in activation of glia and the respiratory burst [56, 57]. In the subacute phase after 72 hours, monocytes differentiated into macrophages and expressed mediators worsening secondary injury [58]. Microglia proliferation continued to persist for weeks [31]. Redox imbalance due to ROS and nitric oxide (NO) species, genes - HspBl and HspA4 were found to be important in the pathophysiology of SCI [16]. Microglia and macrophages produced both species [59 - 62]. Decreasing NO production was neuroprotective [59 - 62]. Moreover, peroxynitrite radical formation, a product of NO and superoxide were involved in neuronal apoptosis [63]. Calcium regulation was crucial in preventing cell death [64, 65]. Its dysregulation was responsible for production of free radicals and activation of calpains [64, 65]. Excitotoxic cell death has been postulated to be due to disruption of osmotic balance [66 - 68]. This cascade was due to disruption of active cell membrane function maintaining glutamate homeostasis [69]. This process leads to overstimulation of glutamate receptors and increased influx of sodium and calcium *via* the N-Methyl-D-aspartate and α-amino-3-hydroxy-5-methyl-4-isoxazolepropionic acid receptors [66 - 68]. Subacute phase was marked by the proliferation and hypertrophy of astrocytes with formation of the glial scar releasing inhibitory substances like chondroitin sulfate proteoglycans [18, 70 - 75] (*Refer to* Fig. (**4**)).

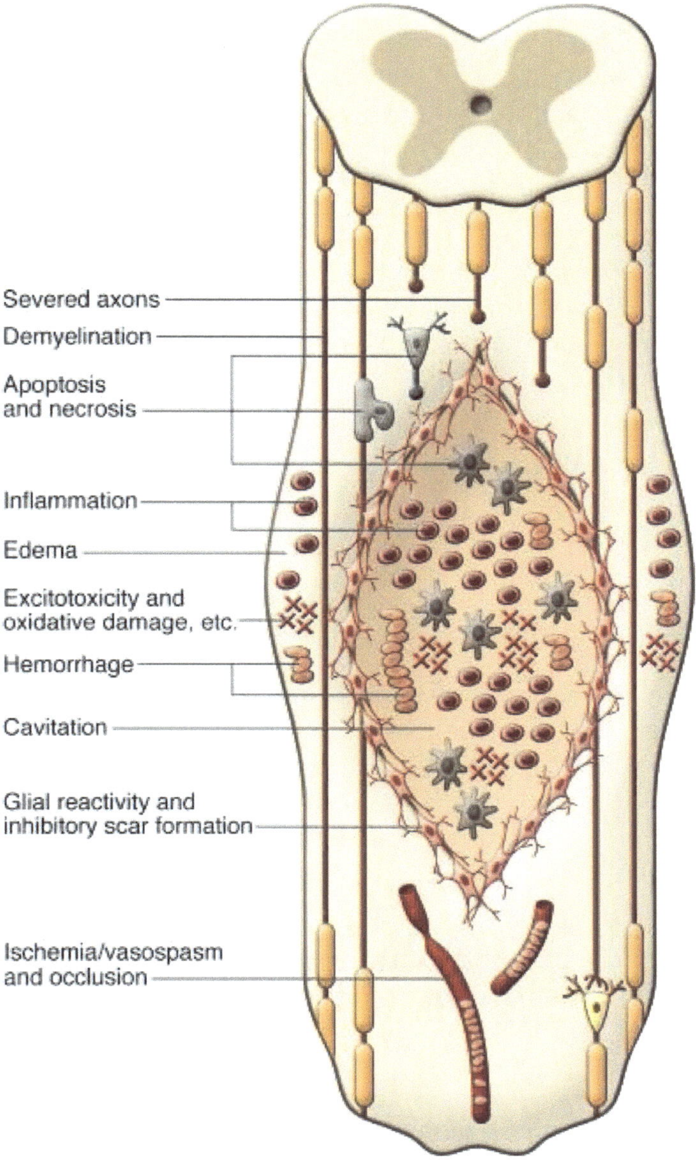

Fig. (1). Spinal cord injury pathophysiology.
This figure demonstrated the acute, subacute and chronic phases of SCI. As a result of primary and secondary injury mechanisms there has been a presence of edema, hemorrhage, inflammation, apoptosis, necrosis, excitotoxicity, lipid peroxidation, electrolyte imbalance, ischemia/vasospasm and blood vessel occlusion. Axonal demyelination and disruption of synaptic transmission occurred secondary to oligodendrocyte and neuronal death. Fluid-filled lentiform-shaped cavities or cysts in the center of the cord with surrounding hypertrophic astrocytes and macrophages in the subacute and chronic phases were also depicted.
J Clin Invest. 2012 Nov 1; 122(11):3824-3834; Copyright © 2012, American Society for Clinical Investigation

Intermediate Phase

Intermediate phase was marked by maturation of the glial scar (*Refer to* Fig. (**4**)). Although not clinically significant, there was axonal sprouting of the corticospinal and reticulospinal tract fibers [16, 76].

Chronic Phase

Chronic phase was marked by further stabilization of the glial scar and syrinx development [18, 77] (*Refer to* Fig. (**4**)). The pathophysiology consisted of clearance of debris of neural tissue by microglia and macrophages with some sparing at the syrinx rim [28, 78, 79]. This stage negatively affected neuroregeneration [80].

CLINICAL MANAGEMENT

Management for SCI has been discussed along the surgical, medical and NSC axis.

Surgical

Early surgical decompression, less than 24 hours post injury has constantly been shown to improve neurological recovery through various trials such as the Surgical Timing in Acute Spinal Cord Injury study (STASCIS). Neurosurgical practices at this time used for SCIs have been notably more empirical than evidence based [81].

Medical

Medical management has been discussed in terms of supportive therapy, pharmacotherapy and rehabilitation.

Supportive

Supportive medical management depended on whether the injury involved the upper spinal cord or the lower spinal cord. Injury to the upper spinal cord lead to hypotension due to disruption of the central supraspinal sympathetic control [4, 82]. Minimum arterial pressures of 85 mmHg should be maintained to prevent spinal cord hypoperfusion [83]. First line management in SCI above the sixth thoracic level involved administrating intravenous crystalloids fluids followed by colloid fluids and vasopressors with inotropic, chronotropic and vasoconstrictive properties [84, 85]. Management of lower spinal cord injuries involved administrating vasopressors with only peripheral vasoconstriction [84, 86]. Both types of SCI required subcutaneous heparin prophylaxis for venous thromboembolism for three months [4, 87].

Pharmacotherapy

Pharmacotherapy consisted of methylprednisolone sodium succinate (MPSS) [17]. MPSS has shown to have an immunomodulatory effect by reducing peroxidation of membranes and reducing inflammation [88]. National Acute Spinal Cord Injury Study (NASCIS) I, II, III have established dosage, timing and adverse effects of MPSS. 30mg/kg bolus with 5.4mg/kg/h for 23 hours was thought to be neuroprotective as per NASCIS II [89]. Optimal timing for administration of MPSS was between 3-8 hours. As per NASCIS III there was some additional motor improvement if given for 48 hours [89 - 91]. Adverse effects included increased risk of infections such as pneumonia [92, 93]. STASCIS showed that early decompression and MPSS had a synergistic effect in cervical cord injury [81, 93].

Exercise Rehabilitation

Exercise rehabilitation was complementary to the medical and surgical strategies [17]. Studies have shown that longer stays in rehabilitation centers improve functional recovery [94, 95]. Ginis *et al.* review of literature recommended that patient should engage in 20 minutes of moderate to heavy aerobic activities with major muscle group strengthening exercises 2 times a week [96].

STEM CELL BASED TREATMENT FOR SCI

Stem cells have capabilities to self-renew and generate into more specialized cells. Therefore, they have been potentially useful for regenerative medicine [97, 98]. Furthermore, totipotent stem cells such as zygotes and some early stage cleavage-stage blastomeres have the potential to fully form an organism with a central nervous system (CNS) and peripheral nervous system (PNS) [99, 100]. Pluripotent stem cells such as embryonic stem cells (ESCs) have the potential to form all cells within the nervous system, but cannot form trophoblasts of the placenta [99].

The CNS has the potential to regenerate with NSCs. A NSC has been defined as a self-renewing cell with the ability to generate neural tissue [97, 99, 101]. Initially, NSCs were isolated through dissecting brain regions such as the subventricular zone (SVZ) and the hippocampus [98]. The tissue isolated was dissociated into component cells and exposed to growth factors such as fibroblast growth factor (FGF)-2 and epidermal growth factor (EGF). The differentiated cells were further isolated and selective cell lines were induced through withdrawing exposure to various factors [98]. Analysis was performed through staining with antibodies specific for astrocytes with glial fibrillary acidic protein (GFAP), oligodendrocytes with Ga1C and neurons with TUJ1 (neuronal beta III tubulin expressed in the CNS and PNS) [97].

Exogenous Transplantation Methods for SCI

Many types of stem cells have been considered for enhancing repair after SCI. Human ESCs were pluripotent, telomerase-positive and immortal cells that could self-renew and differentiate into multiple cell lineages [102 - 104]. *In-vitro* use of a 42 day differentiation protocol with H1 and H7 lines of human ESCs were shown to differentiate into oligodendrocyte progenitor cells (OPCs) [105, 106]. In rat models these OPCs enhanced axon remyelination and reestablished motor function within the injury site [107]. An important consideration in the clinical utility of human ESCs were the immunogenic consequences of allogenic cell transplantation [108]. OPCs were shown to be resistant to human natural killer (NK) cell destruction and the lytic effects of allogeneic and xenogeneic antibodies. Although human ESCs were associated with ethical dilemmas they have an advantage over bone marrow mesenchymal stem cells (BMMSCs). Conversely, the utilization of patient's own BMMSCs was found to be beneficial for immune regulation, histocompatibility and repeated harvestability [109].

Stem Cells

Bone Marrow Stem Cells

Bone marrow stem cells (BMSCs) were shown to be relatively easy to isolate and grow in tissues cultures. They have been approved for use in hematopoietic diseases [110]. BMSCs consist of stromal cells and bone marrow (BM) containing nucleated cells [111 - 113]. Intravenous injection of both these types of cells demonstrated improved recovery of hind limb motor function after spinal cord compression in rats. Sykova *et al.* through their preclinical trials have demonstrated a benefit with different BMSC populations in both behavioral outcome and histopathological assessment [110].

Mesenchymal Stem Cells

Mesenchymal stromal cells were mesodermal, non-hematopoietic and easy to obtain. They were cultured *ex-vivo* and differentiated into neuronal lineages [114] (*Refer to* Fig. (**2**)). They were a target in acute, subacute and chronic SCI, and have been shown to migrate and integrate with injured spinal tissue, reduce cyst size and improve functional recovery [115].

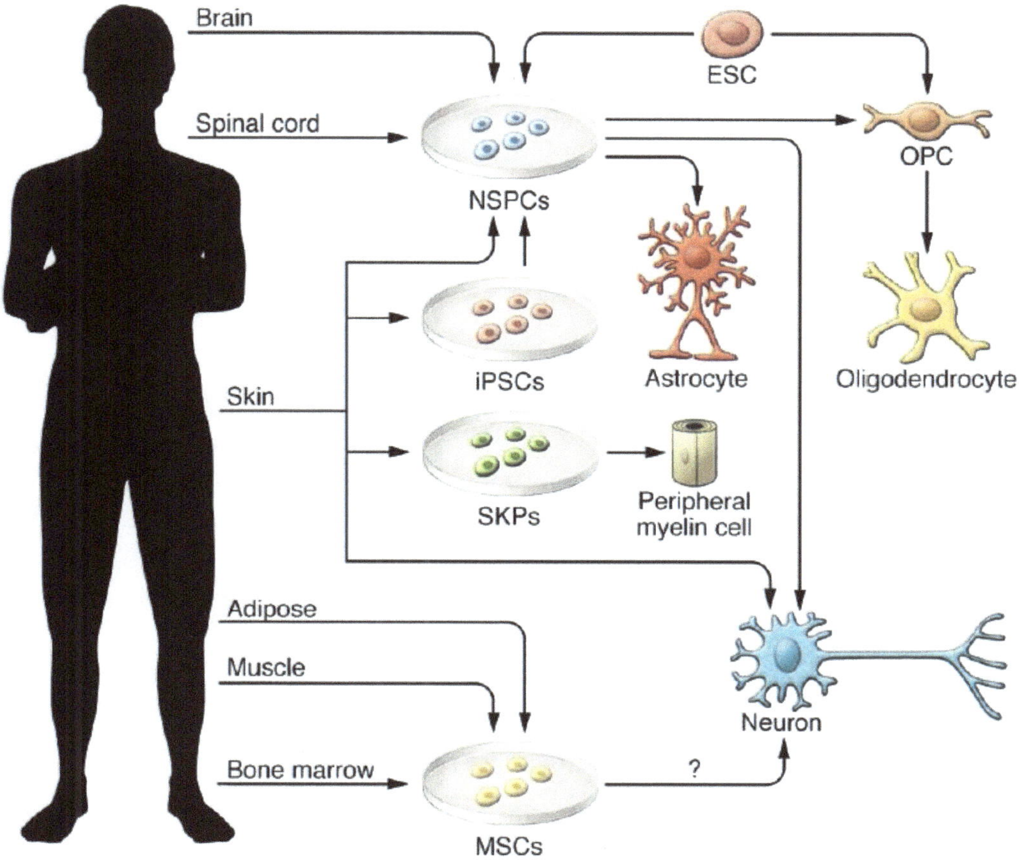

Fig. (2). Exogenous stem cell transplantation - Various sources of stem cell transplantation post spinal cord injury.
Stem cells demonstrated in figure were neural stem/progenitor cells, (iPSCs), skin-derived precursors and MSCs. MSCs were obtained from the bone marrow, umbilical cord, adipose tissue, muscle and dental pulp from deciduous baby teeth. Dependent on culture conditions and growth factor exposure, differentiation of neural stem/progenitor cells can lead to OPCs, mature oligodendrocytes, astrocytes or neurons. Fibroblasts from the skin were reprogrammed into iPSCs and conversion of them directly to neurons and NSPCs without the need to pass into the pluripotent stage.
J Clin Invest. 2012 Nov 1; 122(11):3824-3834; Copyright © 2012, American Society for Clinical Investigation

Embryonic Stem Cells

Several groups have demonstrated in-situ oligodendrocyte differentiation and neurologic improvement after ESC implantation [116, 117]. Keirstead *et al.* through transplantation of human ESC-derived OPCs in rats continued to demonstrate oligodendrocyte differentiation, remyelination and locomotor improvement in acute injury within seven days [107]. Manley *et al.* supported the clinical use of OPC therapy for cervical SCI based on their preclinical efficacy and safety studies. This study used nude rats with cervical SCI lesions and was treated with AST-OPC1, a cell therapy product comprised of OPCs differentiated from the H7 human ESC line. After treatment with graft migration into the lesion site which suppressed the parenchymal cavitation there was improved motor behavioral recovery after four months. This was measured as locomotor performance. There was no effect on teratoma or tumor formation, development of allodynia or on mortality and morbidity rate [118].

Umbilical Cord Stem Cells

Liu *et al.* in six patients with complete SCI utilized umbilical cord mesenchymal stem cells (MSCs) demonstrated poor treatment response [119]. Tarasenko *et al.* demonstrated that primed (2×10^6 cells maintained for 7 days in Poly- D-lysine/Laminin-coated, 25-cm^2 flasks of 4 ml of medium supplemented with 20 ng/ml basic FGF, 5 μg/ml heparin, and 0.5 μg/ml laminin) grafted human NSCs differentiated into cholinergic neurons in the intact spinal cord. When transplanted in mice 9 days after injury in the thoracic region the grafted NSCs survived for 3 months. They showed functional improvement of truncal stability. This was suggestive of functional improvement being dependent on timing of transplantation after injury, sites of grafting and sustainability of newly differentiated neurons and oligodendrocytes [120].

Adipose Derived Stem Cells

Forostyak *et al.* injected autologous adipose derived (AAD) MSCs intrathecally and found improvement in the American Spinal Injury Association (ASIA) motor and sensory scores after an 8 month follow up with nil magnetic resonance imaging interval changes [121].

Neural Stem Cells

NSCs have demonstrated to be multipotent cells with the ability to differentiate into neurons, oligodendrocytes and astrocytes. They were obtained from the developing and/or the adult CNS or were prepared as differentiated derivatives from the embryonic or induced pluripotent stem cells [122 - 128]. Since, there were ethical issues associated with obtaining fetal tissue, immortalized neuroepithelial stem cell lines have been developed. An example was cMycERTAM which was used to achieve conditional growth-promoting gene (c-myc) [129]. Induced pluripotent stem cells (iPSCs) were obtained from differentiated cells by reprogramming techniques through implanting specific transcription factors responsible for pluripotency [130] (*Refer to* Fig. (**2**)).

Many groups have demonstrated some degree of locomotor improvement after NSC grafting in SCI. Several studies demonstrated grafting of rat neural progenitor (NP) cells, human fetal spinal cord NP cells and human fetal NSCs after SCI [131 - 133]. Other studies have demonstrated transplantation of human brain-derived stem cells and human spinal cord tissue in injured mouse spinal cords [134 - 136]. Most grafted cells were shown to differentiate into glia. However, a small percentage differentiated into neurons [117, 131, 133]. Recovery of function noted in these studies were likely secondary to remyelination or release of trophic factors involved in survival and regeneration of host neurons from the grafted cells [69, 117, 133, 137 - 142].

Cummings *et al.* demonstrated that human CNS stem cells from fetal brain when grown as neurospheres were found to survive, migrate, engraft, differentiate and express markers for neurons and oligodendrocytes after SCI in NOD-scid mice. Conversely, the selective ablation of the grafted human neural cells by diphtheria toxin eliminated the observed benefits. Microscopically, new synapse formation between the grafted human NSCs and the host neurons was observed. Human NSCs when grafted after SCI in mice differentiated into gamma-aminobutyric acid (GABA) neurons [143].

Human NSCs derived from iPSCs have shown strong potential in experimental treatment of SCI [128, 144 - 146]. Ruzicka *et al.* found that use of iPSC-NP cells positively affected spinal cord regeneration by factors such glial scar formation, axonal sprouting, tissue sparing and cytokine levels. This lead to better outcomes in advanced locomotor tests (flat beam test and rotarod) [130].

Tarasenko *et al.* highlighted issues that needed to be addressed in NSC therapy in order to effectively replace lost neurons [120]. First issue was maintenance of these cells in culture. Second issue was methods to enhance neuronal differentiation of grafted stem cells in adults. Third issue was methods to directly stimulate stem cells to specific neuronal phenotypes. The intrinsic properties of grafted cells and microenvironment of host tissue were thought to guide phenotypic differentiation post grafting [147]. Wu *et al.* developed an *in-vitro* priming technology which lead to more than fifty percent of human fetal NSCs obtaining cholinergic phenotype in non-injured spinal cords [101]. Tarasenko *et al.* demonstrated that *in-vitro* priming of the grafted cells promoted differentiation to cholinergic phenotypic neurons in the post injury spinal cord [120]. Notably, the timing of grafting was important for graft survival [120, 148, 149]. Grafting of cells on day 9 was found to be most effective. Earlier grafting was less likely to survive due to detrimental effects of high levels of extracellular concentrations of glutamate and aspartate early after a SCI [150 - 152].

Other Cell Types for Transplantation

Schwann Cells Grafts

Schwann cells (SCs) present in grafts were involved in the growth promoting function when peripheral nerve grafts were transplanted into the CNS [153]. SCs were found to promote nerve fiber growth through secretion of neurotrophic factors, expressed cell adhesion molecules and various extracellular matrix molecules [154]. SCs not contacted by axons were found to release neurotrophic factors. SCs contacted by neurons stopped secreting the neurotrophic growth factors [155 - 158]. Xu *et al.* demonstrated that after placing SC grafts into 4-5mm spaces in the thoracic spinal cord of an adult rat that there was regeneration of more than 17,000 axonal processes. Barakat *et al.* found a three month survival of transplanted SCs in the chronic contusion model of SCI in rats [159]. There was regrowth of the corticospinal axons rostral to the injury graft interface in the descending serotonergic axons and the ascending calcitonin gene-related peptide positive fibers. There was reported modest recovery of locomotor function and other neurobiological indices [159, 160]. Martin *et al.* demonstrated that SC grafts were invaded by axonal profiles originating from the dorsal root ganglion neurons. There was an absence of descending fiber regeneration likely due to poor neurotrophic factor secretion and nil penetration of the supraspinal axons into the graft [161].

Olfactory Ensheathing Cells

Olfactory ensheathing cells (OECs) were not shown to be typical stem cells, but have demonstrated the ability to support neuronal regeneration in the olfactory and other areas of the CNS [162] (*Refer to* Fig. (**3**)). In SC transplantation, OECs were better integrated into the host glia, and had better migration along the white matter tracts [163]. OECs were present in the olfactory nerve layer, which was the outer layer of the olfactory bulb, and around the olfactory nerve layer [164 - 166]. Both areas contained two types of OECs, SC-like and astrocyte-like [167, 168]. SC-like OECs had diffuse glial-specific intermediate filament (IF) and GFAP staining with expression of low affinity growth factor (LAGF) p75 receptors [169, 170]. LAGF p75 receptors were downregulated after remyelination and regeneration of the axons with P0-positive (Schwann type) myelin. The astrocyte-like OECs had fibrous GFAP and expressed the embryonic form of neural cell adhesion molecule [168, 171, 172]. *In-vitro*, they have been found to transform from non-myelinating to myelinating [173]. Ramon-Cueto *et al.* demonstrated long-distance regeneration of spinal axons after transplanting OECs in a completely transected thoracic spinal cord [174]. Moreover, OECs have had a more directed elongation in regeneration of corticospinal axons along the lesion site, and into the distal part of the damaged white matter tract [175]. OECs were also shown to remyelinate axons, which have been chemotoxically demyelinated [176]. In sum, OECs were found to establish a glial environment, promote axonal growth, increase angiogenesis, prevent cavity formation and modify scar formation [177].

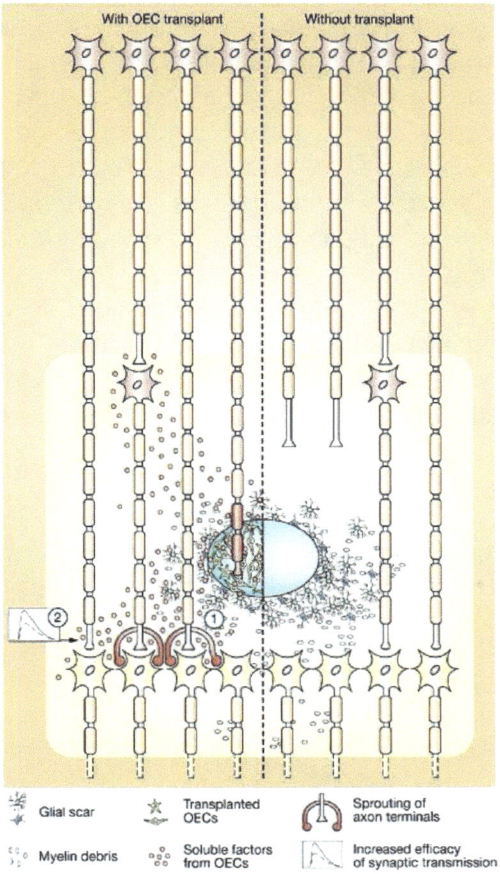

Fig. (3). Olfactory ensheathing cell transplants - Promoting of anatomical and synaptic plasticity through olfactory ensheathing cell transplants.
This figure demonstrated that both anatomical sprouting and synaptic plasticity could improve function within the zone of partial preservation above a complete/partial injury and would maximize the function of spared fibers passing the lesion site.
(1) Shows release of trophic factors by olfactory ensheathing cell (OEC) transplants promoting sprouting of intact fiber systems.
(2) Shows release of neurotrophic factors released from OEC transplants and induce an increase in the efficacy of synaptic transmission of the surviving connections. The inner square (in yellow) represents the area of injury.
Nature Reviews Neurology. 2007 Mar 1; 3:152-161; Copyright © 2007, Springer Nature.

Endogenous Stem Cells for SCI

Neuronal regeneration has become an area of research with potential for clinical application. Although primarily at the animal stage, previous and current studies as discussed below have shown to be promising for eNSCs.

Altman *et al.* introduced the notion of new neurons in the adult rodent brain [178]. SCI research initially emphasized regrowth and reconnection techniques of severed corticospinal tracts through re-innervating alpha motor neurons in the ventral horns [179]. More recently, there has been an up rise in attempts to use nascent local spinal interneurons to form a neuronal relay network. This would serve as a "bridge" to circumvent long-distance axonal growth [180, 181]. Adult neurogenesis was found to be an active process of proliferation, fate specification of NPs, differentiation, maturation, navigation and functional integration of neurons into the existing neuronal circuitry [182]. It was also found to be a dynamic process regulated by stimuli and modulated by the existing neuronal circuitry [183]. Neurogenesis occured in the normal brain in new olfactory interneurons and hippocampal granule cells [184]. 5-ethynyl-2'-deoxyuridine labelling of mice brains has shown that the main proliferative zone has proliferative potential. However, all proliferative cells were not likely to be differentiated into neurons and incorporated into neuronal circuitry [185]. Compared to postnatal development, adult neurogenesis had a more prolonged process. Adult borne dentate granule cell development, formation and maturation lasted more than eight weeks. Postnatally, a similar process occurred within two weeks [186, 187].

Advantages of Endogenous versus Exogenous Stem cells

The utilization of eNSCs has become appealing as they are noninvasive, avoid need for immunosuppression and concern for immune rejection [188]. Moreover, they avoid associated ethical dilemmas and concern for tumor formation secondary to dysregulated cell proliferation [189].

Areas of Neurogenesis in Spinal Cord

eNSCs have shown to exist in the brain and the spinal cord. The adult mammalian spinal cord was a non-neurogenic region. In non-pathological situations the progenitor cells were glial-restricted leading to oligodendrocytes and astrocytes formation, but not neurons. NSCs were located in the white matter parenchyma and close to the ependymal/subependymal layer (SEL) in the central canal [190 - 195]. Two models exist on the location of neural stems cells in the intact spinal cord [190]. The first model postulated that the multipotent stem cells resided in the ependymal layer of central canal, known as ependymal stem/progenitor cells (epSPCs). Here, they divided asymmetrically into daughter cells and NP cells [191, 193]. They further migrated to the outer portion of the spinal cord becoming either proliferative glial progenitor (GP) cells or mature glial cells of the spinal cord [196]. The second model postulated that the stem cells and GP resided independent of the proliferating ependymal cells (ECs) in the spinal cord parenchyma [196, 197]. NP cells in the spinal cord white matter expressed markers such as chondroitin sulphate proteoglycan neuron-glial antigen (NG) 2 and transcription factors such as oligodendrocyte transcription factor (Olig2) and NKx2.2. This represented a spectrum ranging from multipotent to restricted GP cells designated to the oligodendrocyte lineage [190, 194, 198].

Histology of the Central Spinal Cord

The ependymal and periependymal region of the central spinal cord region consisted of tanycytes, ependymocytes and cerebrospinal fluid (CSF)-contacting neurons [199]. Tanycytes, involved in communication between CSF and capillaries radiated from the central canal into the gray matter and terminated on blood vessels [199, 200]. Ependymocytes were ciliated cells and connected with each other through gap junctions [201]. CSF-contacting neurons were found in lower vertebrates. They have been shown to have similar electrophysiological properties similar to immature neurons, and communicate with local neurons at higher spinal segments. They were responsible in maintaining a potential of hydrogen in the CSF through P2X receptors, and express neuronal nuclei (NeuN) and doublecortin neuronal lineage markers [202 - 204]. Studies have demonstrated that cells in the SEL had increased proliferative capacity after a SCI event [205, 206]. Post injury, cells in mice expressed NSC markers, migrated to the site of injury in the spinal cord and differentiated into astrocytes and oligodendrocytes. Therefore they were key components in forming the glial scar and forming the functional neuronal circuits [193, 205 - 209] (*Refer to* Fig. (**4**)).

Forms of Stem Cells in the Spinal Cord

OPCs, ECs and astrocytes were postulated as potential stem cells in the spinal cord [190 - 194]. Studies by several groups indirectly found that OPCs had NSC properties, and formed multiple lineages in the spinal cord [194, 198, 210, 211]. Others found that ECs consisted of the main cell population in the intact spinal cord, and gave rise to differentiated progeny after injury. Although ECs had *in-vitro* NSC properties, their origin after injury was not known [193, 212]. Buffo *et al.* found that astrocytes had minimal proliferative activity with only *in-vitro* NSC properties after injury in the cortical astrocytes [213]. They only had stem cell properties in the SVZ and dentate gyrus [214]. Astroglial activation included two sources: reactivated resting astrocytes and new astrocytes from eNSCs [116].

Shihabuddin *et al.* showed that transplantation of NSCs obtained from the spinal cord into the hippocampal dentate gyrus gave rise to neurons. Furthermore, with re-implantation of NSCs in the spinal cord they continued to differentiate into glial cells. This study had two implications. First, the NSCs isolated from the spinal cord were multipotent. Second, the NSCs obtained from the spinal cord were not fate restricted, but restricted to express their multipotency in the spinal cord. This was likely due to environmental and intrinsic factors regulating proliferation, migration and differentiation in the spinal cord [215].

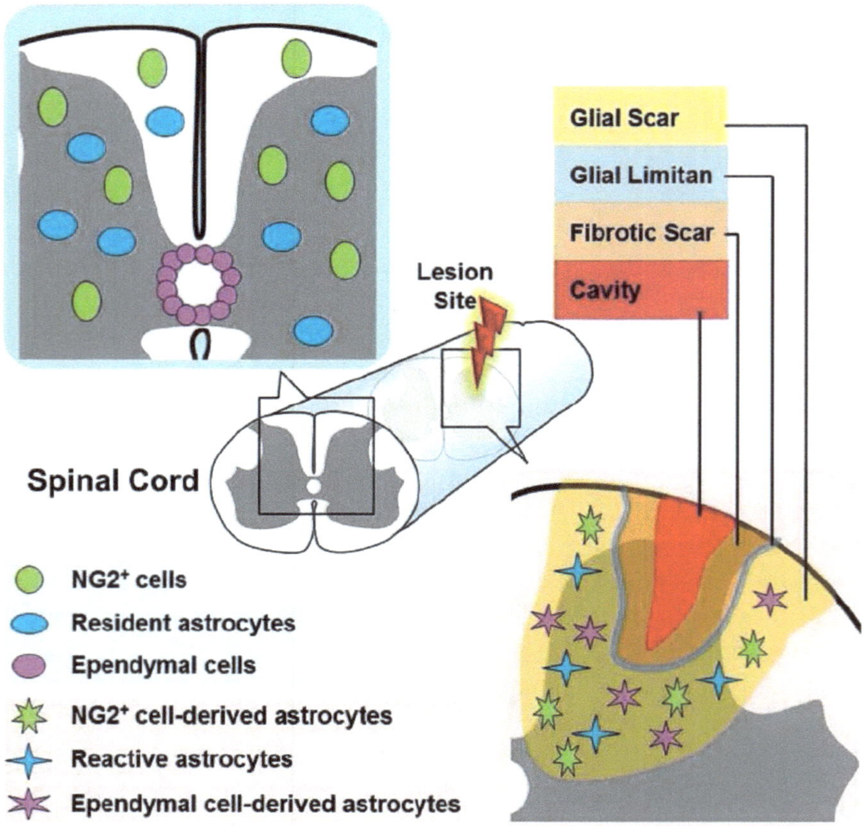

Fig. (4). Glial scar formation - Cellular composition post spinal cord injury. Figure demonstrated the heterogeneous neural cell infiltrate in the glial scar post spinal cord injury.
(Top left) showed location of neural cells in the intact spinal cord
(Bottom right) showed pathology of spinal cord injury with layers of the healing process and differential of the neural cells post injury.
Neurosci Bull. 2013 Jan 1; 29(4):421-435; Copyright © 2013, Springer Nature

Environmental Support for Stem Cells

NSC environments were found to be nurtured with local astrocytes [216 - 219]. They were present throughout the total life span of the organism [220]. Astroglial cell staging was important in supporting or inhibiting NSC migration [221]. Primary astrocytes from neurogenic locations have the ability to promote neurogenic differentiation [222]. Non-neurogenic areas in mice could be induced with injecting subgranular zone astrocytes and sonic hedgehog [217]. Astrocytes were not fate determined, and under specific conditions differentiated accordingly [223 - 228].

IF has been shown to have both structural and non-structural function in eukaryotes [229]. The non-structural function of IF was an important component of signaling pathways for cell survival and migration [230]. Nestin was transiently expressed in NSCs, and down regulation of its expression was related to differentiation of cells into astrocytes, neurons and oligodendrocytes. It has been found that nestin expression of active astrocytes in mice was mediated through EGF-ErbB1. Further intracellular interactions were through the Ras, Raf, MEK and extracellular signal-regulated kinase (ERK) pathways [220, 223, 231]. Nestin was not expressed in mature cells [220]. However, during acute injury, nestin expression was suggestive of active proliferation of progenitor astrocytes [223]. Same gene sequences controlling nestin during CNS development were involved in up-regulating nestin in astrocytes during states of CNS injury [220, 231, 232]. EGF through the EGF receptor - ErbB1 was involved in transformation into active astrocytes [233].

Yang Z *et al.* demonstrated the importance of establishing a nurturing microenvironment for eNSCs and ExSCs to stimulate nascent neural circuitry in the neural repair process [179]. Through a biodegradable material chitosan, a slow release of neurotrophin-3 was achieved in the microenvironment over 14 weeks. This allowed eNSCs to be activated, and later differentiate. The neutrotrophin-3-chitosan model functions through the neutrophin-3 (NT3)-tropomyosin receptor kinase (Trk) C signaling. The three modes of actions in new neuronal growth were: long distance nerve fiber growth through the formation of a nascent neural synaptic network, inhibition of inflammatory immune activity and stimulation in formation of blood vessels [179, 234, 235].

Implications of Spinal Cord Injury and its Effects on Regeneration

Hofstetter *et al.* based on better functional recovery after *in-vitro* transplantation of NSCs found that the adult spinal cord had stem cells which were activated by injury, but were unable to exert the beneficial effects in normal conditions [236]. SCI activated astrocytes, fibroblasts and macrophages which were involved in forming a scar to hinder further injury [237]. This scar was proposed to be detrimental because it hindered axonal regeneration and acted as a physical barrier [238 - 240]. The molecular and cellular mechanism of scar formation was altered at the messenger ribonucleic acid (RNA) level [192]. After a SCI, MicroRNA-486 was responsible for regulation of inflammation. miR-20a was involved in cell death and miR-21 in astrocyte activity. Although they were "actively" involved in these processes. There was no evidence that they directly acted on NSCs [241 - 243]. Moreover, long non-coding RNAs have also been found to affect posttranscriptional regulation of SCI events [244 - 246]. Wang *et al.* demonstrated that knockdown of IncSCIR1 caused astrocyte proliferation and migration. There was a strong correlation between the downregulation of IncSCIR1, decrease of Wnt3 and increase of bone morphogenetic protein (BMP) 7. Therefore, local overexpression of IncSCIR1 promoted functional recovery with an inhibitory effect on the environment around the location of injury [247].

Glial Scar Debate

Moore *et al.* examined the histochemical changes after a SCI [248]. They found that there was a 3-fold increase in GFAP expression in the SEL at the site of lesion and at proximal sites in the canine spinal cord. The basis of these findings was multifold. First, there was increased astrocytic differentiation of neural precursor cells (NPCs) in SEL. Second, astrocytes were derived from the EC layer and migrated to the injury site with regulatory effects of the inflammation. Third, the glial scar formation process involved astrocyte migration into the injured region [209, 248 - 250] (*Refer to* Fig. (**4**)). Moore *et al.* findings suggested that this process was diffuse and occurred throughout the spinal cord. NSCs transforming into astroglia was possibly a natural protective mechanism to confine injury size, but later contributed to the glial scar and hindrance in regrowth of axons. In addition, neurons and oligodendrocytes were the main cells dying after SCI. NSCs were the potential source to replace them. This enhanced neuronal and oligodendrocyte (ODC) differentiation towards eNSC-based recovery [248].

Anderson *et al.* assessed the role of reactive astrocytes in axon regeneration in post-experimental SCI mouse models. They prevented the formation of reactive astrocytes through selectively killing proliferating scar-forming astrocytes, deleting STAT3 transcription factor, genetically altering the astrocytes to express a diphtheria toxin and killing astrocytes through ultralow doses of diphtheria toxin [251]. In animals, lacking the glial scar yielded poor spontaneous regrowth of damaged axons and increased degeneration of axons from the distal to the proximal end of injury. When compared with controls, areas with nil astrocyte scar formation did not contain any axons. This suggested the importance of chronic astrocyte scar formation in preserving the integrity of the neuronal tissue. They also demonstrated that chondroitin sulfate proteoglycans were significantly higher in the injured spinal cord compared to not injured controls. Further removal of the scar did not decrease the chondroitin sulfate proteoglycan levels. Aggrecan, a growth inhibiting chondroitin sulfate proteoglycan was not detected in scar forming astrocytes. It was also demonstrated that injecting neurotrophins and brain-derived neurotrophic factor (BDNF) into the injury site stimulated axonal growth through the astrocytic scar. Removing the scar inhibited axonal regeneration [251].

Hub Gene Network Analysis

Duan *et al.* used the Weighted Gene Coexpression Analysis to establish hub network genes which regulate microenvironment functions. Hub gene network analysis demonstrated modules (Cm) that were positively and negatively correlated with NT3-chitosan treatment. Cm1 and 2 were involved in neurogenesis. Cm3 was involved in pain and temperature sensory processing. Cm4 was involved in muscle contraction. Cm5 was involved in angiogenesis. Cm6 was involved in immune response. Cm7 was involved in the wound, stress and inflammatory response. Cm1,2,3,4 and 5 were positively correlated with NT3-chitosan. Cm6 and 7 were negatively correlated with NT3-chitosan [235]. Yang *et al.* showed that corticospinal axons made contact with new neurons. This was through labelling the corticospinal motor neurons with biotin, fluorescein and dextran to Bromodeoxyuridine (BrdU)/microtubule associated protein 2 with the new neurons in the neurotrophin-3-chitosan model. To assess for functional recovery, cytosine beta-D-arabinofuranoside was used to decrease nestin-positive NSCs and new NeuN-positive neurons. The treatment group showed 30-40% decrease in Basso Beattie Bresnahan scores in the bilateral hind limbs at 1 year. Moreover a decrease in amplitude by 30% with nil decrease in latency was noted in somatosensory evoked potentials (SEP) and motor evoked potentials (MEP) after treatment. This suggested that cytosine beta-D-arabinofuranoside did not alter the number of regenerating axons. Change in amplitudes suggested that locomotor functional recovery was due to regeneration of neurons [179].

Purinergic Receptors

Gómez-Villafuertes *et al.* demonstrated the properties of the purinergic receptors altering the intrinsic properties of the epSPCs by enhancing regulation of proliferation, differentiation and lineage specification after a SCI [252, 253]. Nucleotides activated two different types of purinergic receptors called P2X and P2Y. P2X (P2X 1-7) were ion-gated channels involved in fast calcium influx (ionotropic). P2Y (P2Y 1-14) were G-protein-coupled receptors (metabotropic). NP cells found in adult SVZ and subgranular layer of the hippocampal dentate gyrus of the hippocampus had functional purinergic receptors [254]. Moreover, neurospheres obtained from fetal rat brain expressed P2X2-P2X7 and P2Y1,2,4,6.

Adult mouse SVZ expressed P2Y1 and P2Y2 purinergic receptors types [255 - 258].

After a SCI, there were two effects of activation of the purinergic receptors in the area of lesion. First, there was an activation of purinergic receptors due to release of large amounts of adenosine triphosphate (ATP) and nucleotides by the injured tissues. Due to the interaction of various growth factors there was remodeling and repair of the lesion [255, 259]. Studies have shown that the NP cells have been an important source of local ATP. This was supported by purinergic receptors being localized in regions of mitotic progenitor cell expansion and neurogenesis in the adult brain [254, 260, 261]. External ATP supplementation increased the mitotic index and rate of NP cells [260]. Second, purinergic receptor interaction led to decreased gliosis and glial scar formation [262].

Gómez-Villafuertes *et al.* through microfluorimetric techniques using calcium dye Fura-2 demonstrated that nucleotide activation through ionotropic P2X4,7 and metabotropic P2Y1,4 led to epSPC response. P2X7 receptor was shown to have dual functions. First, it had an inhibitory effect on neurogenesis and axon out-growth. Second, it was a cell death inducer [252]. Other studies have found that P2X7 receptor inhibition have led to cultured hippocampal neuron axonal growth and branching [263]. Activation of P2X7 receptors has led to necrosis and apoptosis of embryonic and adult NPCs. This could adversely affect NP cell survival after a SCI secondary to excessive neuro or gliogenesis [264, 265]. P2Y1,2 have been found to stimulate NP cell migration which facilitates movement of NSCs in the neurogenic niches [257, 266]. Moreover, it could have played a crucial role in NP cell expansion. Further, which could have helped maintain NSC/NP cell niches in adult brains [252]. P2Y1 receptor antagonists aided NSC differentiation into neurons and glial cells through decreasing the size and frequency of primary neurospheres [267]. epSPCs were found to express functional P2Y1 and P2Y4 receptors [252]. P2Y receptor antagonists were found to suppress mitotic index of cells, which suppressed neurosphere expansion [260]. Severe spinal cord contusions have led to early and sustained increase in P2X4,7 receptors [252]. de Rivero Vaccari JP *et al.* found that acute transplantation of undifferentiated epSPCs which were derived from rats with SCI reversed the effect [265]. P2X4 knockout mice demonstrated decreased levels of neuroinflammation after SCI leading to improved functional recovery during the first week after injury [268]. Use of P2X7 receptor was controversial after a SCI as some studies demonstrated failed motor and histopathological improvement. Other studies have shown that P2X7 antagonism improved function and reduced the pathological consequences of SCI [252, 269].

Genetic Fate Mapping

Although the findings of adult neurogenesis and presence of NSCs have been encouraging. However, there has been difficulty in determining the identity and in localizing the cells in the spinal cord [190, 270]. Therefore, the utility of genetic fate mapping (GFM) has been used to characterize, identify and aid in localizing potential NSC populations [123]. The principle of GFM involves labeling of precursor cells, and then tracing their lineage both *in-vivo* and *in-vitro*. It was also a technique used in determining cell behaviors such as proliferation, movement and lineage exclusion [270, 271].

Types of Labelling Techniques

Different types of labelling techniques were used to delineate different cell lineages. Fluorescent labeling of neurosphere-initiating cells demonstrated that multipotent NSCs resided close to the central canal of the spinal cord [191, 192]. Several groups showed that NSCs were stimulated from the medial and lateral parts of the spinal cord and were able to expand passages up to two times [195, 272]. Indirect methods to label cells decreased specificity of detection of NSCs [273]. BrdU labeling involved BrdU being incorporated and substituting thymidine into the newly synthesized DNA during the S phase of the cell cycle. BrdU specific antibodies were used to detect the incorporated chemical which was suggestive of active replication of DNA [274].

Labelling with BrdU was achieved through drinking water for 5 weeks which predominantly marked approximately 80% of ECs, astrocytes and oligodendrocytes. Minority cells labelled were microglia and blood vessel associated cells in the spinal cord [116, 190]. These labelled cells proliferated in normal conditions and after injury, but did not provide lineage data [190, 198, 275]. Retroviral infections were limited in their mapping function as they labelled only a small population of dividing cells with low specificity [194, 198].

Cell Structure Markers

Cells of various structures express markers that help identify their origin. Nestin, a VI IF protein was expressed by NSC/NP cell markers *in-vivo*, *in-vitro* and in endothelial cells [231, 276, 277]. Also, endothelial cells were known to produce nestin positive progeny, therefore nestin cannot be used as an absolute NSC marker [278, 279]. Expression of nestin was decreased when the NSC/NP cell differentiated into neurons/glial cells [280, 281]. It was a marker protein at the stage before exit from the cell cycle, and was committed to a mature progeny of a specific lineage [282]. Moreover, cell markers for central canal cells were prominin-1, Musashi1, platelet-derived growth factor receptor (PDGFR)-alpha,

Sox2, Sox3 and Sox 9 [116]. Sox1 and Sox 9 identified oligodendrocyte lineage cells and astrocytes, respectively [193, 283 - 285]. Sox 9 identified ECs based on position. NG2 was expressed by proliferating OPCs [286].

Meletis *et al.* performed a dorsal funiculus incision which demonstrated that majority of ependymal progeny were immunoreactive to Sox9 and vimentin. Astrocytic morphology was most abundant in the core of the glial scar (*Refer to* Fig. (**4**)). A small number of the ependymal progeny expressing GFAP and nestin were most abundant in the surface of the spinal cord. This represented the reactive resident astrocytes of the glial scar. The same group found that post dorsal and lateral funiculus injury after the first month, Olig2 expressing cells derived from ependyma were scattered throughout the injury site. After 10 months, ECs were found to produce myelinating oligodendrocytes as the progeny expressed Olig2 and demonstrated a cell morphology consistent with mature oligodendrocytes. These cells were found in the glial scar and were predominantly found in the intact gray and white matter surrounding the injury site [193].

Tamoxifen-Dependent Cre System

Tamoxifen-dependent Cre has been demonstrated to be an inducible Cre/loxP-based gene expression system [193]. The *in-vivo* use of this system required two sets of transgenic animals. First group expressed tamoxifen-inducible Cre recombinase under a promoter specifically for NSCs. Second group carried a reporter gene such as green fluorescent protein (GFP) that had a Floxed stop signal between the promoter and the transgene. Double-transgenic mice expressed the reporter gene only after the stop signal was excised *via* Cre/loxP recombination. This was activated by tamoxifen administration. In this mouse model, a reporter was expressed only in NSCs upon tamoxifen induction. This specific labeling technique enabled a precise tracking of the fate of eNSCs in spinal cord [116, 193]. The investigators used CreER with different cell type-specific promoters such as Forkhead box protein (FoxJ1) for ECs, Connexin 30 for astrocytes and Olig2 for oligodendrocyte cells. FoxJ1 expression was specific to cells containing motile cilia and flagella [193, 287 - 290]. FoxJ1-CreER showed genetic recombination in ECs of spinal cord [116]. Nestin-CreER and FoxJ1-CreER was exclusively expressed by lumen contacting cells with motile cilia in the central canal. Cells containing nestin and vimentin were suggestive of immature neural cells [193, 231]. Meletis *et al.* found that this lineage of cells could be serially passaged at least 8 times with the number of cells increasing exponentially during each passage. The neurospheres passaged were multipotent, which further differentiated into neurons, astrocytes and oligodendrocytes. GFP positive ECs formed neurospheres. GFP negative cells did not form any neurospheres. BrdU labeled ECs were found in pairs suggestive of self-duplication [193].

Genetic labeling frequency at 10 months after tamoxifen administration demonstrated that ECs were maintained through self-duplication, and were not a progeny of another cell population [193, 291, 292]. Migration of the ECs after a SCI demonstrated loss of immunoreactivity to Sox2, Sox3 and CreER expression from FoxJ1 promotor. Most cells expressed Sox9 and GFAP [193].

Ependymal Cells, Oligodendrocytes, Astrocytes

Barnabe-Heider *et al.* reported that in the intact spinal cord, ECs and astrocytes demonstrated only self-duplication. Oligodendrocyte progenitor cells self-duplicated and produced mature cells [116]. BrdU in the intact spinal cord was incorporated in oligodendrocytes 10 times more than ECs and astrocytes. After injury, ECs and astrocytes produced more progeny than oligodendrocytes. ECs through GFM were found to be the "quiescent stem cell population", which were retrievable after a SCI. These cells lining the lateral ventricle could give progeny to other cell types [116].

Oligodendrocytes

OPCs were found to divide after a SCI, but lacked multilineage potential [193, 293]. Zhu *et al.* demonstrated that dorsal oligodendrocytes similar to ventral oligodendrocytes expressed Olig2, Sox10, adenomatous polyposis coli and myelin proteins. They were capable of migrating to the injury site and forming myelin sheaths in axons [293]. Ventral oligodendrocytes were present in all areas of the spinal cord. Dorsal oligodendrocytes present were Laz and GFP labelled [293 - 296]. Ventral oligodendrocytes were likely to myelinate axons in early development. Dorsal oligodendrocytes myelinated axons in the late stages and post injury. Dorsal oligodendrocytes were comprised of only 20% of the oligodendrocyte population. They had delayed differentiation compared to ventral oligodendrocytes [293]. The mechanism for the delay was the inhibitory effects of BMP signalling on dorsal oligodendrocytes. As BMP levels declined, dorsal oligodendrocytes activity increased [293 - 296]. The Shh signaling increased generation of ventral oligodendrocytes in the early phases of the injury. This decreased the space for the dorsal oligodendrocytes to migrate ventrally [293, 297, 298]. Around the lesion at two weeks post injury, 60% of the oligodendrocyte population was derived dorsally [293, 299, 300]. The restrictor factor around the lesion formed the glial scar preventing oligodendrocyte migration [70, 301, 302].

Astrocytes

Astrocytes were found to proliferate extensively after a SCI. They have not demonstrated multi-lineage potential, but have the potential to produce neurospheres after injury [116, 213]. Astrocytes were the predominant cells found in the glial scar post injury, which were found to prevent axonal regeneration through chondroitin sulfate proteoglycans (*Refer to* Fig. (**4**)). Astrocytes were of two fates. First type was the self-duplication of local astrocytes in the periphery. Second type was derived from ECs, which were GFAP negative and formed the core of the glial scar [116, 237, 303 - 305].

STEM CELL TRANSPLANTATION

Methods for Transplantation

The most effective approach for exogenous transplantation was local transplantation into lesions [306]. Other approaches used were subarachnoid transplantation and intravenous infusion [307, 308]. Intrathecal administration was deemed safe as MSC transplantation was given intrathecally in chronic SCI (mean - 33 months after injury), which demonstrated no treatment related adverse effects [309]. Standard procedure involved open surgical exploration exposing the injured spinal cord with direct intra-spinal cell transplantation. Some disadvantages associated with this procedure were long post-operative recovery, causing extensive trauma and difficulty in repetition. Advanced methods for local stem cell transplantation through computed tomography guided puncture were minimally invasive and had a quicker recovery [310].

Mechanism, Benefits and Adverse Effects

ExSCs have been demonstrated to function through production of growth factors, cytokine production, neurotrophic factors and stimulation of endogenous NP cells [311]. Transplanted NSCs have given rise to astrocytes through restricted number of oligodendrocytes [131, 236, 312 - 314]. Hofstetter *et al.* suggested that astrocytes support recovery through production of neurotrophic factors. This supported the survival of non-injured cells causing axonal local sprouting. Oligodendrocytes were involved in remyelination and axonal ion channel reorganization leading to some recovery [315]. There was a shift in the balance to increase production of oligodendrocytes. Decreased astrocytes showed improvement in sensory and locomotor deficits [236]. The success of grafted cells were based more on the preservation of the local environment than the intrinsic character of the cell [316].

With the limited studies available the benefits of BMMSC therapy were demonstrated to be:

1. Axonal regeneration and remyelination [317, 318]
2. Promotion of vascular regeneration and improvement of local blood supply [319 - 321]
3. Stimulation of eNSC migration [322]
4. Regulation of the local inflammatory environment [323]
5. Secretion of neurotrophic factors [324]
6. Neuronal protection through rewrapping injured nerve fibers suffering demyelization [324 - 327]
7. Replacement of neuronal cells [109, 116, 328]

Adverse effects reported by Dai *et al.* after stem cell transplantation were transient fevers lasting for 72 hours with a maximum temperature of 38.5 celsius, headache, dizziness, nausea and vomiting [109]. Large volume transplantations caused increased edema and could potentially cause a secondary SCI [109].

Clinical Trials

ExSC therapy was approved by the Food and Drug Administration (FDA) of the United States of America and the State FDA in China for therapy of SCI. Clinical trials evaluating stem cell modalities have mainly focused on autologous BMSC transplantation, umbilical MSCs and AAD stem cells in both acute and chronic cervical and thoracic SCI [329]. Results from these clinical trials have been limited and controversial at this time. *For details of current trials refer to Table 1: Current Clinical Trials Using Exogenous Stem Cells for Spinal Cord Injury.*

Clinical Results and Outcomes

Benefit in the Acute Setting

Dai *et al.* demonstrated 50% clinical improvement of motor function after stem cell therapy. Patients with nil muscle strength as per ASIA motor scoring showed some contraction in the brachial muscles (increase of 1 point on ASIA) status post stem cell transplantation (SPSCT). Moreover, patients classified as weak in motor strength improved to strong, with added improvements of physical flexibility and coordination. Electromyographical studies showed increase in wave amplitudes after therapy. They also found significant improvement in sensory function. As per paraspinal SEP examinations, sensory levels decreased 1.8 to 8 levels. SPSCT symptoms of zonesthesia also improved, due to a poorly understood mechanism [109]. Yoon *et al.* used ASIA grade A SCI patients. In the acute stage of SCI (less than two weeks), 29.5% patients improved from ASIA grade A to B or C. In the subacute stage of SCI (2-8 weeks), 33.3% patients improved from ASIA grade A to B or C. There was no improvement noted in patients in the chronic stage (greater than 8 weeks) [310].

Table 1. Current Clinical Trials Using Exogenous Stem Cells for Spinal Cord Injury.

Completion Date of Study	Title of Study	NCT Number	Sponsors/ Collaborators	Study Design	Age Groups (Years)	Interventions	Outcome Measures
Mesenchymal Stem Cells							
Not stated	Safety and Effect of Adipose Tissue Derived Mesenchymal Stem Cell Implantation in Patients With Spinal Cord Injury	NCT01769872	1. Biostar 2. Korea University Anam Hospital	• Endpoint Classification: Safety/Efficacy Study • Intervention Model: Single Group Assignment • Masking: Open Label • Primary Purpose: Treatment	19 to 70	Autologous adipose tissue derived MSCs	• ASIA (American Spinal Injury Association) scale • Magnetic Resonance Imaging (MRI) • MEP/SSEP|ADL (activities of daily living) • SF-36 • ODI (Oswestry Disability Questionnaire) • Frequency of Adverse Events
February 2010	Autologous Adipose Derived MSCs Transplantation in Patient With Spinal Cord Injury.	NCT01274975	Biostar	• Allocation: Randomized • Endpoint Classification: Safety Study • Intervention Model: Single Group Assignment • Masking: Open Label • Primary Purpose: Treatment	19 to 60	Autologous adipose derived mesenchymal stem cells	• Safety evaluation
May 2014	Intrathecal Transplantation Of Autologous Adipose Tissue Derived MSC in the Patients With Spinal Cord Injury	NCT01624779	Bukwang Pharmaceutical	• Endpoint Classification: Safety/Efficacy Study • Intervention Model: Factorial Assignment • Masking: Open Label • Primary Purpose: Treatment	19 to 70	Autologous adipose tissue derived mesenchymal stem cells	• Significant MRI - change before and after intervention • Significant neurologic function - change before and after intervention • Significant electrophysiological change before and after intervention • Adverse event
June 2014	Autologous Mesenchymal Stem Cells in Spinal Cord Injury (SCI) Patients	NCT01694927	1. Clínica Las Condes. LIT INNOVA CORFO 2. Clinica las Condes, Chile	• Endpoint Classification: Safety/Efficacy Study • Intervention Model: Single Group Assignment • Masking: Open Label • Primary Purpose: Treatment	2 to 65	Autologous mesenchymal stem cells	• Safety of autologous expanded mesenchymal stem cells transplantation in SCI patients • Functional improvement in muscle strength • Functional Improvement in sphincters control • Functional improvement in spasticity control

(Table 1) contd.....

Completion Date of Study	Title of Study	NCT Number	Sponsors/ Collaborators	Study Design	Age Groups (Years)	Interventions	Outcome Measures
December 2015	Evaluation of Autologous Mesenchymal Stem Cell Transplantation in Chronic Spinal Cord Injury: a Pilot Study	NCT02152657	Hospital Sao Rafael	• Endpoint Classification: Safety/Efficacy Study • Intervention Model: Single Group Assignment • Masking: Open Label • Primary Purpose: Treatment	18 to 65	Mesenchymal stem cells	• MRI • Sensitivity and motor strength on the inferior limbs • Improvement in urological function • Improvements in sensorial mapping and neuropathic pain
March 2016	Autologous Mesenchymal Stem Cells Transplantation for Spinal Cord Injury A Phase I Clinical Study	NCT02482194	1. Armed Forces Bone Marrow Transplant Center, Rawalpindi, Pakistan 2. Armed Forces Institute of Regenerative Medicine	• Safety Study • Intervention Model: Single Group Assignment • Masking: Open Label • Primary Purpose: Treatment	18 to 50	Mesenchymal stem cells	• Number of adverse events • Sensory and motor strength • Functional independence • Muscle strength assessment
December 2016	Autologous Mesenchymal Stem Cells Transplantation in Cervical Chronic and Complete Spinal Cord Injury	NCT02574572	Hospital Sao Rafael	• Endpoint Classification: Safety/Efficacy Study • Intervention Model: Single Group Assignment • Masking: Open Label • Primary Purpose: Treatment	18 to 65	Autologous mesenchymal cells	• Number of participants with treatment-related adverse events as assessed by MRI • Functional improvement in ASIA grade • Functional improvement regarding the ASIA impairment Scale (AIS) scores (light touch, pin prick and motor force) • Improvements in sensorial mapping and neuropathic pain
December 2017	Autologous Mesenchymal Stem Cells Transplantation in Thoracolumbar Chronic and Complete Spinal Cord Injury Spinal Cord Injury	NCT02574585	Hospital Sao Rafael	• Allocation: Randomized • Endpoint Classification: Efficacy Study • Intervention Model: Parallel Assignment • Masking: Open Label • Primary Purpose: Treatment	18 to 65	Autologous mesenchymal cells	• Number of participants with treatment-related adverse events as assessed by MRI SCI • Functional improvement in ASIA grade • Functional improvement in AIS scores (light touch, pin prick and motor force) • Improvements in sensorial mapping and neuropathic pain

(Table 1) cont.....

Completion Date of Study	Title of Study	NCT Number	Sponsors/ Collaborators	Study Design	Age Groups (Years)	Interventions	Outcome Measures
December 2018	Umbilical Cord Mesenchymal Stem Cells Transplantation to Patients With Spinal Cord Injury	NCT02481440	1. Limin Rong 2. Third Affiliated Hospital, Sun Yat-Sen University	• Allocation: Non-Randomized • Endpoint Classification: Safety/Efficacy Study • Intervention Model: Factorial Assignment • Masking: Open Label • Primary Purpose: Treatment	18 to 60	Umbilical cord mesenchymal stem cells	• Changes in motor and sensory assessment by the ASIA score • Number of participants with adverse events • Changes in electromyogram and electroneurophysiologic test
December 2020	Safety and Efficacy of Autologous Mesenchymal Stem Cells in Chronic Spinal Cord Injury	NCT01676441	Pharmicell Co., Ltd	• Endpoint Classification: Safety/Efficacy Study • Intervention Model: Single Group Assignment • Masking: Open Label • Primary Purpose: Treatment	16 to 65	Mesenchymal stem cell transplantation	• Motor score of the ASIA scale • Sensory score of the ASIA scale • Electromyogram and electroneurophysiologic test • MRI and diffusion tensor imaging of spinal cord • Number of adverse events
Bone Marrow Stem Cells							
December 2008	Cell Transplant in Spinal Cord Injury Patients	NCT00816803	1. Cairo University 2. Al-Azhar University 3. Medical Military Academy, Egypt 4. Alexandria University	• Endpoint Classification: Safety/Efficacy Study • Intervention Model: Parallel Assignment • Masking: Single Blind (Outcomes Assessor) • Primary Purpose: Treatment	10 to 36	• Autologous bone marrow (BM) transplant • Procedure: Physical therapy	• Safety of autologous BM transplant - measured by absence of neuronal changes, infections or increased intracranial tension and monitoring for any abnormal growth or tumor formation by MRI • Efficacy of BM cell transplant in improving neurological functions in patients with chronic SCI. • Improvement in motor, sensory, sphincteric functions and quality of life using ASIA scores and MRI
August 2010	Safety and Efficacy of Autologous Bone Marrow Stem Cells in Treating Spinal Cord Injury	NCT01186679	International Stem cell Services Limited	• Allocation: Non-Randomized • Endpoint Classification: Safety/Efficacy Study • Intervention Model: Parallel Assignment • Masking: Open Label • Primary Purpose: Treatment	20 to 55	• Procedure: laminectomy • Procedure: intrathecal	• Number of participants with adverse events as a measure of safety and tolerability. • Significant clinical improvement in AIS and general condition • Changes in the MRI, neurological improvement (cranial/spinal reflexes) and evoked potentials study

(Table 1) cont.....

Completion Date of Study	Title of Study	NCT Number	Sponsors/ Collaborators	Study Design	Age Groups (Years)	Interventions	Outcome Measures
December 2010	Autologous Bone Marrow Stem Cells in Pressure Ulcer Treatment	NCT01572376	Hospital Universitario Central de Asturias	• Allocation: Non-Randomized • Endpoint Classification: Safety/Efficacy Study • Intervention Model: Single Group Assignment • Masking: Single Blind (Outcomes Assessor) • Primary Purpose: Treatment	18 to 75	Infusion of autologous BM stem cells after wound debridement	• Absence of adverse effects during timeframe as infections or complications related with the intervention • Improvement and closure of the pressure ulcer
December 2012	Autologous Bone Marrow Stem Cell Transplantation in Patients With Spinal Cord Injury	NCT01325103	1. Hospital Sao Rafael 2. Oswaldo Cruz Foundation 3. Irep Sociedade de Ensino Superior Médio e Fundamental Limitada 4. Hospital Espanhol	• Allocation: Non-Randomized • Endpoint Classification: Safety/Efficacy Study • Intervention Model: Single Group Assignment • Masking: Open Label • Primary Purpose: Treatment	18 to 50	Stem cell transplantation	• Feasibility and safety of BM stem cell transplantation in patients with spinal cord injury • Functional improvement in muscle strength • Improvement of sphincter control
February 2013	Safety and Efficacy of Stem Cell Therapy in Spinal Cord Injury	NCT02027246	Neurogen Brain and Spine Institute	• Intervention Model: Single Group Assignment • Masking: Open Label • Primary Purpose: Treatment	8 months to 63	Autologous BM mononuclear cell transplantation	• Change in clinical symptoms • Functional independence measure
May 2014	Transfer of Bone Marrow Derived Stem Cells for the Treatment of Spinal Cord Injury	NCT01162915	TCA Cellular Therapy	• Endpoint Classification: Safety Study • Intervention Model: Single Group Assignment • Masking: Open Label • Primary Purpose: Treatment	18 to 65	Autologous BM-derived mesenchymal stem cells	• Safety

(Table 1) cont.....

Completion Date of Study	Title of Study	NCT Number	Sponsors/ Collaborators	Study Design	Age Groups (Years)	Interventions	Outcome Measures
March 2015	Safety Study of Local Administration of Autologous Bone Marrow Stromal Cells in Chronic Paraplegia	NCT01909154	Puerto de Hierro University Hospital	• Endpoint Classification: Safety/Efficacy Study • Intervention Model: Single Group Assignment • Masking: Open Label • Primary Purpose: Treatment	18 to 60	Mesenchymal stromal cell	• Safety • Efficacy
June 2016	Autologous Stem Cells for Spinal Cord Injury (SCI) in Children	NCT01328860	1. James E.Baumgartner, MD 2. The Institute for Rehabilitation and Research Foundation 3. The University of Texas Health Science Center, Houston 4. M.D. Anderson Cancer Center 5. Baylor College of Medicine 6. Florida Hospital for Children 7. Memorial Hermann Health System	• Endpoint Classification: Safety/Efficacy Study • Intervention Model: Single Group Assignment • Masking: Open Label • Primary Purpose: Treatment	1 to 15	BM progenitor cells (BMPC) autologous transplantation	• ASIA - Standard neurological classification of spinal cord injury • Standard neuropathic pain rating scale
September 2016	Study the Safety and Efficacy of Bone Marrow Derived Autologous Cells for the Treatment of Spinal Cord Injury	NCT01833975	Chaitanya Hospital, Pune	• Endpoint Classification: Safety/Efficacy Study • Intervention Model: Single Group Assignment • Masking: Open Label • Primary Purpose: Treatment	18 to 55	Autologous stem cell	• Improvement in overall sensory for motor control using Frankel score • Improvement in pain sensation and significant changes in muscle tones from base line • Improvement in sensory and motor dysfunction using ASIA scale • Significant changes in muscle tones and improvement in pain sensation from base line

(Table 1) cont.....

Completion Date of Study	Title of Study	NCT Number	Sponsors/ Collaborators	Study Design	Age Groups (Years)	Interventions	Outcome Measures
May 2017	Administration of Expanded Autologous Adult Bone Marrow Mesenchymal Cells in Established Chronic Spinal Cord Injuries	NCT02570932	Puerta de Hierro University Hospital	• Endpoint Classification: Safety/Efficacy Study • Intervention Model: Single Group Assignment • Masking: Open Label • Primary Purpose: Treatment	18 to 70	Autologous mesenchymal BM cell	• Changes in IANR-SCIFRS score are considerate for the motor and sensory functions evaluation • Changes in PENN score are considerate for the motor and sensory functions evaluation • Changes in VAS score were considerate for quantification of pain • Changes in BDS score were considerate for the assessment of the functional status • Changes in GEFFNER score were considerate for the assessment of the functional status • Neurotrophic factors levels in cerebrospinal fluid (CSF) and/or number of adverse events related to treatment
Central Nervous System Stem Cells							
April 2015	Study of Human Central Nervous System Stem Cells (HuCNS-SC) in Patients With Thoracic Spinal Cord Injury	NCT01321333	StemCells, Inc.	• Endpoint Classification: Safety/Efficacy Study • Intervention Model: Single Group Assignment • Masking: Open Label • Primary Purpose: Treatment	18 to 60	HuCNS-SC cells	• Types and frequencies of adverse events and serious adverse events (SAE)
March 2016	Safety Study of Human Spinal Cord-derived Neural Stem Cell Transplantation for the Treatment of Chronic SCI	NCT01772810	Neuralstem Inc.	• Endpoint Classification: Safety Study • Intervention Model: Single Group Assignment • Masking: Open Label • Primary Purpose: Treatment	18 to 65	Human spinal cord stem cells	• The primary outcome measure of the study is to determine the incidence of adverse events including SAEs and clinically significant laboratory abnormalities • The secondary outcome measure of the study was to evaluate the graft survival in the transplant site as determined by MRI (with and without gadolinium contrast) and *via* autopsy, if one is completed

(Table 1) cont.....

Completion Date of Study	Title of Study	NCT Number	Sponsors/ Collaborators	Study Design	Age Groups (Years)	Interventions	Outcome Measures
May 2016	Study of Human Central Nervous System (CNS) Stem Cell Transplantation in Cervical Spinal Cord Injury	NCT02163876	StemCells, Inc.	• Allocation: Randomized • Endpoint Classification: Safety/Efficacy Study • Intervention Model: Parallel Assignment • Masking: Single Blind (Outcomes Assessor) • Primary Purpose: Treatment	18 to 60	HuCNS-SC cells	• Change from baseline in ISNCSCI upper extremity motor scores • Number of participants with serious and non-serious adverse events
May 2016	Long-Term Follow-Up of Transplanted Human Central Nervous System Stem Cells (HuCNS-SC) in Spinal Cord Trauma Subjects	NCT01725880	StemCells, Inc.	• Observational Model: Cohort • Time Perspective: Prospective	18 to 65	Observation	• ASIA impairment scale improvement
October 2018	Neural Stem Cell Transplantation in Traumatic Spinal Cord Injury	NCT02326662	1. Federal Research Clinical Center of Federal Medical & Biological Agency, Russia 2. Novagenesis Foundation 3. Ophiuchus Technologies AG	• Allocation: Non-Randomized • Endpoint Classification: Safety/Efficacy Study • Intervention Model: Factorial Assignment • Masking: Open Label • Primary Purpose: Treatment	18 to 50	Autologous Stem Cell	• Feasibility and safety of autologous neural stem cell transplantation in patients with complete traumatic spinal cord injury as measured by reported side effects and surgery procedure • 324-point ASIA and other scoring including muscle control and voluntary movement and improvement of sphincter control • MRI spinal cord
Stem Cell Combination/Other Intervention							
December 2016	NeuroRegen Scaffold™ With Stem Cells for Chronic Spinal Cord Injury Repair	NCT02352077	1. Chinese Academy of Sciences 2. Affiliated Hospital of Logistics University of CAPF 3. The First Affiliated Hospital of Soochow University 4. First Hospitals affiliated to the China PLA General Hospital	• Endpoint Classification: Safety/Efficacy Study • Intervention Model: Single Group Assignment • Masking: Open Label • Primary Purpose: Treatment	18 to 65	NeuroRegen scaffold with BMMCs or MSCs	• Safety evaluation number of patients with adverse events • Improvements in neurophysiological measures somatosensory evoked potentials (SSEP) and motor evoked potentials (MEP) monitoring was assessed

(Table 1) cont.....

Completion Date of Study	Title of Study	NCT Number	Sponsors/ Collaborators	Study Design	Age Groups (Years)	Interventions	Outcome Measures
							Improvements in ASIA impairment scale • Improvements in independence measures and quality of life • Improvements in pain assessed based on visual analog scale • Improvements in urinary and bowel function assessed based on bladder pressure monitor before and after transplantation • Changes at the transplantation site in spinal cord by MRI
December 2017	NeuroRegen Scaffold™ Combined With Stem Cells for Chronic Spinal Cord Injury Repair	NCT02688049	1. Chinese Academy of Sciences 2. Affiliated Hospital of Logistics University of CAPF	• Allocation: Randomized • Endpoint Classification: Safety/Efficacy Study • Intervention Model: Parallel Assignment • Masking: Double Blind (Subject, Outcomes Assessor) • Primary Purpose: Treatment	18 to 65	• NeuroRegen scaffold/mesenchymal stem cells • NeuroRegen scaffold/neural stem cells	• Improvements in AIS • Improvements in SSEP • Improvements in MEP • Improvements in independence measures • Changes at the transplantation site in spinal cord by MRI • Improvements in urinary and bowel function • Safety and tolerability assessed by adverse events
March 2018	Safety and Feasibility Study of Cell Therapy in Treatment of Spinal Cord Injury	NCT02237547	Translational Biosciences	• Endpoint Classification: Safety/Efficacy Study • Intervention Model: Single Group Assignment • Masking: Open Label • Primary Purpose: Treatment	18 to 50	Intravenous and intrathecal human umbilical cord tissue-derived mesenchymal stem cells and BM mononuclear cells	• Number of patients with adverse events • Number of subjects with a change in ASIA score from baseline • Number of subjects with a change in Frankel scale score from baseline
September 2018	Dose Escalation Study of AST-OPC1 in Spinal Cord Injury	NCT02302157	Asterias Biotherapeutics, Inc.	• Endpoint Classification: Safety Study • Intervention Model: Single Group Assignment • Masking: Open Label • Primary Purpose: Treatment	18 to 69	AST-OPC1	• Number of adverse events within 1 year (365 days) that are related to AST-OPC1 injection • Neurological function as measured by upper extremity motor scores and motor level on international standards for neurological classification of spinal cord • Injury examinations at 30, 60, 90, 180, and 365 days after injection of AST-OPC1

(Table 1) cont.....

Completion Date of Study	Title of Study	NCT Number	Sponsors/ Collaborators	Study Design	Age Groups (Years)	Interventions	Outcome Measures
December 2021	Transplantation of Autologous Bone Marrow or Leukapheresis-Derived Stem Cells for Treatment of Spinal Cord Injury	NCT02687672	Stem Cells Arabia	• Allocation: Randomized • Endpoint Classification: Safety/Efficacy Study • Intervention Model: Parallel Assignment • Masking: Open Label • Primary Purpose: Treatment	5 to 50	Stem cells	• Overall clinical improvement in sensory and motor functions using ASIA • Improvement in urine and stool incontinence using a questionnaire • Functional improvement in impotence and previous sexual erection status (male patients) using a questionnaire • Improvement in quality of life using a questionnaire • Improvement in personal independence and productivity using questionnaire
Source: Clinicaltrials.gov – Neural stem cells in spinal cord injury							

Motor Evoked Potentials

Dai *et al.* studied MEPs in patients who showed improvement in excitation of nerve impulses and an increase in the number of nerve fibers involved in impulse formation from zero. MEPs also demonstrated shortened latency with no improvement in amplitude. This indicated improvement in the degree of conduction function with no improvement in the degree of excitement of the nerve fibers [328]. Jiang *et al.* found that there was 75% improvement in ASIA grading [324].

Effects on Muscle Tone

Jiang *et al.* study showed that abnormal muscle tone, either hypertonia or hypotonia were associated with the location of the lesion. Posterior funicular lesions caused hypotonia. Pyramidal tract lesions caused hypertonia. With limitations of a small sample size, they found that all patients with hypertonia benefited from stem cell therapy. There was improvement in 33% with hypotonia [324].

Effects on Sphincteric Tone

Studies were conflicting for sphincteric dysfunction. In patients with urinary dysfunction after SCI and SPSCT, there was nil recovery in urinary function between the treatment and control group [330]. Jiang *et al.* found that two patients experienced full recovery and one patient experienced urinary incontinence [324]. Dai *et al.* found significant improvement in the quantity of residual urine in the treatment group with nil improvement in urine perception and urine control [109]. Studies for bowel dysfunction SPSCT showed that 75% of patients experienced improvement in bowel function. Restoration of intestinal secretions was a significant factor in improving bowel dysfunction [324, 330].

Summary of Findings

Several studies have shown positive results for ExSCs after a SCI. Barnabe-Heider *et al.* demonstrated that there were significant limitations to these studies. Some limitations discussed were varying degrees of characterization of the transplanted cells, use of different injury models, implantation occurring at different times after injury and small sample sizes [116]. This made the data collected difficult to validate, extrapolate and generalize to the population. Several studies have been refuted based on alternative explanations to the expression of neural markers since both hematopoietic stem cells and MSCs had the potential to generate neurons and glial cells [331]. Parr *et al.* demonstrated improvement in neuronal function after replacing MSCs and macrophages which did not generate glial cells or neurons. The exogenous BM derived NSCs did not quantitatively produce neurons. Therefore, it was postulated that the improvement in neuronal function was secondary to a modulating effect of the inflammatory response through trophic factor support. Moreover, transplanted cells were more likely to be indirectly involved in spinal cord repair [311].

ExSC therapy has been at the clinical stage. The short term data is supportive. However, the long term data is limited and controversial. Although, the cumulative data for eNSCs is preclinical, there seems to be more promise long term for SCI patients.

FACTORS INFLUENCING THE EFFICACY OF STEM CELL TRANSPLANTATION

Combination Therapies

Due to the structural barriers of regeneration newer methods such as nervous tissue/synthetic bridges have been developed to bypass the injured area [162, 332, 333]. These bridges were biodegradable polymer scaffolds, which were both structural supports and a guide to axonal regeneration [162] (*Refer to* Fig. (**5**)). 85:15 polyactic acid:polyglycolic acid scaffolds were used most extensively in peripheral nerve regeneration [334, 335]. Alpha-hydroxy acid polymer scaffolds were used in both peripheral nerve injuries and SCI [336 - 341]. Multichannel scaffolds were used to prevent lumen collapse seen with single lumen scaffolds [335, 340, 342]. A unique approach to scaffolds was a matrix composed of peptide amphiphile molecules known to self-assemble from aqueous solution into

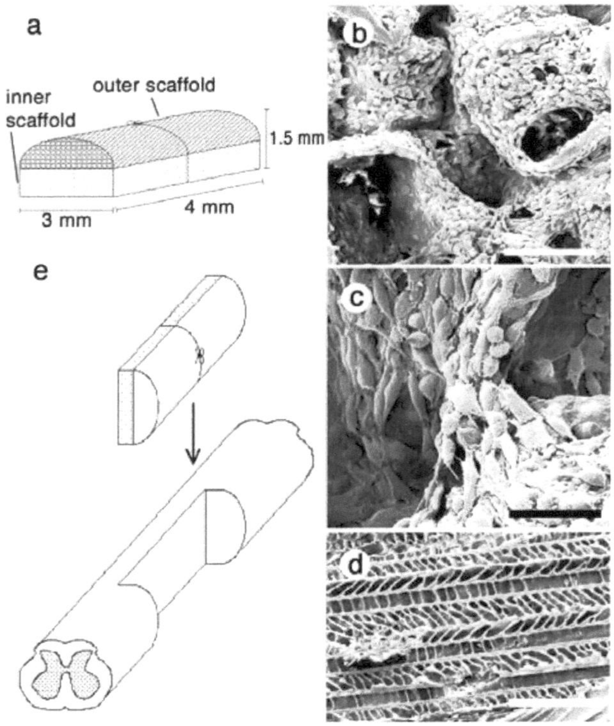

Fig. (5). Polymer scaffolds - Functional recovery post spinal cord injury mediated by a unique polymer scaffold seeded with neural stem cells.
(a) Components of the scaffold - inner and outer scaffolds
(b) Inner scaffold seeded with neural stem cells (scale: Fig. (**b**) - 200μm and Fig. (**c**) - 50μm)
(c) Outer section of scaffold fabricated with solid-liquid phase separation technique producing long, axially oriented pores for axonal guidance as well as radial pores to allow fluid transport and inhibit ingrowth of scar tissue (scale: 100μm)
(e) Demonstrated surgical insertion of implant in spinal cord
Teng YD, Lavik EB, Xianlu Q, Park KI, Ourednik J, Zurakowski D, *et al.* Functional recovery following traumatic spinal cord injury mediated by a unique polymer scaffold seeded with neural stem cells. Proceedings of the National Academy of Sciences. 2002 Mar; 99(5):3024-3029; Copyright © 2002, National Academy of Sciences, U.S.A.

cylindrical nanofibers [343 - 345]. They had a built in signaling sequence into the backbone which instructed surrounding cells. Therefore different epitopes such as isoleucine-lysine-valine-alanine-valine peptide amphiphile could reduce gliosis and cell death at a distance of 400 micrometers [317]. Olson *et al.* demonstrated that seeding of NSCs and SCs into multichannel biodegradable polymer scaffolds promoted axonal regeneration, especially in the SC group. Immunohistochemical analysis of the scaffold 1 month after transplantation showed extensive nestin and neurofilament staining in the NSC group [162]. Friedman *et al.* demonstrated that implantation of the scaffold with NSCs improved functional recovery compared to NSCs alone [346]. The polymer scaffold used consisted of an inner layer representing the gray matter which was seeded with NSCs. The outer portion represented the white matter which inhibited ingrowth of scar tissue with radial porosity allowing fluid transport. Scaffold plus NSCs maintained for 1 year demonstrated continuing recovery as per the Basso-Beattie-Bresnahan scale. The scaffold alone demonstrated functional recovery through decreasing epidural and glial scar formation. NSCs functioned as trophic support in this process. They were less likely involved in cellular replacement. Functional recovery was defined as regeneration of approximately 5-10% of corticospinal axons [347]. Studies have demonstrated 1% axonal regeneration. This discrepancy was due to the complicated axonal regeneration process. This involved the axons traversing the scaffold, exiting the opposite scaffold-cord interface, reaching the appropriate targets and forming the appropriate functional connections [162].

The expression of chondroitin sulfate proteoglycans post injury have been found to be detrimental to the survival of transplanted NPCs in the chronically injured spinal cord [71]. Fouad *et al.* demonstrated that chondroitin sulfate proteoglycans inhibited migratory outgrowth and the velocity of cells from the neurosphere [348]. Chondroitin sulfate proteoglycan inhibition supported neural plasticity, improved functional recovery and promoted axonal regeneration in acute and subacute SCI [349 - 355]. The addition of NSC transplantation was involved in stimulating neurotrophic factors causing neurite outgrowth, axonal regeneration and repair of the disrupted neuronal circuitry through remyelination [69, 131, 180]. Karimi-Abdolrezaee *et al.* found in a chronic model of compressive SCI in mice that after intrathecal infusion of chondroitinase ABC for 7 days followed by transplantation of NPCs with EGF, FGF and PDGF supplementation there was improved long term survival, migration and integration of NPCs in chronic SCIs [71]. Ikegami *et al.* found that chondroitin sulfate proteoglycan-coated wells that were pretreated with chondroitinase ABC showed that cells recovered their migration, but the survival and differentiation assays did not affect the viability or differentiation of NSCs. Moreover, a significant difference in the GAP-43 positive fiber outgrowth at the site of the lesion was also demonstrated in mice treated with a combination of chondroitinase ABC with NSCs [356].

Cytokines

The immune response cascade after a SCI is overwhelming which can lead to secondary damage with an effect on the functionality of NSCs. There were many factors from different groups that have shown to have stimulatory or inhibitory effects on NSCs. The underlying basis of these mechanisms have been discussed below.

As a result of interferon (IFN)-gamma secretion, a post inflammatory response to SCI and remyelination failure occurred due to apoptosis of vulnerable OPCs in cultured mice [357 - 364]. There have been several mechanisms proposed for IFN-gamma induced damage or death of OPCs in literature. This was likely due to increased expression of chemokine (C-X-C motif) ligand (CXCL)-10 messenger RNA transcripts and protein binding to the C-X-C chemokine receptor (CXCR)-3 on OPCs [365]. In animal models blocking of CXCL10 hindered T-lymphocyte accumulation in the CNS. This lead to decreased IFN-gamma expression and improved clinical outcome [366 - 369]. Tirotta *et al.* found that there was a presence of local CXCL10 in the white matter tracts which induced damage and death in cells of oligodendrocyte lineage [365]. Liu *et al.* confirmed that blocking CXCL10 signaling in mice infected with a neurotropic JHM strain of hepatitis virus lead to preservation of myelin-producing cells and remyelination [367]. The other mechanisms discussed were independent of CXCL10. IFN-gamma induced OPC death which was secondary to ER-related stress lead to dysfunction of myelin formation [370 - 372]. Tirotta *et al.* demonstrated that CXCR2 signaling with addition of CXCL1 in a “tonic manner” protected OPCs from IFN-gamma and CXCL10 apoptosis [365].

T-cells had a promotive effect on neurogenesis through stimulation of microglia. Stimulation by T-helper-1-IFN-gamma lead to neurogenesis in co-cultured NSCs [373]. Stimulation by IL-4 lead to neurogenesis towards oligodendrogenesis. IL-4 was also involved in the expression of IGF-1 in microglia involved in neuroprotection and neurogenesis [374]. Myelin specific T-cells also promoted neuroprotection and neurogenesis in the adult spinal cord [375].

Both glucocorticoid and mineralocorticoid receptors were present on the spinal cord. The use of synthetic glucocorticoids affected the proliferation of macrophages and microglia. This had a deleterious effect on the proliferator effect of eNSCs. It also altered the inflammatory environment through downregulating the amount of NG2-positive precursor cells involved in oligodendrocyte repair [376].

Neuronal differentiation of NSCs was promoted by IL-1b and IL-6. It was inhibited by insulin-like growth factor binding protein 6 and proteoglycan decorin [216]. Cytokine response to SCI has been important in the growth and differentiation of NSCs. There was upregulation of TNF-alpha, IL-6, IL-1a and IL-1b after a SCI [32]. Vela *et al.* demonstrated the inhibitory effects of IL-1b on the proliferation and differentiation of oligodendrocytes [377]. TNF-alpha, IL-1b and IL-6 were the most acutely reacting cytokines (in minutes). *In-vivo*, mice experiments have shown that TNF-alpha stimulate TNF-RI and RII, which in turn stimulate oligodendrocyte proliferation and promote survival [378, 379].

Studies by several groups have investigated the effects of IL-6 on NSCs after a SCI. IL-6 has been found to have an astrogliosis effect in the spinal cord [380, 381]. Improved functional recovery was demonstrated in mice after blocking IL-6 receptors with antibodies [381]. *In-vitro*, IL-6 expression caused more astrocytic differentiation. Blocking of IL-6 lead to its inhibition [380]. *In-vivo* after blocking IL-6 there was a decrease in number of new astrocytes. Leukemia inhibitory factor promoted self-renewal capacity of NSCs [382, 383]. In the spinal cord, increased nestin-positive cells were found after systemic administration of this factor [208].

Many growth factors have been implicated in the stimulation and proliferation of eNSCs in various CNS locations. Although, the full spectrum of effects of cytokines is not fully understood. The current understanding of the growth factors and the inflammatory cascade as outlined above has shown promise in regulating this complex process.

Neurovascular Niche

Parallels have shown to exist between the development of the nervous system and the vascular system [384]. The nervous system has shown to consist of neurons and surrounding glia. The blood vessels have shown to consist of endothelial cells and supporting smooth muscle cells or pericytes. NSCs and endothelial cells were present in similar locations in the germinal zones [385, 386]. There has been evidence of synchronized development and dependence of the nervous and vascular systems. These areas have demonstrated active neurogenesis during adult life with coexistence of angioblasts and neuroblasts. In regions of the CNS which were going through post-natal neurogenesis. Endothelial precursor cells, capillaries and NSCs were in close proximity forming the vascular niche [387, 388]. Therefore, the neurovascular niche described the bidirectional interaction between the vessels and the NSCs. Numerous vascular growth factors implicated in neuronal growth and differentiation were the vascular endothelial growth factor (VEGF), ephrins, neurotrophins and angiopoietins. In songbirds, NSCs proliferated in clusters surrounding the dividing capillaries [389].

Growth Factors

Vascular Endothelial Growth Factor

VEGF was involved in maintaining and regulating the integrity of both the central and peripheral nervous systems [384]. Also, it was an important biphasic mediator of the "oligovascular niche" through the endothelial-oligodendrocyte trophic coupling [390]. VEGF has shown to acutely worsen blood-brain-barrier permeability [391, 392]. Delayed VEGF action was shown to regulate endothelium-oligodendrocyte interaction to help preserve and alter white matter function [390]. Moreover, VEGF164/5 bonded and exerted effects through the neuropilin receptor (Nrp) 1-57-59 (semaphorin family of neural ligands). These receptors were involved in the formation of both the vascular and nervous systems [393 - 396]. VEGF family molecules have a neurotrophic, neuroprotective and neurogenic function influencing cell survival and axonal growth. VEGF and Semaphorin 3A (Nrp1 ligand in the semaphorin family) antagonistically affected NSCs, endothelial cells and the expression of Nrp1 receptors in the arterial beds [397].

Ephrins

Ephrins were involved in the patterning process of the nervous and vascular systems through the tyrosine kinase receptor family [384, 398 - 400]. Neurotrophins consisted of nerve growth factors, BDNF and neurotrophin-3,4/5,6,7 [346]. The growth factors stimulated neural growth, survival and regeneration in the adult and injured nervous system [401]. Endothelial cells expressed various neurotrophins such as BDNF, NGF, p75 receptors and TrkA. BDNF stimulated OPC proliferation and maturation into oligodendrocytes through the TrkB receptor and ERK pathway [390, 402 - 405]. BDNF regulated the oligodendrocyte lineage cells after demyelination. It had an effect on NSC differentiation into oligodendrocyte lineage through activation of the Wnt/beta catenin signaling pathway [406, 407]. McTigue *et al.* found that BDNF secreting fibroblast grafts enhanced BrdU positive oligodendrocytes through a growth-permissive substrate for local, peripheral and descending axons demonstrating significantly more axonal ingrowth [404]. This type of graft uniquely consisted of enhanced cholinergic fiber growth which in response to BDNF demonstrated extensive growth of injured spinal motor neurons [408].

Fibroblast Growth Factor-2

FGF-2 induced proliferation of late-stage OPCs. They blocked the process of terminal differentiation into mature oligodendrocytes [409, 410]. FGF-2 interacted with PDGF to increase OPC proliferation. FGF-2 also enhanced migration of pre-OPC from oligospheres, but not of late stage OPCs [411]. Although still controversial, FGF-2 has been shown to prevent OPCs from succumbing to apoptotic stress [412].

Adrenomedullin

Adrenomedullin (AM) is a product of pheochromocytoma cells that has shown to modulate and have a protective effect towards oligodendrocytes in white matter through AM receptors [413, 414]. AM has been found to be an important factor in modulating the differentiation and proliferation of NSCs [415]. AM membrane receptors and the phosphatidylinositol 3-kinase pathway could alter the proportion of neurons, astrocytes and oligodendrocytes from undifferentiated progenitor cells. AM was involved in further increasing growth factors such as VEGF and FGF [416 - 419]. AM was found to increase the messenger RNA expression of VEGF and FGF in human umbilical vein endothelial cells [416]. The AM receptor and the phosphatidylinositol 3-kinase pathway were involved in the upregulation of VEGF and FGF growth factors. VEGF in the CNS was known to directly activate NP cells. This promoted adult neurogenesis [420]. Both VEGF and FGF acted synergistically through the enhancement of endogenous PDGF-B–PDGF--beta in the formation of new non-leaky vessels [421].

Endothelial Growth Factor

EGF has been shown to promote NSCs in the CNS. In the spinal cord both EGF and basic FGF were required (*Refer to* Fig. (**6**)). It was proposed that thoracic

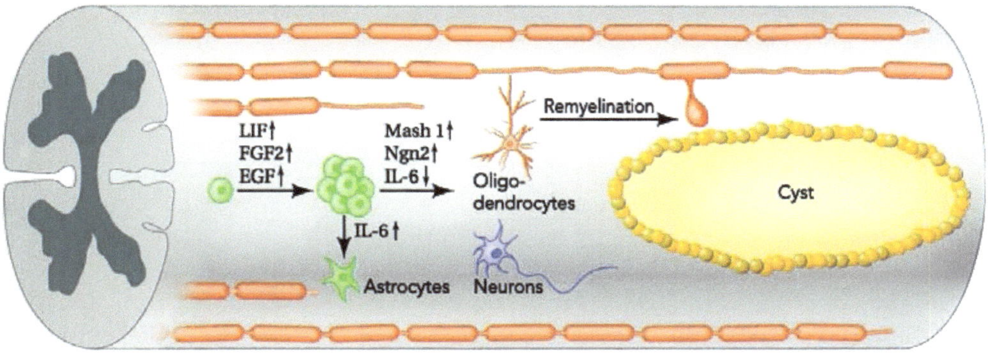

Fig. (6). Endogenous progenitor cells - Post spinal cord injury and experimental manipulations influencing endogenous progenitor cells.
This figure demonstrated various points of intervention influencing endogenous progenitors to help repair and recover from spinal cord injury through cell replacement, remyelination and supporting trophic factors. Physiology. 2008 Oct 1; 23(5):296-304; Copyright © 2008, The American Physiological Society

spinal cord cell division was stimulated by activation of the basic FGF receptor with adequate heparin. However, the proliferation of secondary stem cells could be stimulated with either basic FGF or EGF [422, 423]. Moreover, the intrathecal administration consisted of EGF, FGF-2, EGF plus FGF-2 *versus* artificial CSF. 42 days post moderate SCI, EGF plus FGF-2 group showed a greater functional recovery than the artificial CSF. This group at 14 days showed increased counts of astrocytes and ECs with increased migration of ECs into the adjacent spinal cord. Histologically, EGF and FGF-2 did not improve nestin expression, neuron-specific marker NeuN, OPCs, microglia or macrophage counts [424]. Martens *et al.* demonstrated that infusions of EGF plus FGF-2 plus heparin into the fourth ventricle resulted in 33.5-fold increase in the number of BrdU positive cells around the central canal of the spinal cord [192]. Seven weeks after infusion, only new astrocytes and oligodendrocytes were present in the spinal cord. After administration of exogenous growth factors BrdU positive neurons were found in the spinal cord [192].

Angiopoietin

Angiopoietin consisted of ang1,2 and 3 isoforms in mice that bounded with tyrosine kinase (Tie2) and integrin (B1) group of receptors [425, 426]. Tie2 interactions were involved in organization of neuronal processes of the mouse brain. Beta-1 integrin interactions mediated neurite and axon growth, neuroblast migration, glial development and synapse formation [384, 427]. In the CNS, beta-1 integrin receptors were present on meningeal cells, neurons, astrocytes, oligodendrocytes, radial glia and endothelial cells. *In-vivo*, transforming growth factor - beta was expressed by the spinal cord [428]. It was involved with migration of cultured OPC when maintained on astrocytes. It also modulated the cell adhesion property of OPCs [429, 430].

Transcription Factors

Transcription factors have not yet been successful in producing persistent neuronal progeny. However, after a SCI understanding their role in regulation and differentiation into various lineages has been crucial.

Ohori *et al.* found that transcription factors such as neurogenin2 increased production of new neurons. Mash-1 increased production of oligodendrocytes after a SCI event [194] (*Refer to* Fig. (**6**)). After a few weeks these new neurons and oligodendrocytes were not traceable suggesting that they integrated themselves. Olig2 is important in oligodendrocyte specification throughout the neural axis [431 - 433]. Olig2 and Mash-1 were shown to give rise to GABA neurons, and later to oligodendrocytes regulated by BMP [434, 435]. Yin Yang 1 was shown to regulate early stages of oligodendrocyte differentiation, which could potentially be useful in altering the fate of NSCs towards oligodendrocytes [436].

Electrical Stimulation

The use of endogenous and exogenous electrical currents for CNS regeneration after a SCI through NSC manipulation has been a promising area of research. Endogenous electrical activation has shown to be beneficial in spontaneous regeneration through neural tube formation, axonal regeneration and neural cell migration [437 - 439]. Becker *et al.* showed that after functional electrical stimulation (FES) there was an 82-86% increase in the NP cell formation in rats after chronic SCI [439]. Also, there was a 2-fold increase in cells that expressed nestin and tripotential progenitors. FES upregulated EGF and FGF leading to proliferation of NSCs/NP cells with a peak in NP cells seen in 7 days [440]. This suggested that controlled electrical activation promoted regeneration after injury [439]. Moreover, electroacupuncture was an invasive electrical stimulation technique incorporating traditional acupuncture with electrical currents to incite needle specific acupoints [440]. EA stimulated NSC/NP cell proliferation through the BDNF/VEGF signaling pathways [441].

Electrical current provided "directedness value of migration" which was 2-2.5 times greater than the control group [440]. Directedness value quantified directional migration of neuroblasts towards the cathode. In epilepsy models, abnormal brain electrical activity increased NSC populations by 163% in the dentate subgranular proliferative zone [442]. Exogenous electric currents influenced NSCs and NPCs to migrate to the site of lesion in a voltage or duration approach [443 - 446]. Electrical stimulation augmented the differentiation of NSCs and NP cells [443, 444, 447, 448]. Astrocyte activation was observed *in-vivo* when NSCs and NP cells were present. This led to the formation of a glial scar hindering neural regeneration [238] (*Refer to* Fig. (**4**)). Kobelt *et al.* found that electrical currents stimulated mature neuronal differentiation, which was further amplified by interferon-gamma biochemical mediums [448].

CONCLUSION

As discussed above, the tremendous growth in understanding the utility of ExSCs, eNSCs and combination therapies in acute and chronic spinal cord injuries has given a new direction to SCI research. Before quests for new therapies are undertaken it is imperative to understand the physiology of injury and the repair process of the spinal cord. As highly impactful research has occurred globally by various well established research groups. Knowledge about NSCs in SCI has significantly matured. Some recent studies have contradicted previous beliefs, and have influenced a new paradigm in the field. There needs to be a continuous strive towards taking a holistic approach in understanding that the process and application of NSC therapy is a dynamic balance of numerous exogenous and endogenous factors and micro-environments. ExSCs initially was an exciting area of research, however the emerging data has been controversial. It has shown inconsistent outcomes and many associated ethical issues. Although still at the animal stage, eNSC interventions have a more optimistic outlook. Stimulation of intrinsic cells, growth factors and transcription factors have shown promise in recovery after a SCI. Finally, newer more innovative techniques through electrical stimulation have given insight into alternative techniques to NSC therapy in SCI. In sum, although there have been great leaps in understanding SCIs and their interventions at the basic science level. The translational utility of NSCs from the basic sciences into its clinical application still continues to be an emerging entity.

CONSENT FOR PUBLICATION

Not applicable.

CONFLICT OF INTEREST

The authors confirm that there is no conflict of interest to declare for this publication.

ACKNOWLEDGEMENTS

We are thankful to Dr. Richard E Coggeshall for his review of the chapter.

REFERENCES

[1] Cripps RA, Lee BB, Wing P, Weerts E, Mackay J, Brown D. A global map for traumatic spinal cord injury epidemiology: towards a living data repository for injury prevention. Spinal Cord 2011; 49(4): 493-501.
[http://dx.doi.org/10.1038/sc.2010.146] [PMID: 21102572]

[2] Wyndaele M, Wyndaele JJ. Incidence, prevalence and epidemiology of spinal cord injury: what learns a worldwide literature survey? Spinal Cord 2006; 44(9): 523-9.
[http://dx.doi.org/10.1038/sj.sc.3101893] [PMID: 16389270]

[3] Sebastià-Alcácer V, Alcanyis-Alberola M, Giner-Pascual M, Gomez-Pajares F. Are the characteristics of the patient with a spinal cord injury changing? Spinal Cord 2014; 52(1): 29-33.
[http://dx.doi.org/10.1038/sc.2013.128] [PMID: 24190078]

[4] Furlan JC, Fehlings MG. Cardiovascular complications after acute spinal cord injury: pathophysiology, diagnosis, and management. Neurosurg Focus 2008; 25(5): E13.
[http://dx.doi.org/10.3171/FOC.2008.25.11.E13] [PMID: 18980473]

[5] DeVivo MJ, Chen Y. Trends in new injuries, prevalent cases, and aging with spinal cord injury. Arch Phys Med Rehabil 2011; 92(3): 332-8.
[http://dx.doi.org/10.1016/j.apmr.2010.08.031] [PMID: 21353817]

[6] Burney RE, Maio RF, Maynard F, Karunas R. Incidence, characteristics, and outcome of spinal cord injury at trauma centers in North America. Arch Surg 1993; 128(5): 596-9.
[http://dx.doi.org/10.1001/archsurg.1993.01420170132021] [PMID: 8489395]

[7] DeVivo MJ, Biering-Sørensen F, New P, Chen Y. Standardization of data analysis and reporting of results from the International Spinal Cord Injury Core Data Set. Spinal Cord 2011; 49(5): 596-9.
[http://dx.doi.org/10.1038/sc.2010.172] [PMID: 21135863]

[8] Sekhon LH, Fehlings MG. Epidemiology, demographics, and pathophysiology of acute spinal cord injury. Spine 2001; 26(24) (Suppl.): S2-S12.
[http://dx.doi.org/10.1097/00007632-200112151-00002] [PMID: 11805601]

[9] Albert T, Ravaud JF. Rehabilitation of spinal cord injury in France: a nationwide multicentre study of incidence and regional disparities. Spinal Cord 2005; 43(6): 357-65.
[http://dx.doi.org/10.1038/sj.sc.3101717] [PMID: 15741980]

[10] Pickett W, Simpson K, Walker J, Brison RJ. Traumatic spinal cord injury in Ontario, Canada. J Trauma 2003; 55(6): 1070-6. [http://dx.doi.org/10.1097/01.TA.0000034228.18541.D1] [PMID: 14676653]

[11] Acton PA, Farley T, Freni LW, Ilegbodu VA, Sniezek JE, Wohlleb JC. Traumatic spinal cord injury in Arkansas, 1980 to 1989. Arch Phys Med Rehabil 1993; 74(10): 1035-40. [http://dx.doi.org/10.1016/0003-9993(93)90058-I] [PMID: 8215853]

[12] Devivo MJ. Epidemiology of traumatic spinal cord injury: trends and future implications. Spinal Cord 2012; 50(5): 365-72. [http://dx.doi.org/10.1038/sc.2011.178] [PMID: 22270188]

[13] New PW, Rawicki HB, Bailey MJ. Nontraumatic spinal cord injury: demographic characteristics and complications. Arch Phys Med Rehabil 2002; 83(7): 996-1001. [http://dx.doi.org/10.1053/apmr.2002.33100] [PMID: 12098161]

[14] McDonald JW, Sadowsky C. Spinal-cord injury. Lancet 2002; 359(9304): 417-25. [http://dx.doi.org/10.1016/S0140-6736(02)07603-1] [PMID: 11844532]

[15] Tator CH. Update on the pathophysiology and pathology of acute spinal cord injury. Brain Pathol 1995; 5(4): 407-13. [http://dx.doi.org/10.1111/j.1750-3639.1995.tb00619.x] [PMID: 8974623]

[16] Sabapathy V, Tharion G, Kumar S. Cell Therapy Augments Functional Recovery Subsequent to Spinal Cord Injury under Experimental Conditions. Stem Cells Int 2015; 2015: 132172. [http://dx.doi.org/10.1155/2015/132172] [PMID: 26240569]

[17] Siddiqui AM, Khazaei M, Fehlings MG. Translating mechanisms of neuroprotection, regeneration, and repair to treatment of spinal cord injury. Prog Brain Res 2015; 218: 15-54. [http://dx.doi.org/10.1016/bs.pbr.2014.12.007] [PMID: 25890131]

[18] Rowland JW, Hawryluk GW, Kwon B, Fehlings MG. Current status of acute spinal cord injury pathophysiology and emerging therapies: promise on the horizon. Neurosurg Focus 2008; 25(5): E2. [http://dx.doi.org/10.3171/FOC.2008.25.11.E2] [PMID: 18980476]

[19] McDonald JW, Belegu V. Demyelination and remyelination after spinal cord injury. J Neurotrauma 2006; 23(3-4): 345-59. [http://dx.doi.org/10.1089/neu.2006.23.345] [PMID: 16629621]

[20] Radojicic M, Reier PJ, Steward O, Keirstead HS. Septations in chronic spinal cord injury cavities contain axons. Exp Neurol 2005; 196(2): 339-41. [http://dx.doi.org/10.1016/j.expneurol.2005.08.009] [PMID: 16153640]

[21] Nashmi R, Fehlings MG. Mechanisms of axonal dysfunction after spinal cord injury: with an emphasis on the role of voltage-gated potassium channels. Brain Res Brain Res Rev 2001; 38(1-2): 165-91. [http://dx.doi.org/10.1016/S0165-0173(01)00134-5] [PMID: 11750932]

[22] Fehlings MG, Agrawal S. Role of sodium in the pathophysiology of secondary spinal cord injury. Spine 1995; 20(20): 2187-91. [http://dx.doi.org/10.1097/00007632-199510001-00002] [PMID: 8545710]

[23] Kakulas BA. Neuropathology: the foundation for new treatments in spinal cord injury. Spinal Cord 2004; 42(10): 549-63. [http://dx.doi.org/10.1038/sj.sc.3101670] [PMID: 15346131]

[24] Braughler JM, Duncan LA, Chase RL. Interaction of lipid peroxidation and calcium in the pathogenesis of neuronal injury. Cent Nerv Syst Trauma 1985; 2(4): 269-83. [http://dx.doi.org/10.1089/cns.1985.2.269] [PMID: 2424624]

[25] Wagner FC Jr, Stewart WB. Effect of trauma dose on spinal cord edema. J Neurosurg 1981; 54(6): 802-6. [http://dx.doi.org/10.3171/jns.1981.54.6.0802] [PMID: 6165810]

[26] Vawda R, Fehlings MG. Mesenchymal cells in the treatment of spinal cord injury: current & future perspectives. Curr Stem Cell Res Ther 2013; 8(1): 25-38.
[http://dx.doi.org/10.2174/1574888X11308010005] [PMID: 23270635]

[27] Kakulas BA. A review of the neuropathology of human spinal cord injury with emphasis on special features. J Spinal Cord Med 1999; 22(2): 119-24.
[http://dx.doi.org/10.1080/10790268.1999.11719557] [PMID: 10826269]

[28] Norenberg MD, Smith J, Marcillo A. The pathology of human spinal cord injury: defining the problems. J Neurotrauma 2004; 21(4): 429-40.
[http://dx.doi.org/10.1089/089771504323004575] [PMID: 15115592]

[29] Davalos D, Grutzendler J, Yang G, *et al.* ATP mediates rapid microglial response to local brain injury *in vivo.* Nat Neurosci 2005; 8(6): 752-8.
[http://dx.doi.org/10.1038/nn1472] [PMID: 15895084]

[30] David S, Kroner A. Repertoire of microglial and macrophage responses after spinal cord injury. Nat Rev Neurosci 2011; 12(7): 388-99.
[http://dx.doi.org/10.1038/nrn3053] [PMID: 21673720]

[31] Donnelly DJ, Popovich PG. Inflammation and its role in neuroprotection, axonal regeneration and functional recovery after spinal cord injury. Exp Neurol 2008; 209(2): 378-88.
[http://dx.doi.org/10.1016/j.expneurol.2007.06.009] [PMID: 17662717]

[32] Pineau I, Lacroix S. Proinflammatory cytokine synthesis in the injured mouse spinal cord: multiphasic expression pattern and identification of the cell types involved. J Comp Neurol 2007; 500(2): 267-85.
[http://dx.doi.org/10.1002/cne.21149] [PMID: 17111361]

[33] Tator CH, Fehlings MG. Review of the secondary injury theory of acute spinal cord trauma with emphasis on vascular mechanisms. J Neurosurg 1991; 75(1): 15-26.
[http://dx.doi.org/10.3171/jns.1991.75.1.0015] [PMID: 2045903]

[34] Boland RA, Lin CS, Engel S, Kiernan MC. Adaptation of motor function after spinal cord injury: novel insights into spinal shock. Brain 2011; 134(Pt 2): 495-505.
[http://dx.doi.org/10.1093/brain/awq289] [PMID: 20952380]

[35] Yip PK, Malaspina A. Spinal cord trauma and the molecular point of no return. Mol Neurodegener 2012; 7: 6.
[http://dx.doi.org/10.1186/1750-1326-7-6] [PMID: 22315999]

[36] Beattie MS, Hermann GE, Rogers RC, Bresnahan JC. Cell death in models of spinal cord injury. Prog Brain Res 2002; 137: 37-47.
[http://dx.doi.org/10.1016/S0079-6123(02)37006-7] [PMID: 12440358]

[37] Keane RW, Kraydieh S, Lotocki G, *et al.* Apoptotic and anti-apoptotic mechanisms following spinal cord injury. J Neuropathol Exp Neurol 2001; 60(5): 422-9.
[http://dx.doi.org/10.1093/jnen/60.5.422] [PMID: 11379817]

[38] Yu WR, Liu T, Fehlings TK, Fehlings MG. Involvement of mitochondrial signaling pathways in the mechanism of Fas-mediated apoptosis after spinal cord injury. Eur J Neurosci 2009; 29(1): 114-31.
[http://dx.doi.org/10.1111/j.1460-9568.2008.06555.x] [PMID: 19120440]

[39] Ackery A, Robins S, Fehlings MG. Inhibition of Fas-mediated apoptosis through administration of soluble Fas receptor improves functional outcome and reduces posttraumatic axonal degeneration after acute spinal cord injury. J Neurotrauma 2006; 23(5): 604-16.
[http://dx.doi.org/10.1089/neu.2006.23.604] [PMID: 16689665]

[40] Casha S, Yu WR, Fehlings MG. Oligodendroglial apoptosis occurs along degenerating axons and is associated with FAS and p75 expression following spinal cord injury in the rat. Neuroscience 2001; 103(1): 203-18.
[http://dx.doi.org/10.1016/S0306-4522(00)00538-8] [PMID: 11311801]

[41] Casha S, Yu WR, Fehlings MG. FAS deficiency reduces apoptosis, spares axons and improves function after spinal cord injury. Exp Neurol 2005; 196(2): 390-400. [http://dx.doi.org/10.1016/j.expneurol.2005.08.020] [PMID: 16202410]

[42] Crowe MJ, Bresnahan JC, Shuman SL, Masters JN, Beattie MS. Apoptosis and delayed degeneration after spinal cord injury in rats and monkeys. Nat Med 1997; 3(1): 73-6. [http://dx.doi.org/10.1038/nm0197-73] [PMID: 8986744]

[43] Yu WR, Fehlings MG. Fas/FasL-mediated apoptosis and inflammation are key features of acute human spinal cord injury: implications for translational, clinical application. Acta Neuropathol 2011; 122(6): 747-61. [http://dx.doi.org/10.1007/s00401-011-0882-3] [PMID: 22038545]

[44] Chu GK, Yu W, Fehlings MG. The p75 neurotrophin receptor is essential for neuronal cell survival and improvement of functional recovery after spinal cord injury. Neuroscience 2007; 148(3): 668-82. [http://dx.doi.org/10.1016/j.neuroscience.2007.05.028] [PMID: 17706365]

[45] Demjen D, Klussmann S, Kleber S, *et al.* Neutralization of CD95 ligand promotes regeneration and functional recovery after spinal cord injury. Nat Med 2004; 10(4): 389-95. [http://dx.doi.org/10.1038/nm1007] [PMID: 15004554]

[46] Emery E, Aldana P, Bunge MB, *et al.* Apoptosis after traumatic human spinal cord injury. J Neurosurg 1998; 89(6): 911-20. [http://dx.doi.org/10.3171/jns.1998.89.6.0911] [PMID: 9833815]

[47] Takagi T, Takayasu M, Mizuno M, Yoshimoto M, Yoshida J. Caspase activation in neuronal and glial apoptosis following spinal cord injury in mice. Neurol Med Chir (Tokyo) 2003; 43(1): 20-9. [http://dx.doi.org/10.2176/nmc.43.20] [PMID: 12568318]

[48] Yoshino O, Matsuno H, Nakamura H, *et al.* The role of Fas-mediated apoptosis after traumatic spinal cord injury. Spine 2004; 29(13): 1394-404. [http://dx.doi.org/10.1097/01.BRS.0000129894.34550.48] [PMID: 15223929]

[49] Austin JW, Fehlings MG. Molecular mechanisms of Fas-mediated cell death in oligodendrocytes. J Neurotrauma 2008; 25(5): 411-26. [http://dx.doi.org/10.1089/neu.2007.0436] [PMID: 18435595]

[50] Robins-Steele S, Nguyen DH, Fehlings MG. The delayed post-injury administration of soluble fas receptor attenuates post-traumatic neural degeneration and enhances functional recovery after traumatic cervical spinal cord injury. J Neurotrauma 2012; 29(8): 1586-99. [http://dx.doi.org/10.1089/neu.2011.2005] [PMID: 22260324]

[51] Kwon BK, Tetzlaff W, Grauer JN, Beiner J, Vaccaro AR. Pathophysiology and pharmacologic treatment of acute spinal cord injury. Spine J 2004; 4(4): 451-64. [http://dx.doi.org/10.1016/j.spinee.2003.07.007] [PMID: 15246307]

[52] Mautes AE, Weinzierl MR, Donovan F, Noble LJ. Vascular events after spinal cord injury: contribution to secondary pathogenesis. Phys Ther 2000; 80(7): 673-87. [PMID: 10869130]

[53] Ng MT, Stammers AT, Kwon BK. Vascular disruption and the role of angiogenic proteins after spinal cord injury. Transl Stroke Res 2011; 2(4): 474-91. [http://dx.doi.org/10.1007/s12975-011-0109-x] [PMID: 22448202]

[54] Zhang N, Yin Y, Xu SJ, Wu YP, Chen WS. Inflammation & apoptosis in spinal cord injury. Indian J Med Res 2012; 135: 287-96. [PMID: 22561613]

[55] Mueller CA, Schluesener HJ, Conrad S, Pietsch T, Schwab JM. Spinal cord injury-induced expression of the immune-regulatory chemokine interleukin-16 caused by activated microglia/macrophages and CD8+ cells. J Neurosurg Spine 2006; 4(3): 233-40. [http://dx.doi.org/10.3171/spi.2006.4.3.233] [PMID: 16572623]

[56] Tzekou A, Fehlings MG. Treatment of spinal cord injury with intravenous immunoglobulin G: preliminary evidence and future perspectives. J Clin Immunol 2014; 34 (Suppl. 1): S132-8. [http://dx.doi.org/10.1007/s10875-014-0021-8] [PMID: 24722853]

[57] Carlson SL, Parrish ME, Springer JE, Doty K, Dossett L. Acute inflammatory response in spinal cord following impact injury. Exp Neurol 1998; 151(1): 77-88. [http://dx.doi.org/10.1006/exnr.1998.6785] [PMID: 9582256]

[58] Leskovar A, Moriarty LJ, Turek JJ, Schoenlein IA, Borgens RB. The macrophage in acute neural injury: changes in cell numbers over time and levels of cytokine production in mammalian central and peripheral nervous systems. J Exp Biol 2000; 203(Pt 12): 1783-95. [PMID: 10821736]

[59] Chatzipanteli K, Garcia R, Marcillo AE, Loor KE, Kraydieh S, Dietrich WD. Temporal and segmental distribution of constitutive and inducible nitric oxide synthases after traumatic spinal cord injury: effect of aminoguanidine treatment. J Neurotrauma 2002; 19(5): 639-51. [http://dx.doi.org/10.1089/089771502753754109] [PMID: 12042098]

[60] Koeberle PD, Ball AK. Nitric oxide synthase inhibition delays axonal degeneration and promotes the survival of axotomized retinal ganglion cells. Exp Neurol 1999; 158(2): 366-81. [http://dx.doi.org/10.1006/exnr.1999.7113] [PMID: 10415143]

[61] López-Vales R, García-Alías G, Forés J, Navarro X, Verdú E. Increased expression of cyclo-oxygenase 2 and vascular endothelial growth factor in lesioned spinal cord by transplanted olfactory ensheathing cells. J Neurotrauma 2004; 21(8): 1031-43. [http://dx.doi.org/10.1089/0897715041651105] [PMID: 15319002]

[62] Pearse DD, Chatzipanteli K, Marcillo AE, Bunge MB, Dietrich WD. Comparison of iNOS inhibition by antisense and pharmacological inhibitors after spinal cord injury. J Neuropathol Exp Neurol 2003; 62(11): 1096-107. [http://dx.doi.org/10.1093/jnen/62.11.1096] [PMID: 14656068]

[63] Xiong Y, Rabchevsky AG, Hall ED. Role of peroxynitrite in secondary oxidative damage after spinal cord injury. J Neurochem 2007; 100(3): 639-49. [http://dx.doi.org/10.1111/j.1471-4159.2006.04312.x] [PMID: 17181549]

[64] Schanne FA, Kane AB, Young EE, Farber JL. Calcium dependence of toxic cell death: a final common pathway. Science 1979; 206(4419): 700-2. [http://dx.doi.org/10.1126/science.386513] [PMID: 386513]

[65] Vosler PS, Sun D, Wang S, *et al.* Calcium dysregulation induces apoptosis-inducing factor release: cross-talk between PARP-1- and calpain-signaling pathways. Exp Neurol 2009; 218(2): 213-20. [http://dx.doi.org/10.1016/j.expneurol.2009.04.032] [PMID: 19427306]

[66] Agrawal SK, Fehlings MG. Role of NMDA and non-NMDA ionotropic glutamate receptors in traumatic spinal cord axonal injury. J Neurosci 1997; 17(3): 1055-63. [http://dx.doi.org/10.1523/JNEUROSCI.17-03-01055.1997] [PMID: 8994060]

[67] Gerardo-Nava J, Mayorenko II, Grehl T, Steinbusch HW, Weis J, Brook GA. Differential pattern of neuroprotection in lumbar, cervical and thoracic spinal cord segments in an organotypic rat model of glutamate-induced excitotoxicity. J Chem Neuroanat 2013; 53: 11-7. [http://dx.doi.org/10.1016/j.jchemneu.2013.09.007] [PMID: 24126226]

[68] Park E, Velumian AA, Fehlings MG. The role of excitotoxicity in secondary mechanisms of spinal cord injury: a review with an emphasis on the implications for white matter degeneration. J Neurotrauma 2004; 21(6): 754-74. [http://dx.doi.org/10.1089/0897715041269641] [PMID: 15253803]

[69] Lladó J, Haenggeli C, Maragakis NJ, Snyder EY, Rothstein JD. Neural stem cells protect against glutamate-induced excitotoxicity and promote survival of injured motor neurons through the secretion of neurotrophic factors. Mol Cell Neurosci 2004; 27(3): 322-31.

[http://dx.doi.org/10.1016/j.mcn.2004.07.010] [PMID: 15519246]

[70] Fawcett JW, Asher RA. The glial scar and central nervous system repair. Brain Res Bull 1999; 49(6): 377-91.
[http://dx.doi.org/10.1016/S0361-9230(99)00072-6] [PMID: 10483914]

[71] Karimi-Abdolrezaee S, Eftekharpour E, Wang J, Schut D, Fehlings MG. Synergistic effects of transplanted adult neural stem/progenitor cells, chondroitinase, and growth factors promote functional repair and plasticity of the chronically injured spinal cord. J Neurosci 2010; 30(5): 1657-76.
[http://dx.doi.org/10.1523/JNEUROSCI.3111-09.2010] [PMID: 20130176]

[72] Reier PJ, Houle JD. The glial scar: its bearing on axonal elongation and transplantation approaches to CNS repair. Adv Neurol 1988; 47: 87-138.
[PMID: 3278533]

[73] Young W. Spinal cord regeneration. Cell Transplant 2014; 23(4-5): 573-611.
[http://dx.doi.org/10.3727/096368914X678427] [PMID: 24816452]

[74] Fitch MT, Silver J. CNS injury, glial scars, and inflammation: Inhibitory extracellular matrices and regeneration failure. Exp Neurol 2008; 209(2): 294-301.
[http://dx.doi.org/10.1016/j.expneurol.2007.05.014] [PMID: 17617407]

[75] Hagg T, Oudega M. Degenerative and spontaneous regenerative processes after spinal cord injury. J Neurotrauma 2006; 23(3-4): 264-80.
[http://dx.doi.org/10.1089/neu.2006.23.263] [PMID: 16629615]

[76] Hill CE, Beattie MS, Bresnahan JC. Degeneration and sprouting of identified descending supraspinal axons after contusive spinal cord injury in the rat. Exp Neurol 2001; 171(1): 153-69.
[http://dx.doi.org/10.1006/exnr.2001.7734] [PMID: 11520130]

[77] Li J, Lepski G. Cell transplantation for spinal cord injury: a systematic review. BioMed Res Int 2013; 2013: 786475.
[PMID: 23484157]

[78] Basso DM, Beattie MS, Bresnahan JC. Graded histological and locomotor outcomes after spinal cord contusion using the NYU weight-drop device *versus* transection. Exp Neurol 1996; 139(2): 244-56.
[http://dx.doi.org/10.1006/exnr.1996.0098] [PMID: 8654527]

[79] Fleming JC, Norenberg MD, Ramsay DA, *et al.* The cellular inflammatory response in human spinal cords after injury. Brain 2006; 129(Pt 12): 3249-69.
[http://dx.doi.org/10.1093/brain/awl296] [PMID: 17071951]

[80] Kramer AS, Harvey AR, Plant GW, Hodgetts SI. Systematic review of induced pluripotent stem cell technology as a potential clinical therapy for spinal cord injury. Cell Transplant 2013; 22(4): 571-617.
[http://dx.doi.org/10.3727/096368912X655208] [PMID: 22944020]

[81] Fehlings MG, Vaccaro A, Wilson JR, *et al.* Early *versus* delayed decompression for traumatic cervical spinal cord injury: results of the Surgical Timing in Acute Spinal Cord Injury Study (STASCIS). PLoS One 2012; 7(2): e32037.
[http://dx.doi.org/10.1371/journal.pone.0032037] [PMID: 22384132]

[82] Raw DA, Beattie JK, Hunter JM. Anaesthesia for spinal surgery in adults. Br J Anaesth 2003; 91(6): 886-904.
[http://dx.doi.org/10.1093/bja/aeg253] [PMID: 14633762]

[83] Stratman RC, Wiesner AM, Smith KM, Cook AM. Hemodynamic management after spinal cord injury. Orthopedics 2008; 31(3): 252-5.
[http://dx.doi.org/10.3928/01477447-20080301-39] [PMID: 18351045]

[84] Nockels RP. Nonoperative management of acute spinal cord injury. Spine 2001; 26(24) (Suppl.): S31-7.
[http://dx.doi.org/10.1097/00007632-200112151-00007] [PMID: 11805606]

[85] Wing PC. Early acute management in adults with spinal cord injury: a clinical practice guideline for health-care providers. Who should read it? J Spinal Cord Med 2008; 31(4): 360. [http://dx.doi.org/10.1080/10790268.2008.11760737] [PMID: 18959352]

[86] Hadley MN, Walters BC, Grabb PA, *et al.* Guidelines for the management of acute cervical spine and spinal cord injuries. Clin Neurosurg 2002; 49: 407-98. [PMID: 12506565]

[87] Winemiller MH, Stolp-Smith KA, Silverstein MD, Therneau TM. Prevention of venous thromboembolism in patients with spinal cord injury: effects of sequential pneumatic compression and heparin. J Spinal Cord Med 1999; 22(3): 182-91. [http://dx.doi.org/10.1080/10790268.1999.11719568] [PMID: 10685383]

[88] Bartholdi D, Schwab ME. Methylprednisolone inhibits early inflammatory processes but not ischemic cell death after experimental spinal cord lesion in the rat. Brain Res 1995; 672(1-2): 177-86. [http://dx.doi.org/10.1016/0006-8993(94)01410-J] [PMID: 7749740]

[89] Bracken MB, Shepard MJ, Collins WF, *et al.* A randomized, controlled trial of methylprednisolone or naloxone in the treatment of acute spinal-cord injury. Results of the Second National Acute Spinal Cord Injury Study. N Engl J Med 1990; 322(20): 1405-11. [http://dx.doi.org/10.1056/NEJM199005173222001] [PMID: 2278545]

[90] Bracken MB, Shepard MJ, Holford TR, *et al.* Methylprednisolone or tirilazad mesylate administration after acute spinal cord injury: 1-year follow up. Results of the third National Acute Spinal Cord Injury randomized controlled trial. J Neurosurg 1998; 89(5): 699-706. [http://dx.doi.org/10.3171/jns.1998.89.5.0699] [PMID: 9817404]

[91] Bracken MB. Methylprednisolone and acute spinal cord injury: an update of the randomized evidence. Spine 2001; 26(24) (Suppl.): S47-54. [http://dx.doi.org/10.1097/00007632-200112151-00010] [PMID: 11805609]

[92] Hall ED, Springer JE. Neuroprotection and acute spinal cord injury: a reappraisal. NeuroRx 2004; 1(1): 80-100. [http://dx.doi.org/10.1602/neurorx.1.1.80] [PMID: 15717009]

[93] Wilson JR, Forgione N, Fehlings MG. Emerging therapies for acute traumatic spinal cord injury. CMAJ 2013; 185(6): 485-92. [http://dx.doi.org/10.1503/cmaj.121206] [PMID: 23228995]

[94] Spivak JM, Weiss MA, Cotler JM, Call M. Cervical spine injuries in patients 65 and older. Spine 1994; 19(20): 2302-6. [http://dx.doi.org/10.1097/00007632-199410150-00009] [PMID: 7846575]

[95] Sumida M, Fujimoto M, Tokuhiro A, Tominaga T, Magara A, Uchida R. Early rehabilitation effect for traumatic spinal cord injury. Arch Phys Med Rehabil 2001; 82(3): 391-5. [http://dx.doi.org/10.1053/apmr.2001.19780] [PMID: 11245763]

[96] Ginis KAMLA, McKechnie K, Ditor DS, McCartney N, Hicks AL. Using exercise to enhance subjective well-being among people with spinal cord injury: The mediating influences of stress and pain. Rehabil Psychol 2003; 48(3): 157-64. [http://dx.doi.org/10.1037/0090-5550.48.3.157]

[97] Ma DK, Bonaguidi MA, Ming GL, Song H. Adult neural stem cells in the mammalian central nervous system. Cell Res 2009; 19(6): 672-82. [http://dx.doi.org/10.1038/cr.2009.56] [PMID: 19436263]

[98] Guo W, Patzlaff NE, Jobe EM, Zhao X. Isolation of multipotent neural stem or progenitor cells from both the dentate gyrus and subventricular zone of a single adult mouse. Nat Protoc 2012; 7(11): 2005-12. [http://dx.doi.org/10.1038/nprot.2012.123] [PMID: 23080272]

[99] Mariano ED, Teixeira MJ, Marie SK, Lepski G. Adult stem cells in neural repair: Current options,

limitations and perspectives. World J Stem Cells 2015; 7(2): 477-82.
[http://dx.doi.org/10.4252/wjsc.v7.i2.477] [PMID: 25815131]

[100] Condic ML. Totipotency: what it is and what it is not. Stem Cells Dev 2014; 23(8): 796-812.
[http://dx.doi.org/10.1089/scd.2013.0364] [PMID: 24368070]

[101] Wu P, Tarasenko YI, Gu Y, Huang LY, Coggeshall RE, Yu Y. Region-specific generation of cholinergic neurons from fetal human neural stem cells grafted in adult rat. Nat Neurosci 2002; 5(12): 1271-8.
[http://dx.doi.org/10.1038/nn974] [PMID: 12426573]

[102] Hoffman LM, Carpenter MK. Characterization and culture of human embryonic stem cells. Nat Biotechnol 2005; 23(6): 699-708.
[http://dx.doi.org/10.1038/nbt1102] [PMID: 15940242]

[103] Odorico JS, Kaufman DS, Thomson JA. Multilineage differentiation from human embryonic stem cell lines. Stem Cells 2001; 19(3): 193-204.
[http://dx.doi.org/10.1634/stemcells.19-3-193] [PMID: 11359944]

[104] Schuldiner M, Eiges R, Eden A, *et al.* Induced neuronal differentiation of human embryonic stem cells. Brain Res 2001; 913(2): 201-5.
[http://dx.doi.org/10.1016/S0006-8993(01)02776-7] [PMID: 11549388]

[105] Nistor GI, Totoiu MO, Haque N, Carpenter MK, Keirstead HS. Human embryonic stem cells differentiate into oligodendrocytes in high purity and myelinate after spinal cord transplantation. Glia 2005; 49(3): 385-96.
[http://dx.doi.org/10.1002/glia.20127] [PMID: 15538751]

[106] Zhang SC, Wernig M, Duncan ID, Brüstle O, Thomson JA. *in vitro* differentiation of transplantable neural precursors from human embryonic stem cells. Nat Biotechnol 2001; 19(12): 1129-33.
[http://dx.doi.org/10.1038/nbt1201-1129] [PMID: 11731781]

[107] Keirstead HS, Nistor G, Bernal G, *et al.* Human embryonic stem cell-derived oligodendrocyte progenitor cell transplants remyelinate and restore locomotion after spinal cord injury. J Neurosci 2005; 25(19): 4694-705.
[http://dx.doi.org/10.1523/JNEUROSCI.0311-05.2005] [PMID: 15888645]

[108] Okamura RM, Lebkowski J, Au M, Priest CA, Denham J, Majumdar AS. Immunological properties of human embryonic stem cell-derived oligodendrocyte progenitor cells. J Neuroimmunol 2007; 192(1-2): 134-44.
[http://dx.doi.org/10.1016/j.jneuroim.2007.09.030] [PMID: 17996308]

[109] Dai G, Liu X, Zhang Z, Yang Z, Dai Y, Xu R. Transplantation of autologous bone marrow mesenchymal stem cells in the treatment of complete and chronic cervical spinal cord injury. Brain Res 2013; 1533: 73-9.
[http://dx.doi.org/10.1016/j.brainres.2013.08.016] [PMID: 23948102]

[110] Syková E, Jendelová P, Urdzíková L, Lesný P, Hejcl A. Bone marrow stem cells and polymer hydrogels--two strategies for spinal cord injury repair. Cell Mol Neurobiol 2006; 26(7-8): 1113-29.
[http://dx.doi.org/10.1007/s10571-006-9007-2] [PMID: 16633897]

[111] Sasaki M, Honmou O, Akiyama Y, Uede T, Hashi K, Kocsis JD. Transplantation of an acutely isolated bone marrow fraction repairs demyelinated adult rat spinal cord axons. Glia 2001; 35(1): 26-34.
[http://dx.doi.org/10.1002/glia.1067] [PMID: 11424189]

[112] Akiyama Y, Radtke C, Honmou O, Kocsis JD. Remyelination of the spinal cord following intravenous delivery of bone marrow cells. Glia 2002; 39(3): 229-36.
[http://dx.doi.org/10.1002/glia.10102] [PMID: 12203389]

[113] Inoue M, Honmou O, Oka S, Houkin K, Hashi K, Kocsis JD. Comparative analysis of remyelinating potential of focal and intravenous administration of autologous bone marrow cells into the rat demyelinated spinal cord. Glia 2003; 44(2): 111-8.

[http://dx.doi.org/10.1002/glia.10285] [PMID: 14515327]

[114] de Girolamo L, Lucarelli E, Alessandri G, *et al.* Mesenchymal stem/stromal cells: a new "cells as drugs" paradigm. Efficacy and critical aspects in cell therapy. Curr Pharm Des 2013; 19(13): 2459-73. [http://dx.doi.org/10.2174/1381612811319130015] [PMID: 23278600]

[115] Wright KT, El Masri W, Osman A, Chowdhury J, Johnson WE. Concise review: Bone marrow for the treatment of spinal cord injury: mechanisms and clinical applications. Stem Cells 2011; 29(2): 169-78. [http://dx.doi.org/10.1002/stem.570] [PMID: 21732476]

[116] Barnabé-Heider F, Göritz C, Sabelström H, *et al.* Origin of new glial cells in intact and injured adult spinal cord. Cell Stem Cell 2010; 7(4): 470-82. [http://dx.doi.org/10.1016/j.stem.2010.07.014] [PMID: 20887953]

[117] McDonald JW, Liu XZ, Qu Y, *et al.* Transplanted embryonic stem cells survive, differentiate and promote recovery in injured rat spinal cord. Nat Med 1999; 5(12): 1410-2. [http://dx.doi.org/10.1038/70986] [PMID: 10581084]

[118] Manley NC, Priest CA, Denham J, Wirth ED III, Lebkowski JS. Human Embryonic Stem Cell-Derived Oligodendrocyte Progenitor Cells: Preclinical Efficacy and Safety in Cervical Spinal Cord Injury. Stem Cells Transl Med 2017; 6(10): 1917-29. [http://dx.doi.org/10.1002/sctm.17-0065] [PMID: 28834391]

[119] Liu J, Han D, Wang Z, *et al.* Clinical analysis of the treatment of spinal cord injury with umbilical cord mesenchymal stem cells. Cytotherapy 2013; 15(2): 185-91. [http://dx.doi.org/10.1016/j.jcyt.2012.09.005] [PMID: 23321330]

[120] Tarasenko YI, Gao J, Nie L, *et al.* Human fetal neural stem cells grafted into contusion-injured rat spinal cords improve behavior. J Neurosci Res 2007; 85(1): 47-57. [http://dx.doi.org/10.1002/jnr.21098] [PMID: 17075895]

[121] Forostyak S, Jendelova P, Sykova E. The role of mesenchymal stromal cells in spinal cord injury, regenerative medicine and possible clinical applications. Biochimie 2013; 95(12): 2257-70. [http://dx.doi.org/10.1016/j.biochi.2013.08.004] [PMID: 23994163]

[122] Cocks G, Romanyuk N, Amemori T, *et al.* Conditionally immortalized stem cell lines from human spinal cord retain regional identity and generate functional V2a interneurons and motorneurons. Stem Cell Res Ther 2013; 4(3): 69. [http://dx.doi.org/10.1186/scrt220] [PMID: 23759128]

[123] Cheng I, Mayle RE, Cox CA, *et al.* Functional assessment of the acute local and distal transplantation of human neural stem cells after spinal cord injury. Spine J 2012; 12(11): 1040-4. [http://dx.doi.org/10.1016/j.spinee.2012.09.005] [PMID: 23063425]

[124] Růžička J, Romanyuk N, Hejčl A, *et al.* Treating spinal cord injury in rats with a combination of human fetal neural stem cells and hydrogels modified with serotonin. Acta Neurobiol Exp (Warsz) 2013; 73(1): 102-15. [PMID: 23595287]

[125] Parr AM, Kulbatski I, Zahir T, *et al.* Transplanted adult spinal cord-derived neural stem/progenitor cells promote early functional recovery after rat spinal cord injury. Neuroscience 2008; 155(3): 760-70. [http://dx.doi.org/10.1016/j.neuroscience.2008.05.042] [PMID: 18588947]

[126] Salewski RP, Mitchell RA, Shen C, Fehlings MG. Transplantation of neural stem cells clonally derived from embryonic stem cells promotes recovery after murine spinal cord injury. Stem Cells Dev 2015; 24(1): 36-50. [http://dx.doi.org/10.1089/scd.2014.0096] [PMID: 25119334]

[127] Romanyuk N, Amemori T, Turnovcova K, *et al.* Beneficial Effect of Human Induced Pluripotent Stem Cell-Derived Neural Precursors in Spinal Cord Injury Repair. Cell Transplant 2015; 24(9): 1781-97. [http://dx.doi.org/10.3727/096368914X684042] [PMID: 25259685]

[128] Sareen D, Gowing G, Sahabian A, *et al.* Human induced pluripotent stem cells are a novel source of neural progenitor cells (iNPCs) that migrate and integrate in the rodent spinal cord. J Comp Neurol 2014; 522(12): 2707-28.
[http://dx.doi.org/10.1002/cne.23578] [PMID: 24610630]

[129] Amemori T, Romanyuk N, Jendelova P, *et al.* Human conditionally immortalized neural stem cells improve locomotor function after spinal cord injury in the rat. Stem Cell Res Ther 2013; 4(3): 68.
[http://dx.doi.org/10.1186/scrt219] [PMID: 23759119]

[130] Ruzicka J, Machova-Urdzikova L, Gillick J, *et al.* A Comparative Study of Three Different Types of Stem Cells for Treatment of Rat Spinal Cord Injury. Cell Transplant 2017; 26(4): 585-603.
[http://dx.doi.org/10.3727/096368916X693671] [PMID: 27938489]

[131] Ogawa Y, Sawamoto K, Miyata T, *et al.* Transplantation of *in vitro*-expanded fetal neural progenitor cells results in neurogenesis and functional recovery after spinal cord contusion injury in adult rats. J Neurosci Res 2002; 69(6): 925-33.
[http://dx.doi.org/10.1002/jnr.10341] [PMID: 12205685]

[132] Zompa EA, Cain LD, Everhart AW, Moyer MP, Hulsebosch CE. Transplant therapy: recovery of function after spinal cord injury. J Neurotrauma 1997; 14(8): 479-506.
[http://dx.doi.org/10.1089/neu.1997.14.479] [PMID: 9300561]

[133] Iwanami A, Kaneko S, Nakamura M, *et al.* Transplantation of human neural stem cells for spinal cord injury in primates. J Neurosci Res 2005; 80(2): 182-90.
[http://dx.doi.org/10.1002/jnr.20436] [PMID: 15772979]

[134] Wictorin K, Björklund A. Axon outgrowth from grafts of human embryonic spinal cord in the lesioned adult rat spinal cord. Neuroreport 1992; 3(12): 1045-8.
[http://dx.doi.org/10.1097/00001756-199212000-00003] [PMID: 1493215]

[135] Stepanov GA, Karpenko DO, Aleksandrova MA, *et al.* Xenotransplantation of stem/progenitor cells from human fetal brain to adult rats with spinal trauma. Bull Exp Biol Med 2003; 135(4): 397-400.
[http://dx.doi.org/10.1023/A:1024633403155] [PMID: 12910319]

[136] Akesson E, Holmberg L, Jönhagen ME, *et al.* Solid human embryonic spinal cord xenografts in acute and chronic spinal cord cavities: a morphological and functional study. Exp Neurol 2001; 170(2): 305-16.
[http://dx.doi.org/10.1006/exnr.2001.7707] [PMID: 11476597]

[137] Liu S, Qu Y, Stewart TJ, *et al.* Embryonic stem cells differentiate into oligodendrocytes and myelinate in culture and after spinal cord transplantation. Proc Natl Acad Sci USA 2000; 97(11): 6126-31.
[http://dx.doi.org/10.1073/pnas.97.11.6126] [PMID: 10823956]

[138] Nieto-Sampedro M, Kesslak JP, Gibbs R, Cotman CW. Effects of conditioning lesions on transplant survival, connectivity, and function. Role of neurotrophic factors. Ann N Y Acad Sci 1987; 495: 108-19.
[http://dx.doi.org/10.1111/j.1749-6632.1987.tb23670.x] [PMID: 2440364]

[139] McKay R. Stem cells in the central nervous system. Science 1997; 276(5309): 66-71.
[http://dx.doi.org/10.1126/science.276.5309.66] [PMID: 9082987]

[140] Ourednik J, Ourednik V, Lynch WP, Schachner M, Snyder EY. Neural stem cells display an inherent mechanism for rescuing dysfunctional neurons. Nat Biotechnol 2002; 20(11): 1103-10.
[http://dx.doi.org/10.1038/nbt750] [PMID: 12379867]

[141] Kerr DA, Lladó J, Shamblott MJ, *et al.* Human embryonic germ cell derivatives facilitate motor recovery of rats with diffuse motor neuron injury. J Neurosci 2003; 23(12): 5131-40.
[http://dx.doi.org/10.1523/JNEUROSCI.23-12-05131.2003] [PMID: 12832537]

[142] Shear DA, Tate MC, Archer DR, *et al.* Neural progenitor cell transplants promote long-term functional recovery after traumatic brain injury. Brain Res 2004; 1026(1): 11-22.
[http://dx.doi.org/10.1016/j.brainres.2004.07.087] [PMID: 15476693]

[143] Cummings BJ, Uchida N, Tamaki SJ, *et al.* Human neural stem cells differentiate and promote locomotor recovery in spinal cord-injured mice. Proc Natl Acad Sci USA 2005; 102(39): 14069-74. [http://dx.doi.org/10.1073/pnas.0507063102] [PMID: 16172374]

[144] Kobayashi Y, Okada Y, Itakura G, *et al.* Pre-evaluated safe human iPSC-derived neural stem cells promote functional recovery after spinal cord injury in common marmoset without tumorigenicity. PLoS One 2012; 7(12): e52787. [http://dx.doi.org/10.1371/journal.pone.0052787] [PMID: 23300777]

[145] Nutt SE, Chang EA, Suhr ST, *et al.* Caudalized human iPSC-derived neural progenitor cells produce neurons and glia but fail to restore function in an early chronic spinal cord injury model. Exp Neurol 2013; 248: 491-503. [http://dx.doi.org/10.1016/j.expneurol.2013.07.010] [PMID: 23891888]

[146] Tsuji O, Miura K, Fujiyoshi K, Momoshima S, Nakamura M, Okano H. Cell therapy for spinal cord injury by neural stem/progenitor cells derived from iPS/ES cells. Neurotherapeutics 2011; 8(4): 668-76. [http://dx.doi.org/10.1007/s13311-011-0063-z] [PMID: 21909829]

[147] White PM, Morrison SJ, Orimoto K, Kubu CJ, Verdi JM, Anderson DJ. Neural crest stem cells undergo cell-intrinsic developmental changes in sensitivity to instructive differentiation signals. Neuron 2001; 29(1): 57-71. [http://dx.doi.org/10.1016/S0896-6273(01)00180-5] [PMID: 11182081]

[148] Kelly S, Bliss TM, Shah AK, *et al.* Transplanted human fetal neural stem cells survive, migrate, and differentiate in ischemic rat cerebral cortex. Proc Natl Acad Sci USA 2004; 101(32): 11839-44. [http://dx.doi.org/10.1073/pnas.0404474101] [PMID: 15280535]

[149] Lindvall O, Kokaia Z, Martinez-Serrano A. Stem cell therapy for human neurodegenerative disorders-how to make it work. Nat Med 2004; 10 (Suppl.): S42-50. [http://dx.doi.org/10.1038/nm1064] [PMID: 15272269]

[150] Liu D, Xu GY, Pan E, McAdoo DJ. Neurotoxicity of glutamate at the concentration released upon spinal cord injury. Neuroscience 1999; 93(4): 1383-9. [http://dx.doi.org/10.1016/S0306-4522(99)00278-X] [PMID: 10501463]

[151] Xu GY, Hughes MG, Ye Z, Hulsebosch CE, McAdoo DJ. Concentrations of glutamate released following spinal cord injury kill oligodendrocytes in the spinal cord. Exp Neurol 2004; 187(2): 329-36. [http://dx.doi.org/10.1016/j.expneurol.2004.01.029] [PMID: 15144859]

[152] Bunge MB. Bridging areas of injury in the spinal cord. Neuroscientist 2001; 7(4): 325-39. [http://dx.doi.org/10.1177/107385840100700409] [PMID: 11488398]

[153] Guénard VX, Bunge XM. M.B. use of schwann cell transplantation to foster central nervous system repair. Semin Neurosci 1993; 5(6): 401-11. [http://dx.doi.org/10.1016/S1044-5765(05)80012-0]

[154] Xu XM, Chen A, Guénard V, Kleitman N, Bunge MB. Bridging Schwann cell transplants promote axonal regeneration from both the rostral and caudal stumps of transected adult rat spinal cord. J Neurocytol 1997; 26(1): 1-16. [http://dx.doi.org/10.1023/A:1018557923309] [PMID: 9154524]

[155] Johnson EM Jr, Taniuchi M, DiStefano PS. Expression and possible function of nerve growth factor receptors on Schwann cells. Trends Neurosci 1988; 11(7): 299-304. [http://dx.doi.org/10.1016/0166-2236(88)90090-2] [PMID: 2465632]

[156] Matsuoka I, Meyer M, Thoenen H. Cell-type-specific regulation of nerve growth factor (NGF) synthesis in non-neuronal cells: comparison of Schwann cells with other cell types. J Neurosci 1991; 11(10): 3165-77. [http://dx.doi.org/10.1523/JNEUROSCI.11-10-03165.1991] [PMID: 1658245]

[157] DiStefano PS, Johnson EM Jr. Nerve growth factor receptors on cultured rat Schwann cells. J Neurosci

1988; 8(1): 231-41.
[http://dx.doi.org/10.1523/JNEUROSCI.08-01-00231.1988] [PMID: 2828562]

[158] Heumann R. Regulation of the synthesis of nerve growth factor. J Exp Biol 1987; 132: 133-50. [PMID: 3323396]

[159] Barakat DJ, Gaglani SM, Neravetla SR, *et al.* Survival, integration, and axon growth support of glia transplanted into the chronically contused spinal cord. Cell Transplant 2005; 14(4): 225-40. [http://dx.doi.org/10.3727/000000005783983106] [PMID: 15929557]

[160] Saberi H, Moshayedi P, Aghayan HR, *et al.* Treatment of chronic thoracic spinal cord injury patients with autologous Schwann cell transplantation: an interim report on safety considerations and possible outcomes. Neurosci Lett 2008; 443(1): 46-50. [http://dx.doi.org/10.1016/j.neulet.2008.07.041] [PMID: 18662744]

[161] Martin D, Robe P, Franzen R, *et al.* Effects of Schwann cell transplantation in a contusion model of rat spinal cord injury. J Neurosci Res 1996; 45(5): 588-97. [http://dx.doi.org/10.1002/(SICI)1097-4547(19960901)45:5<588::AID-JNR8>3.0.CO;2-8] [PMID: 8875323]

[162] Olson HE, Rooney GE, Gross L, *et al.* Neural stem cell- and Schwann cell-loaded biodegradable polymer scaffolds support axonal regeneration in the transected spinal cord. Tissue Eng Part A 2009; 15(7): 1797-805. [http://dx.doi.org/10.1089/ten.tea.2008.0364] [PMID: 19191513]

[163] Raisman G. Olfactory ensheathing cells - another miracle cure for spinal cord injury? Nat Rev Neurosci 2001; 2(5): 369-75. [http://dx.doi.org/10.1038/35072576] [PMID: 11331921]

[164] Raisman G. Specialized neuroglial arrangement may explain the capacity of vomeronasal axons to reinnervate central neurons. Neuroscience 1985; 14(1): 237-54. [http://dx.doi.org/10.1016/0306-4522(85)90176-9] [PMID: 3974880]

[165] Doucette R. Development of the nerve fiber layer in the olfactory bulb of mouse embryos. J Comp Neurol 1989; 285(4): 514-27. [http://dx.doi.org/10.1002/cne.902850407] [PMID: 2760269]

[166] Valverde F, Santacana M, Heredia M. Formation of an olfactory glomerulus: morphological aspects of development and organization. Neuroscience 1992; 49(2): 255-75. [http://dx.doi.org/10.1016/0306-4522(92)90094-I] [PMID: 1436469]

[167] Pixley SK. The olfactory nerve contains two populations of glia, identified both *in vivo* and *in vitro*. Glia 1992; 5(4): 269-84. [http://dx.doi.org/10.1002/glia.440050405] [PMID: 1534071]

[168] Franceschini IA, Barnett SC. Low-affinity NGF-receptor and E-N-CAM expression define two types of olfactory nerve ensheathing cells that share a common lineage. Dev Biol 1996; 173(1): 327-43. [http://dx.doi.org/10.1006/dbio.1996.0027] [PMID: 8575633]

[169] Yan Q, Johnson EM Jr. An immunohistochemical study of the nerve growth factor receptor in developing rats. J Neurosci 1988; 8(9): 3481-98. [http://dx.doi.org/10.1523/JNEUROSCI.08-09-03481.1988] [PMID: 2845023]

[170] Bignami A, Eng LF, Dahl D, Uyeda CT. Localization of the glial fibrillary acidic protein in astrocytes by immunofluorescence. Brain Res 1972; 43(2): 429-35. [http://dx.doi.org/10.1016/0006-8993(72)90398-8] [PMID: 4559710]

[171] Figarella-Branger D, Nedelec J, Pellissier JF, Boucraut J, Bianco N, Rougon G. Expression of various isoforms of neural cell adhesive molecules and their highly polysialylated counterparts in diseased human muscles. J Neurol Sci 1990; 98(1): 21-36. [http://dx.doi.org/10.1016/0022-510X(90)90179-Q] [PMID: 2230829]

[172] Häyrinen J, Jennings H, Raff HV, *et al.* Antibodies to polysialic acid and its N-propyl derivative:

binding properties and interaction with human embryonal brain glycopeptides. J Infect Dis 1995; 171(6): 1481-90.
[http://dx.doi.org/10.1093/infdis/171.6.1481] [PMID: 7769282]

[173] Barnett SCR. J. Olfactory ensheathing cells – isolation and culture from the rat olfactory bulbNeural stem cells methods and protocol 2008; pp. 85-94.

[174] Ramón-Cueto A, Plant GW, Avila J, Bunge MB. Long-distance axonal regeneration in the transected adult rat spinal cord is promoted by olfactory ensheathing glia transplants. J Neurosci 1998; 18(10): 3803-15.
[http://dx.doi.org/10.1523/JNEUROSCI.18-10-03803.1998] [PMID: 9570810]

[175] Li Y, Field PM, Raisman G. Regeneration of adult rat corticospinal axons induced by transplanted olfactory ensheathing cells. J Neurosci 1998; 18(24): 10514-24.
[http://dx.doi.org/10.1523/JNEUROSCI.18-24-10514.1998] [PMID: 9852589]

[176] Imaizumi T, Lankford KL, Waxman SG, Greer CA, Kocsis JD. Transplanted olfactory ensheathing cells remyelinate and enhance axonal conduction in the demyelinated dorsal columns of the rat spinal cord. J Neurosci 1998; 18(16): 6176-85.
[http://dx.doi.org/10.1523/JNEUROSCI.18-16-06176.1998] [PMID: 9698311]

[177] Ramer LM, Au E, Richter MW, Liu J, Tetzlaff W, Roskams AJ. Peripheral olfactory ensheathing cells reduce scar and cavity formation and promote regeneration after spinal cord injury. J Comp Neurol 2004; 473(1): 1-15.
[http://dx.doi.org/10.1002/cne.20049] [PMID: 15067714]

[178] Altman J, Das GD. Autoradiographic and histological evidence of postnatal hippocampal neurogenesis in rats. J Comp Neurol 1965; 124(3): 319-35.
[http://dx.doi.org/10.1002/cne.901240303] [PMID: 5861717]

[179] Yang Z, Zhang A, Duan H, *et al.* NT3-chitosan elicits robust endogenous neurogenesis to enable functional recovery after spinal cord injury. Proc Natl Acad Sci USA 2015; 112(43): 13354-9.
[http://dx.doi.org/10.1073/pnas.1510194112] [PMID: 26460015]

[180] Lu P, Wang Y, Graham L, *et al.* Long-distance growth and connectivity of neural stem cells after severe spinal cord injury. Cell 2012; 150(6): 1264-73.
[http://dx.doi.org/10.1016/j.cell.2012.08.020] [PMID: 22980985]

[181] Abematsu M, Tsujimura K, Yamano M, *et al.* Neurons derived from transplanted neural stem cells restore disrupted neuronal circuitry in a mouse model of spinal cord injury. J Clin Invest 2010; 120(9): 3255-66.
[http://dx.doi.org/10.1172/JCI42957] [PMID: 20714104]

[182] Ming GL, Song H. Adult neurogenesis in the mammalian central nervous system. Annu Rev Neurosci 2005; 28: 223-50.
[http://dx.doi.org/10.1146/annurev.neuro.28.051804.101459] [PMID: 16022595]

[183] Duan X, Kang E, Liu CY, Ming GL, Song H. Development of neural stem cell in the adult brain. Curr Opin Neurobiol 2008; 18(1): 108-15.
[http://dx.doi.org/10.1016/j.conb.2008.04.001] [PMID: 18514504]

[184] Bordiuk OL, Smith K, Morin PJ, Semënov MV. Cell proliferation and neurogenesis in adult mouse brain. PLoS One 2014; 9(11): e111453.
[http://dx.doi.org/10.1371/journal.pone.0111453] [PMID: 25375658]

[185] Kim WR, Sun W. Programmed cell death during postnatal development of the rodent nervous system. Dev Growth Differ 2011; 53(2): 225-35.
[http://dx.doi.org/10.1111/j.1440-169X.2010.01226.x] [PMID: 21338348]

[186] Ge S, Goh EL, Sailor KA, Kitabatake Y, Ming GL, Song H. GABA regulates synaptic integration of newly generated neurons in the adult brain. Nature 2006; 439(7076): 589-93.
[http://dx.doi.org/10.1038/nature04404] [PMID: 16341203]

[187] Koizumi H, Higginbotham H, Poon T, Tanaka T, Brinkman BC, Gleeson JG. Doublecortin maintains bipolar shape and nuclear translocation during migration in the adult forebrain. Nat Neurosci 2006; 9(6): 779-86.
[http://dx.doi.org/10.1038/nn1704] [PMID: 16699506]

[188] Talib S, Millan MT, Jorgenson RL, Shepard KA. Proceedings: Immune Tolerance and Stem Cell Transplantation: A CIRM Mini-Symposium and Workshop Report. Stem Cells Transl Med 2015; 4(1): 4-9.
[http://dx.doi.org/10.5966/sctm.2014-0262] [PMID: 25473085]

[189] Strulovici Y, Leopold PL, O'Connor TP, Pergolizzi RG, Crystal RG. Human embryonic stem cells and gene therapy. Mol Ther 2007; 15(5): 850-66.
[http://dx.doi.org/10.1038/mt.sj.6300125] [PMID: 17356540]

[190] Horner PJ, Power AE, Kempermann G, *et al.* Proliferation and differentiation of progenitor cells throughout the intact adult rat spinal cord. J Neurosci 2000; 20(6): 2218-28.
[http://dx.doi.org/10.1523/JNEUROSCI.20-06-02218.2000] [PMID: 10704497]

[191] Johansson CB, Momma S, Clarke DL, Risling M, Lendahl U, Frisén J. Identification of a neural stem cell in the adult mammalian central nervous system. Cell 1999; 96(1): 25-34.
[http://dx.doi.org/10.1016/S0092-8674(00)80956-3] [PMID: 9989494]

[192] Martens DJ, Seaberg RM, van der Kooy D. *In vivo* infusions of exogenous growth factors into the fourth ventricle of the adult mouse brain increase the proliferation of neural progenitors around the fourth ventricle and the central canal of the spinal cord. Eur J Neurosci 2002; 16(6): 1045-57.
[http://dx.doi.org/10.1046/j.1460-9568.2002.02181.x] [PMID: 12383233]

[193] Meletis K, Barnabé-Heider F, Carlén M, *et al.* Spinal cord injury reveals multilineage differentiation of ependymal cells. PLoS Biol 2008; 6(7): e182.
[http://dx.doi.org/10.1371/journal.pbio.0060182] [PMID: 18651793]

[194] Ohori Y, Yamamoto S, Nagao M, *et al.* Growth factor treatment and genetic manipulation stimulate neurogenesis and oligodendrogenesis by endogenous neural progenitors in the injured adult spinal cord. J Neurosci 2006; 26(46): 11948-60.
[http://dx.doi.org/10.1523/JNEUROSCI.3127-06.2006] [PMID: 17108169]

[195] Yamamoto S, Nagao M, Sugimori M, *et al.* Transcription factor expression and Notch-dependent regulation of neural progenitors in the adult rat spinal cord. J Neurosci 2001; 21(24): 9814-23.
[http://dx.doi.org/10.1523/JNEUROSCI.21-24-09814.2001] [PMID: 11739589]

[196] Obermair FJ, Schröter A, Thallmair M. Endogenous neural progenitor cells as therapeutic target after spinal cord injury. Physiology (Bethesda) 2008; 23: 296-304.
[http://dx.doi.org/10.1152/physiol.00017.2008] [PMID: 18927205]

[197] Decimo I, Bifari F, Krampera M, Fumagalli G. Neural stem cell niches in health and diseases. Curr Pharm Des 2012; 18(13): 1755-83.
[http://dx.doi.org/10.2174/138161212799859611] [PMID: 22394166]

[198] Horky LL, Galimi F, Gage FH, Horner PJ. Fate of endogenous stem/progenitor cells following spinal cord injury. J Comp Neurol 2006; 498(4): 525-38.
[http://dx.doi.org/10.1002/cne.21065] [PMID: 16874803]

[199] Hugnot JP, Franzen R. The spinal cord ependymal region: a stem cell niche in the caudal central nervous system. Front Biosci 2011; 16: 1044-59.
[http://dx.doi.org/10.2741/3734] [PMID: 21196217]

[200] Sato K, Kubota T, Ishida M, Handa Y. Spinal tanycytic ependymoma with hematomyelia-case report. Neurol Med Chir (Tokyo) 2005; 45(3): 168-71.
[http://dx.doi.org/10.2176/nmc.45.168] [PMID: 15782011]

[201] Corns LF, Deuchars J, Deuchars SA. GABAergic responses of mammalian ependymal cells in the central canal neurogenic niche of the postnatal spinal cord. Neurosci Lett 2013; 553: 57-62.

[http://dx.doi.org/10.1016/j.neulet.2013.07.007] [PMID: 23872091]

[202] Marichal N, García G, Radmilovich M, Trujillo-Cenóz O, Russo RE. Enigmatic central canal contacting cells: immature neurons in "standby mode"? J Neurosci 2009; 29(32): 10010-24. [http://dx.doi.org/10.1523/JNEUROSCI.6183-08.2009] [PMID: 19675235]

[203] Stoeckel ME, Uhl-Bronner S, Hugel S, *et al.* Cerebrospinal fluid-contacting neurons in the rat spinal cord, a gamma-aminobutyric acidergic system expressing the P2X2 subunit of purinergic receptors, PSA-NCAM, and GAP-43 immunoreactivities: light and electron microscopic study. J Comp Neurol 2003; 457(2): 159-74. [http://dx.doi.org/10.1002/cne.10565] [PMID: 12541316]

[204] Vigh B, Vigh-Teichmann I. Actual problems of the cerebrospinal fluid-contacting neurons. Microsc Res Tech 1998; 41(1): 57-83. [http://dx.doi.org/10.1002/(SICI)1097-0029(19980401)41:1<57::AID-JEMT6>3.0.CO;2-R] [PMID: 9550137]

[205] Beattie MS, Bresnahan JC, Komon J, *et al.* Endogenous repair after spinal cord contusion injuries in the rat. Exp Neurol 1997; 148(2): 453-63. [http://dx.doi.org/10.1006/exnr.1997.6695] [PMID: 9417825]

[206] Lacroix S, Hamilton LK, Vaugeois A, *et al.* Central canal ependymal cells proliferate extensively in response to traumatic spinal cord injury but not demyelinating lesions. PLoS One 2014; 9(1): e85916. [http://dx.doi.org/10.1371/journal.pone.0085916] [PMID: 24475059]

[207] Mothe AJ, Tator CH. Proliferation, migration, and differentiation of endogenous ependymal region stem/progenitor cells following minimal spinal cord injury in the adult rat. Neuroscience 2005; 131(1): 177-87. [http://dx.doi.org/10.1016/j.neuroscience.2004.10.011] [PMID: 15680701]

[208] Namiki J, Tator CH. Cell proliferation and nestin expression in the ependyma of the adult rat spinal cord after injury. J Neuropathol Exp Neurol 1999; 58(5): 489-98. [http://dx.doi.org/10.1097/00005072-199905000-00008] [PMID: 10331437]

[209] Sabelström H, Stenudd M, Réu P, *et al.* Resident neural stem cells restrict tissue damage and neuronal loss after spinal cord injury in mice. Science 2013; 342(6158): 637-40. [http://dx.doi.org/10.1126/science.1242576] [PMID: 24179227]

[210] Zhu X, Bergles DE, Nishiyama A. NG2 cells generate both oligodendrocytes and gray matter astrocytes. Development 2008; 135(1): 145-57. [http://dx.doi.org/10.1242/dev.004895] [PMID: 18045844]

[211] Zhu X, Hill RA, Nishiyama A. NG2 cells generate oligodendrocytes and gray matter astrocytes in the spinal cord. Neuron Glia Biol 2008; 4(1): 19-26. [http://dx.doi.org/10.1017/S1740925X09000015] [PMID: 19006598]

[212] Moreno-Manzano V, Rodríguez-Jiménez FJ, García-Roselló M, *et al.* Activated spinal cord ependymal stem cells rescue neurological function. Stem Cells 2009; 27(3): 733-43. [http://dx.doi.org/10.1002/stem.24] [PMID: 19259940]

[213] Buffo A, Rite I, Tripathi P, *et al.* Origin and progeny of reactive gliosis: A source of multipotent cells in the injured brain. Proc Natl Acad Sci USA 2008; 105(9): 3581-6. [http://dx.doi.org/10.1073/pnas.0709002105] [PMID: 18299565]

[214] Doetsch F, Caillé I, Lim DA, García-Verdugo JM, Alvarez-Buylla A. Subventricular zone astrocytes are neural stem cells in the adult mammalian brain. Cell 1999; 97(6): 703-16. [http://dx.doi.org/10.1016/S0092-8674(00)80783-7] [PMID: 10380923]

[215] Shihabuddin LS, Horner PJ, Ray J, Gage FH. Adult spinal cord stem cells generate neurons after transplantation in the adult dentate gyrus. J Neurosci 2000; 20(23): 8727-35. [http://dx.doi.org/10.1523/JNEUROSCI.20-23-08727.2000] [PMID: 11102479]

[216] Barkho BZ, Song H, Aimone JB, *et al.* Identification of astrocyte-expressed factors that modulate

neural stem/progenitor cell differentiation. Stem Cells Dev 2006; 15(3): 407-21. [http://dx.doi.org/10.1089/scd.2006.15.407] [PMID: 16846377]

[217] Jiao J, Chen DF. Induction of neurogenesis in nonconventional neurogenic regions of the adult central nervous system by niche astrocyte-produced signals. Stem Cells 2008; 26(5): 1221-30. [http://dx.doi.org/10.1634/stemcells.2007-0513] [PMID: 18323412]

[218] Lie DC, Colamarino SA, Song HJ, *et al.* Wnt signalling regulates adult hippocampal neurogenesis. Nature 2005; 437(7063): 1370-5. [http://dx.doi.org/10.1038/nature04108] [PMID: 16251967]

[219] Shihabuddin LS, Ray J, Gage FH. FGF-2 is sufficient to isolate progenitors found in the adult mammalian spinal cord. Exp Neurol 1997; 148(2): 577-86. [http://dx.doi.org/10.1006/exnr.1997.6697] [PMID: 9417834]

[220] Gao WL, Tian F, Zhang SQ, Zhang H, Yin ZS. Epidermal growth factor increases the expression of Nestin in rat reactive astrocytes through the Ras-Raf-ERK pathway. Neurosci Lett 2014; 562: 54-9. [http://dx.doi.org/10.1016/j.neulet.2014.01.018] [PMID: 24462842]

[221] Környei Z, Szlávik V, Szabó B, Gócza E, Czirók A, Madarász E. Humoral and contact interactions in astroglia/stem cell co-cultures in the course of glia-induced neurogenesis. Glia 2005; 49(3): 430-44. [http://dx.doi.org/10.1002/glia.20123] [PMID: 15546152]

[222] Song H, Stevens CF, Gage FH. Astroglia induce neurogenesis from adult neural stem cells. Nature 2002; 417(6884): 39-44. [http://dx.doi.org/10.1038/417039a] [PMID: 11986659]

[223] Selvaraj V, Plane JM, Williams AJ, Deng W. Switching cell fate: the remarkable rise of induced pluripotent stem cells and lineage reprogramming technologies. Trends Biotechnol 2010; 28(4): 214-23. [http://dx.doi.org/10.1016/j.tibtech.2010.01.002] [PMID: 20149468]

[224] Puschmann TB, Dixon KJ, Turnley AM. Species differences in reactivity of mouse and rat astrocytes *in vitro*. Neurosignals 2010; 18(3): 152-63. [http://dx.doi.org/10.1159/000321494] [PMID: 21109727]

[225] Zhou R, Wu X, Skalli O. TGF-alpha induces a stationary, radial-glia like phenotype in cultured astrocytes. Brain Res Bull 2001; 56(1): 37-42. [http://dx.doi.org/10.1016/S0361-9230(01)00591-3] [PMID: 11604246]

[226] Itoh T, Satou T, Nishida S, Hashimoto S, Ito H. Cultured rat astrocytes give rise to neural stem cells. Neurochem Res 2006; 31(11): 1381-7. [http://dx.doi.org/10.1007/s11064-006-9186-8] [PMID: 17053966]

[227] Ghashghaei HT, Weimer JM, Schmid RS, *et al.* Reinduction of ErbB2 in astrocytes promotes radial glial progenitor identity in adult cerebral cortex. Genes Dev 2007; 21(24): 3258-71. [http://dx.doi.org/10.1101/gad.1580407] [PMID: 18079173]

[228] Sharif A, Legendre P, Prévot V, *et al.* Transforming growth factor alpha promotes sequential conversion of mature astrocytes into neural progenitors and stem cells. Oncogene 2007; 26(19): 2695-706. [http://dx.doi.org/10.1038/sj.onc.1210071] [PMID: 17057735]

[229] Coulombe PA, Wong P. Cytoplasmic intermediate filaments revealed as dynamic and multipurpose scaffolds. Nat Cell Biol 2004; 6(8): 699-706. [http://dx.doi.org/10.1038/ncb0804-699] [PMID: 15303099]

[230] Eriksson JE, Dechat T, Grin B, *et al.* Introducing intermediate filaments: from discovery to disease. J Clin Invest 2009; 119(7): 1763-71. [http://dx.doi.org/10.1172/JCI38339] [PMID: 19587451]

[231] Lendahl U, Zimmerman LB, McKay RD. CNS stem cells express a new class of intermediate filament protein. Cell 1990; 60(4): 585-95.

[http://dx.doi.org/10.1016/0092-8674(90)90662-X] [PMID: 1689217]

[232] Blaiss CA, Yu TS, Zhang G, *et al.* Temporally specified genetic ablation of neurogenesis impairs cognitive recovery after traumatic brain injury. J Neurosci 2011; 31(13): 4906-16. [http://dx.doi.org/10.1523/JNEUROSCI.5265-10.2011] [PMID: 21451029]

[233] Codeluppi S, Svensson CI, Hefferan MP, *et al.* The Rheb-mTOR pathway is upregulated in reactive astrocytes of the injured spinal cord. J Neurosci 2009; 29(4): 1093-104. [http://dx.doi.org/10.1523/JNEUROSCI.4103-08.2009] [PMID: 19176818]

[234] Luo Y, Coskun V, Liang A, *et al.* Single-cell transcriptome analyses reveal signals to activate dormant neural stem cells. Cell 2015; 161(5): 1175-86. [http://dx.doi.org/10.1016/j.cell.2015.04.001] [PMID: 26000486]

[235] Duan H, Ge W, Zhang A, *et al.* Transcriptome analyses reveal molecular mechanisms underlying functional recovery after spinal cord injury. Proc Natl Acad Sci USA 2015; 112(43): 13360-5. [http://dx.doi.org/10.1073/pnas.1510176112] [PMID: 26460053]

[236] Hofstetter CP, Holmström NA, Lilja JA, *et al.* Allodynia limits the usefulness of intraspinal neural stem cell grafts; directed differentiation improves outcome. Nat Neurosci 2005; 8(3): 346-53. [http://dx.doi.org/10.1038/nn1405] [PMID: 15711542]

[237] Silver J, Miller JH. Regeneration beyond the glial scar. Nat Rev Neurosci 2004; 5(2): 146-56. [http://dx.doi.org/10.1038/nrn1326] [PMID: 14735117]

[238] Yiu G, He Z. Glial inhibition of CNS axon regeneration. Nat Rev Neurosci 2006; 7(8): 617-27. [http://dx.doi.org/10.1038/nrn1956] [PMID: 16858390]

[239] Liu Y, Young KA, Curtis JT, Aragona BJ, Wang Z. Social bonding decreases the rewarding properties of amphetamine through a dopamine D1 receptor-mediated mechanism. J Neurosci 2011; 31(22): 7960-6. [http://dx.doi.org/10.1523/JNEUROSCI.1006-11.2011] [PMID: 21632917]

[240] Cregg JM, DePaul MA, Filous AR, Lang BT, Tran A, Silver J. Functional regeneration beyond the glial scar. Exp Neurol 2014; 253: 197-207. [http://dx.doi.org/10.1016/j.expneurol.2013.12.024] [PMID: 24424280]

[241] Jee MK, Jung JS, Choi JI, *et al.* MicroRNA 486 is a potentially novel target for the treatment of spinal cord injury. Brain 2012; 135(Pt 4): 1237-52. [http://dx.doi.org/10.1093/brain/aws047] [PMID: 22466292]

[242] Jee MK, Jung JS, Im YB, Jung SJ, Kang SK. Silencing of miR20a is crucial for Ngn1-mediated neuroprotection in injured spinal cord. Hum Gene Ther 2012; 23(5): 508-20. [http://dx.doi.org/10.1089/hum.2011.121] [PMID: 22182208]

[243] Bhalala OG, Pan L, Sahni V, *et al.* microRNA-21 regulates astrocytic response following spinal cord injury. J Neurosci 2012; 32(50): 17935-47. [http://dx.doi.org/10.1523/JNEUROSCI.3860-12.2012] [PMID: 23238710]

[244] Guttman M, Rinn JL. Modular regulatory principles of large non-coding RNAs. Nature 2012; 482(7385): 339-46. [http://dx.doi.org/10.1038/nature10887] [PMID: 22337053]

[245] Geisler S, Coller J. RNA in unexpected places: long non-coding RNA functions in diverse cellular contexts. Nat Rev Mol Cell Biol 2013; 14(11): 699-712. [http://dx.doi.org/10.1038/nrm3679] [PMID: 24105322]

[246] Ulitsky I, Bartel DP. lincRNAs: genomics, evolution, and mechanisms. Cell 2013; 154(1): 26-46. [http://dx.doi.org/10.1016/j.cell.2013.06.020] [PMID: 23827673]

[247] Wang J, Hu B, Cao F, Sun S, Zhang Y, Zhu Q. Down regulation of lncSCIR1 after spinal cord contusion injury in rat. Brain Res 2015; 1624: 314-20. [http://dx.doi.org/10.1016/j.brainres.2015.07.052] [PMID: 26254726]

[248] Moore SA, Oglesbee MJ. Spinal Cord Ependymal Responses to Naturally Occurring Traumatic Spinal Cord Injury in Dogs. Vet Pathol 2015; 52(6): 1108-17. [http://dx.doi.org/10.1177/0300985814560235] [PMID: 25445323]

[249] Hamilton LK, Truong MK, Bednarczyk MR, Aumont A, Fernandes KJ. Cellular organization of the central canal ependymal zone, a niche of latent neural stem cells in the adult mammalian spinal cord. Neuroscience 2009; 164(3): 1044-56. [http://dx.doi.org/10.1016/j.neuroscience.2009.09.006] [PMID: 19747531]

[250] Takahashi M, Arai Y, Kurosawa H, Sueyoshi N, Shirai S. Ependymal cell reactions in spinal cord segments after compression injury in adult rat. J Neuropathol Exp Neurol 2003; 62(2): 185-94. [http://dx.doi.org/10.1093/jnen/62.2.185] [PMID: 12578228]

[251] Anderson MA, Burda JE, Ren Y, *et al.* Astrocyte scar formation aids central nervous system axon regeneration. Nature 2016; 532(7598): 195-200. [http://dx.doi.org/10.1038/nature17623] [PMID: 27027288]

[252] Gómez-Villafuertes R, Rodríguez-Jiménez FJ, Alastrue-Agudo A, Stojkovic M, Miras-Portugal MT, Moreno-Manzano V. Purinergic Receptors in Spinal Cord-Derived Ependymal Stem/Progenitor Cells and Their Potential Role in Cell-Based Therapy for Spinal Cord Injury. Cell Transplant 2015; 24(8): 1493-509. [http://dx.doi.org/10.3727/096368914X682828] [PMID: 25198194]

[253] Trujillo CA, Schwindt TT, Martins AH, Alves JM, Mello LE, Ulrich H. Novel perspectives of neural stem cell differentiation: from neurotransmitters to therapeutics. Cytometry A 2009; 75(1): 38-53. [http://dx.doi.org/10.1002/cyto.a.20666] [PMID: 18988295]

[254] Shukla V, Zimmermann H, Wang L, *et al.* Functional expression of the ecto-ATPase NTPDase2 and of nucleotide receptors by neuronal progenitor cells in the adult murine hippocampus. J Neurosci Res 2005; 80(5): 600-10. [http://dx.doi.org/10.1002/jnr.20508] [PMID: 15884037]

[255] Burnstock G, Ulrich H. Purinergic signaling in embryonic and stem cell development. Cell Mol Life Sci 2011; 68(8): 1369-94. [http://dx.doi.org/10.1007/s00018-010-0614-1] [PMID: 21222015]

[256] Schwindt TT, Trujillo CA, Negraes PD, Lameu C, Ulrich H. Directed differentiation of neural progenitors into neurons is accompanied by altered expression of P2X purinergic receptors. J Mol Neurosci 2011; 44(3): 141-6. [http://dx.doi.org/10.1007/s12031-010-9417-y] [PMID: 20617399]

[257] Grimm I, Ullsperger SN, Zimmermann H. Nucleotides and epidermal growth factor induce parallel cytoskeletal rearrangements and migration in cultured adult murine neural stem cells. Acta Physiol (Oxf) 2010; 199(2): 181-9. [http://dx.doi.org/10.1111/j.1748-1716.2010.02092.x] [PMID: 20121711]

[258] Mishra SK, Braun N, Shukla V, *et al.* Extracellular nucleotide signaling in adult neural stem cells: synergism with growth factor-mediated cellular proliferation. Development 2006; 133(4): 675-84. [http://dx.doi.org/10.1242/dev.02233] [PMID: 16436623]

[259] Glaser T, Cappellari AR, Pillat MM, *et al.* Perspectives of purinergic signaling in stem cell differentiation and tissue regeneration. Purinergic Signal 2012; 8(3): 523-37. [http://dx.doi.org/10.1007/s11302-011-9282-3] [PMID: 22143354]

[260] Lin JH, Takano T, Arcuino G, *et al.* Purinergic signaling regulates neural progenitor cell expansion and neurogenesis. Dev Biol 2007; 302(1): 356-66. [http://dx.doi.org/10.1016/j.ydbio.2006.09.017] [PMID: 17188262]

[261] Braun N, Sévigny J, Mishra SK, *et al.* Expression of the ecto-ATPase NTPDase2 in the germinal zones of the developing and adult rat brain. Eur J Neurosci 2003; 17(7): 1355-64. [http://dx.doi.org/10.1046/j.1460-9568.2003.02567.x] [PMID: 12713638]

[262] Rodríguez-Zayas AE, Torrado AI, Rosas OR, Santiago JM, Figueroa JD, Miranda JD. Blockade of P2 nucleotide receptors after spinal cord injury reduced the gliotic response and spared tissue. J Mol Neurosci 2012; 46(1): 167-76.
[http://dx.doi.org/10.1007/s12031-011-9567-6] [PMID: 21647706]

[263] Díaz-Hernandez M, del Puerto A, Díaz-Hernandez JI, *et al.* Inhibition of the ATP-gated P2X7 receptor promotes axonal growth and branching in cultured hippocampal neurons. J Cell Sci 2008; 121(Pt 22): 3717-28.
[http://dx.doi.org/10.1242/jcs.034082] [PMID: 18987356]

[264] Delarasse C, Gonnord P, Galante M, *et al.* Neural progenitor cell death is induced by extracellular ATP *via* ligation of P2X7 receptor. J Neurochem 2009; 109(3): 846-57.
[http://dx.doi.org/10.1111/j.1471-4159.2009.06008.x] [PMID: 19250337]

[265] Messemer N, Kunert C, Grohmann M, *et al.* P2X7 receptors at adult neural progenitor cells of the mouse subventricular zone. Neuropharmacology 2013; 73: 122-37.
[http://dx.doi.org/10.1016/j.neuropharm.2013.05.017] [PMID: 23727220]

[266] Meda P. Reduced expression of P2Y1 receptors in connexin43-null mice alters calcium signaling and migration of neural progenitor cells. J Neurosci 2004; 23(36): 11444-52.

[267] Stafford MR, Bartlett PF, Adams DJ. Purinergic receptor activation inhibits mitogen-stimulated proliferation in primary neurospheres from the adult mouse subventricular zone. Mol Cell Neurosci 2007; 35(4): 535-48.
[http://dx.doi.org/10.1016/j.mcn.2007.04.013] [PMID: 17553694]

[268] de Rivero Vaccari JP, Bastien D, Yurcisin G, *et al.* P2X4 receptors influence inflammasome activation after spinal cord injury. J Neurosci 2012; 32(9): 3058-66.
[http://dx.doi.org/10.1523/JNEUROSCI.4930-11.2012] [PMID: 22378878]

[269] Marcillo A, Frydel B, Bramlett HM, Dietrich WD. A reassessment of P2X7 receptor inhibition as a neuroprotective strategy in rat models of contusion injury. Exp Neurol 2012; 233(2): 687-92.
[http://dx.doi.org/10.1016/j.expneurol.2011.06.008] [PMID: 22078760]

[270] Beckervordersandforth R, Tripathi P, Ninkovic J, *et al. In vivo* fate mapping and expression analysis reveals molecular hallmarks of prospectively isolated adult neural stem cells. Cell Stem Cell 2010; 7(6): 744-58.
[http://dx.doi.org/10.1016/j.stem.2010.11.017] [PMID: 21112568]

[271] Legué E, Joyner AL. Genetic fate mapping using site-specific recombinases. Methods Enzymol 2010; 477: 153-81.
[http://dx.doi.org/10.1016/S0076-6879(10)77010-5] [PMID: 20699142]

[272] Yoo S, Wrathall JR. Mixed primary culture and clonal analysis provide evidence that NG2 proteoglycan-expressing cells after spinal cord injury are glial progenitors. Dev Neurobiol 2007; 67(7): 860-74.
[http://dx.doi.org/10.1002/dneu.20369] [PMID: 17506499]

[273] Breunig JJ, Arellano JI, Macklis JD, Rakic P. Everything that glitters isn't gold: a critical review of postnatal neural precursor analyses. Cell Stem Cell 2007; 1(6): 612-27.
[http://dx.doi.org/10.1016/j.stem.2007.11.008] [PMID: 18371403]

[274] Konishi T, Takeyasu A, Natsume T, Furusawa Y, Hieda K. Visualization of heavy ion tracks by labeling 3′-OH termini of induced DNA strand breaks. J Radiat Res (Tokyo) 2011; 52(4): 433-40.
[http://dx.doi.org/10.1269/jrr.10097] [PMID: 21785232]

[275] Yang H, Lu P, McKay HM, *et al.* Endogenous neurogenesis replaces oligodendrocytes and astrocytes after primate spinal cord injury. J Neurosci 2006; 26(8): 2157-66.
[http://dx.doi.org/10.1523/JNEUROSCI.4070-05.2005] [PMID: 16495442]

[276] Reynolds BA, Weiss S. Generation of neurons and astrocytes from isolated cells of the adult mammalian central nervous system. Science 1992; 255(5052): 1707-10.

[http://dx.doi.org/10.1126/science.1553558] [PMID: 1553558]

[277] Morshead CM, Reynolds BA, Craig CG, *et al.* Neural stem cells in the adult mammalian forebrain: a relatively quiescent subpopulation of subependymal cells. Neuron 1994; 13(5): 1071-82. [http://dx.doi.org/10.1016/0896-6273(94)90046-9] [PMID: 7946346]

[278] Qian X, Goderie SK, Shen Q, Stern JH, Temple S. Intrinsic programs of patterned cell lineages in isolated vertebrate CNS ventricular zone cells. Development 1998; 125(16): 3143-52. [PMID: 9671587]

[279] Qian X, Shen Q, Goderie SK, *et al.* Timing of CNS cell generation: a programmed sequence of neuron and glial cell production from isolated murine cortical stem cells. Neuron 2000; 28(1): 69-80. [http://dx.doi.org/10.1016/S0896-6273(00)00086-6] [PMID: 11086984]

[280] Frederiksen K, McKay RD. Proliferation and differentiation of rat neuroepithelial precursor cells *in vivo.* J Neurosci 1988; 8(4): 1144-51. [http://dx.doi.org/10.1523/JNEUROSCI.08-04-01144.1988] [PMID: 3357014]

[281] Dahlstrand J, Lardelli M, Lendahl U. Nestin mRNA expression correlates with the central nervous system progenitor cell state in many, but not all, regions of developing central nervous system. Brain Res Dev Brain Res 1995; 84(1): 109-29. [http://dx.doi.org/10.1016/0165-3806(94)00162-S] [PMID: 7720210]

[282] Suzuki S, Namiki J, Shibata S, Mastuzaki Y, Okano H. The neural stem/progenitor cell marker nestin is expressed in proliferative endothelial cells, but not in mature vasculature. J Histochem Cytochem 2010; 58(8): 721-30. [http://dx.doi.org/10.1369/jhc.2010.955609] [PMID: 20421592]

[283] Cahoy JD, Emery B, Kaushal A, *et al.* A transcriptome database for astrocytes, neurons, and oligodendrocytes: a new resource for understanding brain development and function. J Neurosci 2008; 28(1): 264-78. [http://dx.doi.org/10.1523/JNEUROSCI.4178-07.2008] [PMID: 18171944]

[284] Pompolo S, Harley VR. Localisation of the SRY-related HMG box protein, SOX9, in rodent brain. Brain Res 2001; 906(1-2): 143-8. [http://dx.doi.org/10.1016/S0006-8993(01)02574-4] [PMID: 11430871]

[285] Rivers LE, Young KM, Rizzi M, *et al.* PDGFRA/NG2 glia generate myelinating oligodendrocytes and piriform projection neurons in adult mice. Nat Neurosci 2008; 11(12): 1392-401. [http://dx.doi.org/10.1038/nn.2220] [PMID: 18849983]

[286] Nishiyama A, Komitova M, Suzuki R, Zhu X. Polydendrocytes (NG2 cells): multifunctional cells with lineage plasticity. Nat Rev Neurosci 2009; 10(1): 9-22. [http://dx.doi.org/10.1038/nrn2495] [PMID: 19096367]

[287] Lim L, Zhou H, Costa RH. The winged helix transcription factor HFH-4 is expressed during choroid plexus epithelial development in the mouse embryo. Proc Natl Acad Sci USA 1997; 94(7): 3094-9. [http://dx.doi.org/10.1073/pnas.94.7.3094] [PMID: 9096351]

[288] Blatt EN, Yan XH, Wuerffel MK, Hamilos DL, Brody SL. Forkhead transcription factor HFH-4 expression is temporally related to ciliogenesis. Am J Respir Cell Mol Biol 1999; 21(2): 168-76. [http://dx.doi.org/10.1165/ajrcmb.21.2.3691] [PMID: 10423398]

[289] Hackett BP, Brody SL, Liang M, Zeitz ID, Bruns LA, Gitlin JD. Primary structure of hepatocyte nuclear factor/forkhead homologue 4 and characterization of gene expression in the developing respiratory and reproductive epithelium. Proc Natl Acad Sci USA 1995; 92(10): 4249-53. [http://dx.doi.org/10.1073/pnas.92.10.4249] [PMID: 7753791]

[290] Ostrowski LE, Hutchins JR, Zakel K, O'Neal WK. Targeting expression of a transgene to the airway surface epithelium using a ciliated cell-specific promoter. Mol Ther 2003; 8(4): 637-45. [http://dx.doi.org/10.1016/S1525-0016(03)00221-1] [PMID: 14529837]

[291] Dor Y, Brown J, Martinez OI, Melton DA. Adult pancreatic beta-cells are formed by self-duplication

rather than stem-cell differentiation. Nature 2004; 429(6987): 41-6. [http://dx.doi.org/10.1038/nature02520] [PMID: 15129273]

[292] Hsieh PC, Segers VF, Davis ME, *et al.* Evidence from a genetic fate-mapping study that stem cells refresh adult mammalian cardiomyocytes after injury. Nat Med 2007; 13(8): 970-4. [http://dx.doi.org/10.1038/nm1618] [PMID: 17660827]

[293] Zhu Q, Whittemore SR, Devries WH, Zhao X, Kuypers NJ, Qiu M. Dorsally-derived oligodendrocytes in the spinal cord contribute to axonal myelination during development and remyelination following focal demyelination. Glia 2011; 59(11): 1612-21. [http://dx.doi.org/10.1002/glia.21203] [PMID: 21710609]

[294] Cai J, Qi Y, Hu X, *et al.* Generation of oligodendrocyte precursor cells from mouse dorsal spinal cord independent of Nkx6 regulation and Shh signaling. Neuron 2005; 45(1): 41-53. [http://dx.doi.org/10.1016/j.neuron.2004.12.028] [PMID: 15629701]

[295] Fogarty M, Richardson WD, Kessaris N. A subset of oligodendrocytes generated from radial glia in the dorsal spinal cord. Development 2005; 132(8): 1951-9. [http://dx.doi.org/10.1242/dev.01777] [PMID: 15790969]

[296] Vallstedt A, Klos JM, Ericson J. Multiple dorsoventral origins of oligodendrocyte generation in the spinal cord and hindbrain. Neuron 2005; 45(1): 55-67. [http://dx.doi.org/10.1016/j.neuron.2004.12.026] [PMID: 15629702]

[297] Orentas DM, Hayes JE, Dyer KL, Miller RH. Sonic hedgehog signaling is required during the appearance of spinal cord oligodendrocyte precursors. Development 1999; 126(11): 2419-29. [PMID: 10226001]

[298] Pringle NP, Yu WP, Guthrie S, *et al.* Determination of neuroepithelial cell fate: induction of the oligodendrocyte lineage by ventral midline cells and sonic hedgehog. Dev Biol 1996; 177(1): 30-42. [http://dx.doi.org/10.1006/dbio.1996.0142] [PMID: 8660874]

[299] Blakemore WF, Keirstead HS. The origin of remyelinating cells in the central nervous system. J Neuroimmunol 1999; 98(1): 69-76. [http://dx.doi.org/10.1016/S0165-5728(99)00083-1] [PMID: 10426364]

[300] Talbott JF, Loy DN, Liu Y, *et al.* Endogenous Nkx2.2+/Olig2+ oligodendrocyte precursor cells fail to remyelinate the demyelinated adult rat spinal cord in the absence of astrocytes. Exp Neurol 2005; 192(1): 11-24. [http://dx.doi.org/10.1016/j.expneurol.2004.05.038] [PMID: 15698615]

[301] Fok-Seang J, Mathews GA, ffrench-Constant C, Trotter J, Fawcett JW. Migration of oligodendrocyte precursors on astrocytes and meningeal cells. Dev Biol 1995; 171(1): 1-15. [http://dx.doi.org/10.1006/dbio.1995.1255] [PMID: 7556887]

[302] Williams A, Piaton G, Lubetzki C. Astrocytes--friends or foes in multiple sclerosis? Glia 2007; 55(13): 1300-12. [http://dx.doi.org/10.1002/glia.20546] [PMID: 17626262]

[303] Busch SA, Silver J. The role of extracellular matrix in CNS regeneration. Curr Opin Neurobiol 2007; 17(1): 120-7. [http://dx.doi.org/10.1016/j.conb.2006.09.004] [PMID: 17223033]

[304] Faulkner JR, Herrmann JE, Woo MJ, Tansey KE, Doan NB, Sofroniew MV. Reactive astrocytes protect tissue and preserve function after spinal cord injury. J Neurosci 2004; 24(9): 2143-55. [http://dx.doi.org/10.1523/JNEUROSCI.3547-03.2004] [PMID: 14999065]

[305] Okada S, Nakamura M, Katoh H, *et al.* Conditional ablation of Stat3 or Socs3 discloses a dual role for reactive astrocytes after spinal cord injury. Nat Med 2006; 12(7): 829-34. [http://dx.doi.org/10.1038/nm1425] [PMID: 16783372]

[306] Takahashi Y, Tsuji O, Kumagai G, *et al.* Comparative study of methods for administering neural stem/progenitor cells to treat spinal cord injury in mice. Cell Transplant 2011; 20(5): 727-39.

[http://dx.doi.org/10.3727/096368910X536554] [PMID: 21054930]

[307] Cizkova D, Novotna I, Slovinska L, *et al.* Repetitive intrathecal catheter delivery of bone marrow mesenchymal stromal cells improves functional recovery in a rat model of contusive spinal cord injury. J Neurotrauma 2011; 28(9): 1951-61.
[http://dx.doi.org/10.1089/neu.2010.1413] [PMID: 20822464]

[308] Hong EG, Ko HJ, Cho YR, *et al.* Interleukin-10 prevents diet-induced insulin resistance by attenuating macrophage and cytokine response in skeletal muscle. Diabetes 2009; 58(11): 2525-35.
[http://dx.doi.org/10.2337/db08-1261] [PMID: 19690064]

[309] Satti HS, Waheed A, Ahmed P, *et al.* Autologous mesenchymal stromal cell transplantation for spinal cord injury: A Phase I pilot study. Cytotherapy 2016; 18(4): 518-22.
[http://dx.doi.org/10.1016/j.jcyt.2016.01.004] [PMID: 26971680]

[310] Yoon SH, Shim YS, Park YH, *et al.* Complete spinal cord injury treatment using autologous bone marrow cell transplantation and bone marrow stimulation with granulocyte macrophage-colony stimulating factor: Phase I/II clinical trial. Stem Cells 2007; 25(8): 2066-73.
[http://dx.doi.org/10.1634/stemcells.2006-0807] [PMID: 17464087]

[311] Parr AM, Tator CH, Keating A. Bone marrow-derived mesenchymal stromal cells for the repair of central nervous system injury. Bone Marrow Transplant 2007; 40(7): 609-19.
[http://dx.doi.org/10.1038/sj.bmt.1705757] [PMID: 17603514]

[312] Enzmann GU, Benton RL, Talbott JF, Cao Q, Whittemore SR. Functional considerations of stem cell transplantation therapy for spinal cord repair. J Neurotrauma 2006; 23(3-4): 479-95.
[http://dx.doi.org/10.1089/neu.2006.23.479] [PMID: 16629631]

[313] Karimi-Abdolrezaee S, Eftekharpour E, Wang J, Morshead CM, Fehlings MG. Delayed transplantation of adult neural precursor cells promotes remyelination and functional neurological recovery after spinal cord injury. J Neurosci 2006; 26(13): 3377-89.
[http://dx.doi.org/10.1523/JNEUROSCI.4184-05.2006] [PMID: 16571744]

[314] Pfeifer K, Vroemen M, Caioni M, Aigner L, Bogdahn U, Weidner N. Autologous adult rodent neural progenitor cell transplantation represents a feasible strategy to promote structural repair in the chronically injured spinal cord. Regen Med 2006; 1(2): 255-66.
[http://dx.doi.org/10.2217/17460751.1.2.255] [PMID: 17465808]

[315] Eftekharpour E, Karimi-Abdolrezaee S, Wang J, El Beheiry H, Morshead C, Fehlings MG. Myelination of congenitally dysmyelinated spinal cord axons by adult neural precursor cells results in formation of nodes of Ranvier and improved axonal conduction. J Neurosci 2007; 27(13): 3416-28.
[http://dx.doi.org/10.1523/JNEUROSCI.0273-07.2007] [PMID: 17392458]

[316] Gage FH. Mammalian neural stem cells. Science 2000; 287(5457): 1433-8.
[http://dx.doi.org/10.1126/science.287.5457.1433] [PMID: 10688783]

[317] Tysseling-Mattiace VM, Sahni V, Niece KL, *et al.* Self-assembling nanofibers inhibit glial scar formation and promote axon elongation after spinal cord injury. J Neurosci 2008; 28(14): 3814-23.
[http://dx.doi.org/10.1523/JNEUROSCI.0143-08.2008] [PMID: 18385339]

[318] Peru RL, Mandrycky N, Nait-Oumesmar B, Lu QR. Paving the axonal highway: from stem cells to myelin repair. Stem Cell Rev 2008; 4(4): 304-18.
[http://dx.doi.org/10.1007/s12015-008-9043-z] [PMID: 18759012]

[319] Sasaki H, Ishikawa M, Tanaka N, *et al.* Administration of human peripheral blood-derived CD133+ cells accelerates functional recovery in a rat spinal cord injury model. Spine 2009; 34(3): 249-54.
[http://dx.doi.org/10.1097/BRS.0b013e3181913cde] [PMID: 19148043]

[320] Hess DC, Borlongan CV. Stem cells and neurological diseases. Cell Prolif 2008; 41 (Suppl. 1): 94-114.
[http://dx.doi.org/10.1111/j.1365-2184.2008.00486.x] [PMID: 18181951]

[321] Oz Oyar E, Kardeş O, Korkmaz A, Omeroğlu S. Effects of vascular endothelial growth factor on

ischemic spinal cord injury caused by aortic cross-clamping in rabbits. J Surg Res 2009; 151(1): 94-9. [http://dx.doi.org/10.1016/j.jss.2008.01.006] [PMID: 18599078]

[322] Walker TL, White A, Black DM, Wallace RH, Sah P, Bartlett PF. Latent stem and progenitor cells in the hippocampus are activated by neural excitation. J Neurosci 2008; 28(20): 5240-7. [http://dx.doi.org/10.1523/JNEUROSCI.0344-08.2008] [PMID: 18480280]

[323] Walker PA, Harting MT, Shah SK, *et al.* Progenitor cell therapy for the treatment of central nervous system injury: a review of the state of current clinical trials. Stem Cells Int 2010; 2010: 369578. [http://dx.doi.org/10.4061/2010/369578] [PMID: 21048846]

[324] Jiang PC, Xiong WP, Wang G, *et al.* A clinical trial report of autologous bone marrow-derived mesenchymal stem cell transplantation in patients with spinal cord injury. Exp Ther Med 2013; 6(1): 140-6. [http://dx.doi.org/10.3892/etm.2013.1083] [PMID: 23935735]

[325] Alexanian AR, Maiman DJ, Kurpad SN, Gennarelli TA. *in vitro* and *in vivo* characterization of neurally modified mesenchymal stem cells induced by epigenetic modifiers and neural stem cell environment. Stem Cells Dev 2008; 17(6): 1123-30. [http://dx.doi.org/10.1089/scd.2007.0212] [PMID: 18484898]

[326] Saha K, Keung AJ, Irwin EF, *et al.* Substrate modulus directs neural stem cell behavior. Biophys J 2008; 95(9): 4426-38. [http://dx.doi.org/10.1529/biophysj.108.132217] [PMID: 18658232]

[327] Xuan AG, Long DH, Gu HG, Yang DD, Hong LP, Leng SL. BDNF improves the effects of neural stem cells on the rat model of Alzheimer's disease with unilateral lesion of fimbria-fornix. Neurosci Lett 2008; 440(3): 331-5. [http://dx.doi.org/10.1016/j.neulet.2008.05.107] [PMID: 18579298]

[328] Dai G, Liu X, Zhang Z, *et al.* Comparative analysis of curative effect of CT-guided stem cell transplantation and open surgical transplantation for sequelae of spinal cord injury. J Transl Med 2013; 11: 315. [http://dx.doi.org/10.1186/1479-5876-11-315] [PMID: 24355001]

[329] U.S. National Library of Medicine. Neural stem cells and spinal cord injury Retrieved from https://clinicaltrialsgov

[330] Kishk NA, Gabr H, Hamdy S, *et al.* Case control series of intrathecal autologous bone marrow mesenchymal stem cell therapy for chronic spinal cord injury. Neurorehabil Neural Repair 2010; 24(8): 702-8. [http://dx.doi.org/10.1177/1545968310369801] [PMID: 20660620]

[331] Meletis K, Frisén J. Blood on the tracks: A simple twist of fate? Trends Neurosci 2003; 26(6): 292-6. [http://dx.doi.org/10.1016/S0166-2236(03)00125-5] [PMID: 12798597]

[332] Friedman JA, Windebank AJ, Moore MJ, Spinner RJ, Currier BL, Yaszemski MJ. Biodegradable polymer grafts for surgical repair of the injured spinal cord. Neurosurgery 2002; 51(3): 742-51. [http://dx.doi.org/10.1097/00006123-200209000-00024] [PMID: 12188954]

[333] Novikova LN, Novikov LN, Kellerth JO. Biopolymers and biodegradable smart implants for tissue regeneration after spinal cord injury. Curr Opin Neurol 2003; 16(6): 711-5. [http://dx.doi.org/10.1097/00019052-200312000-00011] [PMID: 14624081]

[334] Rokkanen PU, Böstman O, Hirvensalo E, *et al.* Bioabsorbable fixation in orthopaedic surgery and traumatology. Biomaterials 2000; 21(24): 2607-13. [http://dx.doi.org/10.1016/S0142-9612(00)00128-9] [PMID: 11071610]

[335] Weber RA, Breidenbach WC, Brown RE, Jabaley ME, Mass DP. A randomized prospective study of polyglycolic acid conduits for digital nerve reconstruction in humans. Plast Reconstr Surg 2000; 106(5): 1036-45. [http://dx.doi.org/10.1097/00006534-200010000-00013] [PMID: 11039375]

[336] Reid RL, Cutright DE, Garrison JS. Biodegradable cuff an adjunct to peripheral nerve repair: a study in dogs. Hand 1978; 10(3): 259-66. [http://dx.doi.org/10.1016/S0072-968X(78)80047-3] [PMID: 738657]

[337] Meek MF, den Dunnen WF, Robinson PH, Pennings AJ, Schakenraad JM. Evaluation of functional nerve recovery after reconstruction with a new biodegradable poly (DL-lactide-epsilon-caprolactone) nerve guide. Int J Artif Organs 1997; 20(8): 463-8. [http://dx.doi.org/10.1177/039139889702000809] [PMID: 9323510]

[338] Meek MF, Robinson PH, Stokroos I, Blaauw EH, Kors G, den Dunnen WF. Electronmicroscopical evaluation of short-term nerve regeneration through a thin-walled biodegradable poly(DLLA-epsilo--CL) nerve guide filled with modified denatured muscle tissue. Biomaterials 2001; 22(10): 1177-85. [http://dx.doi.org/10.1016/S0142-9612(00)00340-9] [PMID: 11352097]

[339] Gautier SE, Oudega M, Fragoso M, *et al.* Poly(alpha-hydroxyacids) for application in the spinal cord: resorbability and biocompatibility with adult rat Schwann cells and spinal cord. J Biomed Mater Res 1998; 42(4): 642-54. [http://dx.doi.org/10.1002/(SICI)1097-4636(19981215)42:4<642::AID-JBM22>3.0.CO;2-K] [PMID: 9827690]

[340] Oudega M, Gautier SE, Chapon P, *et al.* Axonal regeneration into Schwann cell grafts within resorbable poly(alpha-hydroxyacid) guidance channels in the adult rat spinal cord. Biomaterials 2001; 22(10): 1125-36. [http://dx.doi.org/10.1016/S0142-9612(00)00346-X] [PMID: 11352092]

[341] Pinzon A, Calancie B, Oudega M, Noga BR. Conduction of impulses by axons regenerated in a Schwann cell graft in the transected adult rat thoracic spinal cord. J Neurosci Res 2001; 64(5): 533-41. [http://dx.doi.org/10.1002/jnr.1105] [PMID: 11391708]

[342] Moore MJ, Friedman JA, Lewellyn EB, *et al.* Multiple-channel scaffolds to promote spinal cord axon regeneration. Biomaterials 2006; 27(3): 419-29. [http://dx.doi.org/10.1016/j.biomaterials.2005.07.045] [PMID: 16137759]

[343] Hartgerink JD, Beniash E, Stupp SI. Self-assembly and mineralization of peptide-amphiphile nanofibers. Science 2001; 294(5547): 1684-8. [http://dx.doi.org/10.1126/science.1063187] [PMID: 11721046]

[344] Hartgerink JD, Beniash E, Stupp SI. Peptide-amphiphile nanofibers: A versatile scaffold for the preparation of self-assembling materials. Proc Natl Acad Sci USA 2002; 99(8): 5133-8. [http://dx.doi.org/10.1073/pnas.072699999] [PMID: 11929981]

[345] Silva GA, Czeisler C, Niece KL, *et al.* Selective differentiation of neural progenitor cells by high-epitope density nanofibers. Science 2004; 303(5662): 1352-5. [http://dx.doi.org/10.1126/science.1093783] [PMID: 14739465]

[346] Teng KK, Hempstead BL. Neurotrophins and their receptors: Signaling trios in complex biological systems. Cell Mol Life Sci 2004; 61(1): 35-48. [http://dx.doi.org/10.1007/s00018-003-3099-3] [PMID: 14704852]

[347] Bregman BS, Kunkel-Bagden E, Schnell L, Dai HN, Gao D, Schwab ME. Recovery from spinal cord injury mediated by antibodies to neurite growth inhibitors. Nature 1995; 378(6556): 498-501. [http://dx.doi.org/10.1038/378498a0] [PMID: 7477407]

[348] Kearns SM, Laywell ED, Kukekov VK, Steindler DA. Extracellular matrix effects on neurosphere cell motility. Exp Neurol 2003; 182(1): 240-4. [http://dx.doi.org/10.1016/S0014-4886(03)00124-9] [PMID: 12821394]

[349] Bradbury EJ, Moon LD, Popat RJ, *et al.* Chondroitinase ABC promotes functional recovery after spinal cord injury. Nature 2002; 416(6881): 636-40. [http://dx.doi.org/10.1038/416636a] [PMID: 11948352]

[350] Fouad K, Schnell L, Bunge MB, Schwab ME, Liebscher T, Pearse DD. Combining Schwann cell

bridges and olfactory-ensheathing glia grafts with chondroitinase promotes locomotor recovery after complete transection of the spinal cord. J Neurosci 2005; 25(5): 1169-78. [http://dx.doi.org/10.1523/JNEUROSCI.3562-04.2005] [PMID: 15689553]

[351] Barritt AW, Davies M, Marchand F, *et al.* Chondroitinase ABC promotes sprouting of intact and injured spinal systems after spinal cord injury. J Neurosci 2006; 26(42): 10856-67. [http://dx.doi.org/10.1523/JNEUROSCI.2980-06.2006] [PMID: 17050723]

[352] Massey JM, Amps J, Viapiano MS, *et al.* Increased chondroitin sulfate proteoglycan expression in denervated brainstem targets following spinal cord injury creates a barrier to axonal regeneration overcome by chondroitinase ABC and neurotrophin-3. Exp Neurol 2008; 209(2): 426-45. [http://dx.doi.org/10.1016/j.expneurol.2007.03.029] [PMID: 17540369]

[353] Lemons ML, Howland DR, Anderson DK. Chondroitin sulfate proteoglycan immunoreactivity increases following spinal cord injury and transplantation. Exp Neurol 1999; 160(1): 51-65. [http://dx.doi.org/10.1006/exnr.1999.7184] [PMID: 10630190]

[354] Jones LL, Yamaguchi Y, Stallcup WB, Tuszynski MH. NG2 is a major chondroitin sulfate proteoglycan produced after spinal cord injury and is expressed by macrophages and oligodendrocyte progenitors. J Neurosci 2002; 22(7): 2792-803. [http://dx.doi.org/10.1523/JNEUROSCI.22-07-02792.2002] [PMID: 11923444]

[355] Yick LW, So KF, Cheung PT, Wu WT. Lithium chloride reinforces the regeneration-promoting effect of chondroitinase ABC on rubrospinal neurons after spinal cord injury. J Neurotrauma 2004; 21(7): 932-43. [http://dx.doi.org/10.1089/0897715041526221] [PMID: 15307905]

[356] Ikegami T, Nakamura M, Yamane J, *et al.* Chondroitinase ABC combined with neural stem/progenitor cell transplantation enhances graft cell migration and outgrowth of growth-associated protein-4--positive fibers after rat spinal cord injury. Eur J Neurosci 2005; 22(12): 3036-46. [http://dx.doi.org/10.1111/j.1460-9568.2005.04492.x] [PMID: 16367770]

[357] Baerwald KD, Popko B. Developing and mature oligodendrocytes respond differently to the immune cytokine interferon-gamma. J Neurosci Res 1998; 52(2): 230-9. [http://dx.doi.org/10.1002/(SICI)1097-4547(19980415)52:2<230::AID-JNR11>3.0.CO;2-B] [PMID: 9579413]

[358] Balabanov R, Strand K, Goswami R, *et al.* Interferon-gamma-oligodendrocyte interactions in the regulation of experimental autoimmune encephalomyelitis. J Neurosci 2007; 27(8): 2013-24. [http://dx.doi.org/10.1523/JNEUROSCI.4689-06.2007] [PMID: 17314297]

[359] Buntinx M, Gielen E, Van Hummelen P, *et al.* Cytokine-induced cell death in human oligodendroglial cell lines. II: Alterations in gene expression induced by interferon-gamma and tumor necrosis factor-alpha. J Neurosci Res 2004; 76(6): 846-61. [http://dx.doi.org/10.1002/jnr.20117] [PMID: 15160396]

[360] Buntinx M, Moreels M, Vandenabeele F, *et al.* Cytokine-induced cell death in human oligodendroglial cell lines: I. Synergistic effects of IFN-gamma and TNF-alpha on apoptosis. J Neurosci Res 2004; 76(6): 834-45. [http://dx.doi.org/10.1002/jnr.20118] [PMID: 15160395]

[361] Horiuchi M, Itoh A, Pleasure D, Itoh T. MEK-ERK signaling is involved in interferon-gamma-induced death of oligodendroglial progenitor cells. J Biol Chem 2006; 281(29): 20095-106. [http://dx.doi.org/10.1074/jbc.M603179200] [PMID: 16728393]

[362] Pouly S, Becher B, Blain M, Antel JP. Interferon-gamma modulates human oligodendrocyte susceptibility to Fas-mediated apoptosis. J Neuropathol Exp Neurol 2000; 59(4): 280-6. [http://dx.doi.org/10.1093/jnen/59.4.280] [PMID: 10759183]

[363] Vartanian T, Li Y, Zhao M, Stefansson K. Interferon-gamma-induced oligodendrocyte cell death: implications for the pathogenesis of multiple sclerosis. Mol Med 1995; 1(7): 732-43. [http://dx.doi.org/10.1007/BF03401888] [PMID: 8612196]

[364] Itoh T, Horiuchi M, Wakayama K, Itoh A, Pleasure D. [Demyelination and oligodendrocytes]. Rinsho Shinkeigaku 2011; 51(11): 1036-9. [Demyelination and oligodendrocytes]. [http://dx.doi.org/10.5692/clinicalneurol.51.1036] [PMID: 22277471]

[365] Tirotta E, Kirby LA, Hatch MN, Lane TE. IFN-γ-induced apoptosis of human embryonic stem cell derived oligodendrocyte progenitor cells is restricted by CXCR2 signaling. Stem Cell Res (Amst) 2012; 9(3): 208-17. [http://dx.doi.org/10.1016/j.scr.2012.06.005] [PMID: 22885102]

[366] Fife BT, Kennedy KJ, Paniagua MC, *et al.* CXCL10 (IFN-gamma-inducible protein-10) control of encephalitogenic CD4+ T cell accumulation in the central nervous system during experimental autoimmune encephalomyelitis. J Immunol 2001; 166(12): 7617-24. [http://dx.doi.org/10.4049/jimmunol.166.12.7617] [PMID: 11390519]

[367] Liu MT, Keirstead HS, Lane TE. Neutralization of the chemokine CXCL10 reduces inflammatory cell invasion and demyelination and improves neurological function in a viral model of multiple sclerosis. J Immunol 2001; 167(7): 4091-7. [http://dx.doi.org/10.4049/jimmunol.167.7.4091] [PMID: 11564831]

[368] Stiles LN, Hardison JL, Schaumburg CS, Whitman LM, Lane TE. T cell antiviral effector function is not dependent on CXCL10 following murine coronavirus infection. J Immunol 2006; 177(12): 8372-80. [http://dx.doi.org/10.4049/jimmunol.177.12.8372] [PMID: 17142734]

[369] Wojcik WJ, Swoveland P, Zhang X, Vanguri P. Chronic intrathecal infusion of phosphorothioate or phosphodiester antisense oligonucleotides against cytokine responsive gene-2/IP-10 in experimental allergic encephalomyelitis of lewis rat. J Pharmacol Exp Ther 1996; 278(1): 404-10. [PMID: 8764376]

[370] Lin W, Kemper A, Dupree JL, Harding HP, Ron D, Popko B. Interferon-gamma inhibits central nervous system remyelination through a process modulated by endoplasmic reticulum stress. Brain 2006; 129(Pt 5): 1306-18. [http://dx.doi.org/10.1093/brain/awl044] [PMID: 16504972]

[371] Lin W, Kunkler PE, Harding HP, Ron D, Kraig RP, Popko B. Enhanced integrated stress response promotes myelinating oligodendrocyte survival in response to interferon-gamma. Am J Pathol 2008; 173(5): 1508-17. [http://dx.doi.org/10.2353/ajpath.2008.080449] [PMID: 18818381]

[372] Lin W, Popko B. Endoplasmic reticulum stress in disorders of myelinating cells. Nat Neurosci 2009; 12(4): 379-85. [http://dx.doi.org/10.1038/nn.2273] [PMID: 19287390]

[373] Ziv Y, Schwartz M. Immune-based regulation of adult neurogenesis: Implications for learning and memory. Brain Behav Immun 2008; 22(2): 167-76. [http://dx.doi.org/10.1016/j.bbi.2007.08.006] [PMID: 17905567]

[374] Kohman RA, Rhodes JS. Neurogenesis, inflammation and behavior. Brain Behav Immun 2013; 27(1): 22-32. [http://dx.doi.org/10.1016/j.bbi.2012.09.003] [PMID: 22985767]

[375] Ziv Y, Avidan H, Pluchino S, Martino G, Schwartz M. Synergy between immune cells and adult neural stem/progenitor cells promotes functional recovery from spinal cord injury. Proc Natl Acad Sci USA 2006; 103(35): 13174-9. [http://dx.doi.org/10.1073/pnas.0603747103] [PMID: 16938843]

[376] Schröter A, Lustenberger RM, Obermair FJ, Thallmair M. High-dose corticosteroids after spinal cord injury reduce neural progenitor cell proliferation. Neuroscience 2009; 161(3): 753-63. [http://dx.doi.org/10.1016/j.neuroscience.2009.04.016] [PMID: 19364523]

[377] Vela JM, Molina-Holgado E, Arévalo-Martín A, Almazán G, Guaza C. Interleukin-1 regulates

proliferation and differentiation of oligodendrocyte progenitor cells. Mol Cell Neurosci 2002; 20(3): 489-502.
[http://dx.doi.org/10.1006/mcne.2002.1127] [PMID: 12139924]

[378] Iosif RE, Ekdahl CT, Ahlenius H, *et al.* Tumor necrosis factor receptor 1 is a negative regulator of progenitor proliferation in adult hippocampal neurogenesis. J Neurosci 2006; 26(38): 9703-12.
[http://dx.doi.org/10.1523/JNEUROSCI.2723-06.2006] [PMID: 16988041]

[379] Wu JP, Kuo JS, Liu YL, Tzeng SF. Tumor necrosis factor-alpha modulates the proliferation of neural progenitors in the subventricular/ventricular zone of adult rat brain. Neurosci Lett 2000; 292(3): 203-6.
[http://dx.doi.org/10.1016/S0304-3940(00)01472-5] [PMID: 11018312]

[380] Okada S, Nakamura M, Mikami Y, *et al.* Blockade of interleukin-6 receptor suppresses reactive astrogliosis and ameliorates functional recovery in experimental spinal cord injury. J Neurosci Res 2004; 76(2): 265-76.
[http://dx.doi.org/10.1002/jnr.20044] [PMID: 15048924]

[381] Okano H, Okada S, Nakamura M, Toyama Y. Neural stem cells and regeneration of injured spinal cord. Kidney Int 2005; 68(5): 1927-31.
[http://dx.doi.org/10.1111/j.1523-1755.2005.00621.x] [PMID: 16221167]

[382] Ying QL, Nichols J, Chambers I, Smith A. BMP induction of Id proteins suppresses differentiation and sustains embryonic stem cell self-renewal in collaboration with STAT3. Cell 2003; 115(3): 281-92.
[http://dx.doi.org/10.1016/S0092-8674(03)00847-X] [PMID: 14636556]

[383] Tropepe V, Hitoshi S, Sirard C, Mak TW, Rossant J, van der Kooy D. Direct neural fate specification from embryonic stem cells: A primitive mammalian neural stem cell stage acquired through a default mechanism. Neuron 2001; 30(1): 65-78.
[http://dx.doi.org/10.1016/S0896-6273(01)00263-X] [PMID: 11343645]

[384] Ward NL, Lamanna JC. The neurovascular unit and its growth factors: Coordinated response in the vascular and nervous systems. Neurol Res 2004; 26(8): 870-83.
[http://dx.doi.org/10.1179/016164104X3798] [PMID: 15727271]

[385] Zerlin M, Goldman JE. Interactions between glial progenitors and blood vessels during early postnatal corticogenesis: Blood vessel contact represents an early stage of astrocyte differentiation. J Comp Neurol 1997; 387(4): 537-46.
[http://dx.doi.org/10.1002/(SICI)1096-9861(19971103)387:4<537::AID-CNE5>3.0.CO;2-3] [PMID: 9373012]

[386] Huxlin KR, Sefton AJ, Furby JH. The origin and development of retinal astrocytes in the mouse. J Neurocytol 1992; 21(7): 530-44.
[http://dx.doi.org/10.1007/BF01186955] [PMID: 1500949]

[387] Palmer TD, Willhoite AR, Gage FH. Vascular niche for adult hippocampal neurogenesis. J Comp Neurol 2000; 425(4): 479-94.
[http://dx.doi.org/10.1002/1096-9861(20001002)425:4<479::AID-CNE2>3.0.CO;2-3] [PMID: 10975875]

[388] Palmer TD. Adult neurogenesis and the vascular Nietzsche. Neuron 2002; 34(6): 856-8.
[http://dx.doi.org/10.1016/S0896-6273(02)00738-9] [PMID: 12086632]

[389] Louissaint A Jr, Rao S, Leventhal C, Goldman SA. Coordinated interaction of neurogenesis and angiogenesis in the adult songbird brain. Neuron 2002; 34(6): 945-60.
[http://dx.doi.org/10.1016/S0896-6273(02)00722-5] [PMID: 12086642]

[390] Miyamoto N, Pham LD, Seo JH, Kim KW, Lo EH, Arai K. Crosstalk between cerebral endothelium and oligodendrocyte. Cell Mol Life Sci 2014; 71(6): 1055-66.
[http://dx.doi.org/10.1007/s00018-013-1488-9] [PMID: 24132511]

[391] Fagan SC, Hess DC, Hohnadel EJ, Pollock DM, Ergul A. Targets for vascular protection after acute

ischemic stroke. Stroke 2004; 35(9): 2220-5.
[http://dx.doi.org/10.1161/01.STR.0000138023.60272.9e] [PMID: 15284446]

[392] Hansen TM, Moss AJ, Brindle NP. Vascular endothelial growth factor and angiopoietins in neurovascular regeneration and protection following stroke. Curr Neurovasc Res 2008; 5(4): 236-45.
[http://dx.doi.org/10.2174/156720208786413433] [PMID: 18991658]

[393] Bagnard D, Vaillant C, Khuth ST, *et al.* Semaphorin 3A-vascular endothelial growth factor-165 balance mediates migration and apoptosis of neural progenitor cells by the recruitment of shared receptor. J Neurosci 2001; 21(10): 3332-41.
[http://dx.doi.org/10.1523/JNEUROSCI.21-10-03332.2001] [PMID: 11331362]

[394] Soker S, Miao HQ, Nomi M, Takashima S, Klagsbrun M. VEGF165 mediates formation of complexes containing VEGFR-2 and neuropilin-1 that enhance VEGF165-receptor binding. J Cell Biochem 2002; 85(2): 357-68.
[http://dx.doi.org/10.1002/jcb.10140] [PMID: 11948691]

[395] Bates D, Taylor GI, Minichiello J, *et al.* Neurovascular congruence results from a shared patterning mechanism that utilizes Semaphorin3A and Neuropilin-1. Dev Biol 2003; 255(1): 77-98.
[http://dx.doi.org/10.1016/S0012-1606(02)00045-3] [PMID: 12618135]

[396] Miao HQ, Soker S, Feiner L, Alonso JL, Raper JA, Klagsbrun M. Neuropilin-1 mediates collapsin-1/semaphorin III inhibition of endothelial cell motility: Functional competition of collapsin-1 and vascular endothelial growth factor-165. J Cell Biol 1999; 146(1): 233-42.
[http://dx.doi.org/10.1083/jcb.146.1.233] [PMID: 10402473]

[397] Carmeliet P. Blood vessels and nerves: Common signals, pathways and diseases. Nat Rev Genet 2003; 4(9): 710-20.
[http://dx.doi.org/10.1038/nrg1158] [PMID: 12951572]

[398] Coulthard MG, Duffy S, Down M, *et al.* The role of the Eph-ephrin signalling system in the regulation of developmental patterning. Int J Dev Biol 2002; 46(4): 375-84.
[PMID: 12141423]

[399] Holmberg J, Frisén J. Ephrins are not only unattractive. Trends Neurosci 2002; 25(5): 239-43.
[http://dx.doi.org/10.1016/S0166-2236(02)02149-5] [PMID: 11972959]

[400] Knöll B, Drescher U. Ephrin-As as receptors in topographic projections. Trends Neurosci 2002; 25(3): 145-9.
[http://dx.doi.org/10.1016/S0166-2236(00)02093-2] [PMID: 11852146]

[401] Sokoloff L. Relationships among local functional activity, energy metabolism, and blood flow in the central nervous system. Fed Proc 1981; 40(8): 2311-6.
[PMID: 7238911]

[402] Van't Veer A, Du Y, Fischer TZ, Boetig DR, Wood MR, Dreyfus CF. Brain-derived neurotrophic factor effects on oligodendrocyte progenitors of the basal forebrain are mediated through trkB and the MAP kinase pathway. J Neurosci Res 2009; 87(1): 69-78.
[http://dx.doi.org/10.1002/jnr.21841] [PMID: 18752299]

[403] Vondran MW, Clinton-Luke P, Honeywell JZ, Dreyfus CF. BDNF+/- mice exhibit deficits in oligodendrocyte lineage cells of the basal forebrain. Glia 2010; 58(7): 848-56.
[PMID: 20091777]

[404] Xiao J, Ferner AH, Wong AW, Denham M, Kilpatrick TJ, Murray SS. Extracellular signal-regulated kinase 1/2 signaling promotes oligodendrocyte myelination *in vitro*. J Neurochem 2012; 122(6): 1167-80.
[http://dx.doi.org/10.1111/j.1471-4159.2012.07871.x] [PMID: 22784206]

[405] Xiao J, Wong AW, Willingham MM, van den Buuse M, Kilpatrick TJ, Murray SS. Brain-derived neurotrophic factor promotes central nervous system myelination *via* a direct effect upon oligodendrocytes. Neurosignals 2010; 18(3): 186-202.

[http://dx.doi.org/10.1159/000323170] [PMID: 21242670]

[406] Moyon D, Pardanaud L, Yuan L, Bréant C, Eichmann A. Plasticity of endothelial cells during arterial-venous differentiation in the avian embryo. Development 2001; 128(17): 3359-70. [PMID: 11546752]

[407] de Wit J, Verhaagen J. Role of semaphorins in the adult nervous system. Prog Neurobiol 2003; 71(2-3): 249-67. [http://dx.doi.org/10.1016/j.pneurobio.2003.06.001] [PMID: 14687984]

[408] McTigue DM, Horner PJ, Stokes BT, Gage FH. Neurotrophin-3 and brain-derived neurotrophic factor induce oligodendrocyte proliferation and myelination of regenerating axons in the contused adult rat spinal cord. J Neurosci 1998; 18(14): 5354-65. [http://dx.doi.org/10.1523/JNEUROSCI.18-14-05354.1998] [PMID: 9651218]

[409] McKinnon RD, Matsui T, Dubois-Dalcq M, Aaronson SA. FGF modulates the PDGF-driven pathway of oligodendrocyte development. Neuron 1990; 5(5): 603-14. [http://dx.doi.org/10.1016/0896-6273(90)90215-2] [PMID: 2171589]

[410] Bansal R, Pfeiffer SE. Inhibition of protein and lipid sulfation in oligodendrocytes blocks biological responses to FGF-2 and retards cytoarchitectural maturation, but not developmental lineage progression. Dev Biol 1994; 162(2): 511-24. [http://dx.doi.org/10.1006/dbio.1994.1105] [PMID: 8150210]

[411] Decker L, Avellana-Adalid V, Nait-Oumesmar B, Durbec P, Baron-Van Evercooren A. Oligodendrocyte precursor migration and differentiation: combined effects of PSA residues, growth factors, and substrates. Mol Cell Neurosci 2000; 16(4): 422-39. [http://dx.doi.org/10.1006/mcne.2000.0885] [PMID: 11085879]

[412] Yasuda T, Grinspan J, Stern J, Franceschini B, Bannerman P, Pleasure D. Apoptosis occurs in the oligodendroglial lineage, and is prevented by basic fibroblast growth factor. J Neurosci Res 1995; 40(3): 306-17. [http://dx.doi.org/10.1002/jnr.490400304] [PMID: 7745624]

[413] Maki T, Ihara M, Fujita Y, *et al.* Angiogenic and vasoprotective effects of adrenomedullin on prevention of cognitive decline after chronic cerebral hypoperfusion in mice. Stroke 2011; 42(4): 1122-8. [http://dx.doi.org/10.1161/STROKEAHA.110.603399] [PMID: 21393586]

[414] Uezono Y, Nakamura E, Ueda Y, *et al.* Production of cAMP by adrenomedullin in human oligodendroglial cell line KG1C: Comparison with calcitonin gene-related peptide and amylin. Brain Res Mol Brain Res 2001; 97(1): 59-69. [http://dx.doi.org/10.1016/S0169-328X(01)00288-1] [PMID: 11744163]

[415] Martínez-Herrero S, Larráyoz IM, Ochoa-Callejero L, García-Sanmartín J, Martínez A. Adrenomedullin as a growth and cell fate regulatory factor for adult neural stem cells. Stem Cells Int 2012; 2012: 804717. [http://dx.doi.org/10.1155/2012/804717] [PMID: 23049570]

[416] Maki T, Ihara M, Fujita Y, *et al.* Angiogenic roles of adrenomedullin through vascular endothelial growth factor induction. Neuroreport 2011; 22(9): 442-7. [http://dx.doi.org/10.1097/WNR.0b013e32834757e4] [PMID: 21558971]

[417] Schwarz N, Renshaw D, Kapas S, Hinson JP. Adrenomedullin increases the expression of calcitonin-like receptor and receptor activity modifying protein 2 mRNA in human microvascular endothelial cells. J Endocrinol 2006; 190(2): 505-14. [http://dx.doi.org/10.1677/joe.1.06806] [PMID: 16899583]

[418] Iimuro S, Shindo T, Moriyama N, *et al.* Angiogenic effects of adrenomedullin in ischemia and tumor growth. Circ Res 2004; 95(4): 415-23. [http://dx.doi.org/10.1161/01.RES.0000138018.61065.d1] [PMID: 15242974]

[419] Ichikawa-Shindo Y, Sakurai T, Kamiyoshi A, *et al.* The GPCR modulator protein RAMP2 is essential for angiogenesis and vascular integrity. J Clin Invest 2008; 118(1): 29-39. [http://dx.doi.org/10.1172/JCI33022] [PMID: 18097473]

[420] Jin K, Zhu Y, Sun Y, Mao XO, Xie L, Greenberg DA. Vascular endothelial growth factor (VEGF) stimulates neurogenesis *in vitro* and *in vivo.* Proc Natl Acad Sci USA 2002; 99(18): 11946-50. [http://dx.doi.org/10.1073/pnas.182296499] [PMID: 12181492]

[421] Kano MR, Morishita Y, Iwata C, *et al.* VEGF-A and FGF-2 synergistically promote neoangiogenesis through enhancement of endogenous PDGF-B-PDGFRbeta signaling. J Cell Sci 2005; 118(Pt 16): 3759-68. [http://dx.doi.org/10.1242/jcs.02483] [PMID: 16105884]

[422] Spivak-Kroizman T, Lemmon MA, Dikic I, *et al.* Heparin-induced oligomerization of FGF molecules is responsible for FGF receptor dimerization, activation, and cell proliferation. Cell 1994; 79(6): 1015-24. [http://dx.doi.org/10.1016/0092-8674(94)90032-9] [PMID: 7528103]

[423] Weiss S, Dunne C, Hewson J, *et al.* Multipotent CNS stem cells are present in the adult mammalian spinal cord and ventricular neuroaxis. J Neurosci 1996; 16(23): 7599-609. [http://dx.doi.org/10.1523/JNEUROSCI.16-23-07599.1996] [PMID: 8922416]

[424] Kojima A, Tator CH. Intrathecal administration of epidermal growth factor and fibroblast growth factor 2 promotes ependymal proliferation and functional recovery after spinal cord injury in adult rats. J Neurotrauma 2002; 19(2): 223-38. [http://dx.doi.org/10.1089/08977150252806974] [PMID: 11893024]

[425] Dumont DJ, Yamaguchi TP, Conlon RA, Rossant J, Breitman ML. tek, a novel tyrosine kinase gene located on mouse chromosome 4, is expressed in endothelial cells and their presumptive precursors. Oncogene 1992; 7(8): 1471-80. [PMID: 1630810]

[426] Carlson TR, Feng Y, Maisonpierre PC, Mrksich M, Morla AO. Direct cell adhesion to the angiopoietins mediated by integrins. J Biol Chem 2001; 276(28): 26516-25. [http://dx.doi.org/10.1074/jbc.M100282200] [PMID: 11346644]

[427] Clegg DO. Novel roles for integrins in the nervous system. Mol Cell Biol Res Commun 2000; 3(1): 1-7. [http://dx.doi.org/10.1006/mcbr.1999.0175] [PMID: 10683310]

[428] Hinks GL, Franklin RJ. Distinctive patterns of PDGF-A, FGF-2, IGF-I, and TGF-beta1 gene expression during remyelination of experimentally-induced spinal cord demyelination. Mol Cell Neurosci 1999; 14(2): 153-68. [http://dx.doi.org/10.1006/mcne.1999.0771] [PMID: 10532806]

[429] Fok-Seang J, DiProspero NA, Meiners S, Muir E, Fawcett JW. Cytokine-induced changes in the ability of astrocytes to support migration of oligodendrocyte precursors and axon growth. Eur J Neurosci 1998; 10(7): 2400-15. [http://dx.doi.org/10.1046/j.1460-9568.1998.00251.x] [PMID: 9749768]

[430] Schnädelbach O, Mandl C, Faissner A. Expression of DSD-1-PG in primary neural and glial-derived cell line cultures, upregulation by TGF-beta, and implications for cell-substrate interactions of the glial cell line Oli-neu. Glia 1998; 23(2): 99-119. [http://dx.doi.org/10.1002/(SICI)1098-1136(199806)23:2<99::AID-GLIA2>3.0.CO;2-Z] [PMID: 9600379]

[431] Alberta JA, Park SK, Mora J, *et al.* Sonic hedgehog is required during an early phase of oligodendrocyte development in mammalian brain. Mol Cell Neurosci 2001; 18(4): 434-41. [http://dx.doi.org/10.1006/mcne.2001.1026] [PMID: 11640898]

[432] Lu QR, Sun T, Zhu Z, *et al.* Common developmental requirement for Olig function indicates a motor

neuron/oligodendrocyte connection. Cell 2002; 109(1): 75-86. [http://dx.doi.org/10.1016/S0092-8674(02)00678-5] [PMID: 11955448]

[433] Zhou Q, Anderson DJ. The bHLH transcription factors OLIG2 and OLIG1 couple neuronal and glial subtype specification. Cell 2002; 109(1): 61-73. [http://dx.doi.org/10.1016/S0092-8674(02)00677-3] [PMID: 11955447]

[434] Yung SY, Gokhan S, Jurcsak J, Molero AE, Abrajano JJ, Mehler MF. Differential modulation of BMP signaling promotes the elaboration of cerebral cortical GABAergic neurons or oligodendrocytes from a common sonic hedgehog-responsive ventral forebrain progenitor species. Proc Natl Acad Sci USA 2002; 99(25): 16273-8. [http://dx.doi.org/10.1073/pnas.232586699] [PMID: 12461181]

[435] Gokhan S, Marin-Husstege M, Yung SY, Fontanez D, Casaccia-Bonnefil P, Mehler MF. Combinatorial profiles of oligodendrocyte-selective classes of transcriptional regulators differentially modulate myelin basic protein gene expression. J Neurosci 2005; 25(36): 8311-21. [http://dx.doi.org/10.1523/JNEUROSCI.1850-05.2005] [PMID: 16148239]

[436] He Y, Dupree J, Wang J, *et al.* The transcription factor Yin Yang 1 is essential for oligodendrocyte progenitor differentiation. Neuron 2007; 55(2): 217-30. [http://dx.doi.org/10.1016/j.neuron.2007.06.029] [PMID: 17640524]

[437] Hotary KB, Robinson KR. Endogenous electrical currents and the resultant voltage gradients in the chick embryo. Dev Biol 1990; 140(1): 149-60. [http://dx.doi.org/10.1016/0012-1606(90)90062-N] [PMID: 2358115]

[438] Cao L, Wei D, Reid B, *et al.* Endogenous electric currents might guide rostral migration of neuroblasts. EMBO Rep 2013; 14(2): 184-90. [http://dx.doi.org/10.1038/embor.2012.215] [PMID: 23328740]

[439] Becker D, Gary DS, Rosenzweig ES, Grill WM, McDonald JW. Functional electrical stimulation helps replenish progenitor cells in the injured spinal cord of adult rats. Exp Neurol 2010; 222(2): 211-8. [http://dx.doi.org/10.1016/j.expneurol.2009.12.029] [PMID: 20059998]

[440] Huang Y, Li Y, Chen J, Zhou H, Tan S. Electrical Stimulation Elicits Neural Stem Cells Activation: New Perspectives in CNS Repair. Front Hum Neurosci 2015; 9: 586. [http://dx.doi.org/10.3389/fnhum.2015.00586] [PMID: 26539102]

[441] Kim YR, Kim HN, Ahn SM, Choi YH, Shin HK, Choi BT. Electroacupuncture promotes post-stroke functional recovery *via* enhancing endogenous neurogenesis in mouse focal cerebral ischemia. PLoS One 2014; 9(2): e90000. [http://dx.doi.org/10.1371/journal.pone.0090000] [PMID: 24587178]

[442] Parent JM, Yu TW, Leibowitz RT, Geschwind DH, Sloviter RS, Lowenstein DH. Dentate granule cell neurogenesis is increased by seizures and contributes to aberrant network reorganization in the adult rat hippocampus. J Neurosci 1997; 17(10): 3727-38. [http://dx.doi.org/10.1523/JNEUROSCI.17-10-03727.1997] [PMID: 9133393]

[443] Li L, El-Hayek YH, Liu B, *et al.* Direct-current electrical field guides neuronal stem/progenitor cell migration. Stem Cells 2008; 26(8): 2193-200. [http://dx.doi.org/10.1634/stemcells.2007-1022] [PMID: 18556511]

[444] Ariza CA, Fleury AT, Tormos CJ, *et al.* The influence of electric fields on hippocampal neural progenitor cells. Stem Cell Rev 2010; 6(4): 585-600. [http://dx.doi.org/10.1007/s12015-010-9171-0] [PMID: 20665129]

[445] Meng X, Arocena M, Penninger J, Gage FH, Zhao M, Song B. PI3K mediated electrotaxis of embryonic and adult neural progenitor cells in the presence of growth factors. Exp Neurol 2011; 227(1): 210-7. [http://dx.doi.org/10.1016/j.expneurol.2010.11.002] [PMID: 21092738]

[446] Liu J, Zhu B, Zhang G, *et al.* Electric signals regulate directional migration of ventral midbrain

derived dopaminergic neural progenitor cells *via* Wnt/GSK3β signaling. Exp Neurol 2015; 263: 113-21.
[http://dx.doi.org/10.1016/j.expneurol.2014.09.014] [PMID: 25265211]

[447] Feng JF, Liu J, Zhang XZ, *et al.* Guided migration of neural stem cells derived from human embryonic stem cells by an electric field. Stem Cells 2012; 30(2): 349-55.
[http://dx.doi.org/10.1002/stem.779] [PMID: 22076946]

[448] Kobelt LJ, Wilkinson AE, McCormick AM, Willits RK, Leipzig ND. Short duration electrical stimulation to enhance neurite outgrowth and maturation of adult neural stem progenitor cells. Ann Biomed Eng 2014; 42(10): 2164-76.
[http://dx.doi.org/10.1007/s10439-014-1058-9] [PMID: 24957636]

CHAPTER 2

Toward Induced Pluripotent Stem Cells for Clinical Use: Sources, Methods and Selection

Maria P De Miguel*, **Adrian Moratilla, Alba Cabrera-Fisac** and **P Gonzalez-Molina**

Cell Engineering Laboratory, La Paz University Hospital Health Research Institute, IdiPAZ, Madrid, Spain

Abstract: Since their development by Yamanaka in 2007, much progress has been made in the last decade toward the use of human induced pluripotent stem cells (iPSCs) in clinical practice. In this review, we will focus on the various sources of somatic cells for human iPSC generation, the methods used for generating human iPSCs, their characterization, and the progress on directed differentiation toward several cell types. We will also describe current efforts to prevent culture-driven mutations and the selection of nontumorigenic cells for clinical use. A comprehensive comparison of such methods will aid in the establishment of standardized techniques and highlight areas in which further research is still needed.

Keywords: cMyc, Cell Reprogramming, Cell Therapy, Clinical Grade IPSC, Differentiation, Disease Modelling, Epigenetics, Episomal Vectors, Embryonic Stem Cells, Genetic Stability, Induced Pluripotent Stem Cells, Klf4, Matrigel, Oct4, PBMCs, Pluripotency, Quality Control, Retrovirus, Sox2, SSEA, Transfection, Transplantation.

INTRODUCTION TO INDUCED PLURIPOTENT STEM CELLS

Cell reprogramming has historically been achieved by various methods, such as somatic cell nuclear transfer [1, 2], cell fusion with pluripotent cells [3], incubation with pluripotent cell extracts [4], and derivation of pluripotent embryonic germ cells (EGCs) from primordial germ cells by addition of a specific cocktail of growth factors [5, 6].

More recently, studies on the genes involved in stemness led to the ground-breaking work of Yamanaka and Thomson, showing that forced expression of just

* **Corresponding author Maria P De Miguel:** Cell Engineering Laboratory, La Paz University Hospital Health Research Institute, IdiPAZ, Madrid, Spain; Tel: 34-912071458; Fax: 34-917277050; E-mail: mariapdemiguel@gmail.com

Atta-ur-Rahman and Shazia Anjum (Eds.)

four transcription factors can reprogram mouse and human fibroblasts into a full pluripotent state, the so-called induced pluripotent stem cells (iPSCs) [7 - 9].

Because the use of human embryonic stem cells in the clinic is controversial due to the ethics involved, to be able to reprogram adult somatic cells into pluripotent stem cells, thus avoiding any ethical or moral deterrent, was remarkable; so much so that Yamanaka received the Nobel Prize in Medicine only 6 years after the initial discovery.

SOURCES

The selection of the cell source for reprograming is a very important aspect to take into account, given not every cell type is equally efficiently reprogramed or equally accessible. Here, we will take a look at the beginnings of iPSC derivation and the various cell types that have since been successfully used to create iPSCs. We will also evaluate the various cell types that are actually being used in research, particularly with prospective clinical applications in mind.

In the Beginning: Mouse Embryonic Fibroblasts

Although in 2006 Takahashi and Yamanaka were not the first to achieve or prove cellular reprograming of somatic cells [7], remarkably, they identified a combination of only four transcription factors (Oct4, Sox2, Klf4, and cMyc, named Yamanaka factors or OSKM), present in embryonic stem cells that could completely reverse somatic differentiated cells into an embryonic pluripotent state [7]. The first generation of iPSCs that Yamanaka's group obtained failed to produce viable chimeras when injected in mouse embryos. Upon changing the selection method for the reprogrammed cells, a collaborative effort with other laboratories allowed Yamanaka and colleagues to develop the second generation of iPSCs, which could successfully form chimeras and therefore give rise to the three germ layers of the embryo. The actual establishment of the first iPSC lines was performed on mouse embryonic fibroblasts (MEFs) and later on adult mouse tail tissue fibroblasts (TTFs). Once 24 genes that were deemed important transcripts for the maintenance of embryonic stem cell (ESC) properties were identified, they transduced various combinations of those genes by retroviral transfection in MEFs, and the resulting cells were screened for pluripotency [7]. As previously mentioned, the exact cocktail of transcription factors needed for the reprograming event was found: *Oct3/4*, *Sox2*, *c-Myc* and *Klf4*. When transfected with these four genes, their activity was reactivated and the whole transcription program of the cell changed to that of the pluripotent state in both MEFs and TTFs.

Today, the reprograming of mouse fibroblasts is not the focus of new research on iPSCs, given their clinical application potential is limited. However, mouse fibroblast-derived iPSCs are still used in other types of experiments, for example to test new methods for increased efficiency or to establish new reprograming pathways that differ between cell lineages. MEFs are also widely used as feeder cells in iPSC cultures, although they need to be treated beforehand with mitomycin or irradiation to prevent growth.

Human Fetal and Neonatal Cells

Human Neonatal Fibroblasts

Human neonatal fibroblasts were used as a source for iPSC generation soon after the discovery that younger cells are more easily reprogrammed [10, 11]. These fibroblasts are usually extracted from foreskin samples rather than skin biopsies, which is a safer approach for young individuals. They have the same pros and cons as regular fibroblasts: they require a somewhat long culture time for expansion purposes but then are more easily reprogrammed than other cell types (see below), especially taking into account the young "age" of the cells. They have also been proven to be efficiently reprogrammed even after long-term cryopreservation, which is important in terms of the availability of the samples [12].

Fetal Stem Cells

Fetal stem cells have also been used extensively as a source for reprogrammed cells. Because they are already stem cells, they are easier to reprogram, have better tissue repair capacity, and sometimes have more rapid growth kinetics than adult stem cells. They can be obtained from newborn and fetal tissues, such as blood, liver, and bone marrow, and also from extraembryonic structures, such as the placenta, cord blood, and amniotic fluid [13, 14].

Fetal stem cells are different from adult stem cells in the sense that they are half way between a pluripotent and a stem cell-like phenotype, presenting markers of both states [14]. However, the initial pool of cells can be very heterogeneous, thus affecting the efficiency of the reprogramming between experiments and making it more difficult to pinpoint the exact nature of the original colony-forming cell [15].

Human Umbilical Cord Endothelial Cells

Human umbilical cord endothelial cells (HUVECs) present a very accessible source of good quality cells for reprogramming, given the extraction procedure carries no risks whatsoever. Also, these neonatal cells are young and therefore less prone to carrying somatic mutations at the time of isolation. HUVECs can also be used after long periods of freezing. The most successful attempt to reprogram these cells took place in 2010, in a study by Panopoulos *et al.* [16]. In their study, they successfully reprogrammed HUVECs at 300 times higher efficiency than that already achieved for human fibroblasts and at approximately 100 times higher efficiency than that already obtained in HUVECs using other protocols, such as those by Haase *et al.* [17], or Giorgetti *et al.* [18], Later, protocols eliminating viral vectors were proven, but with lower efficiency (0.1%) [19].

HUVEC-derived iPSCs have been successfully redifferentiated into somatic cell types, such as neurons and astrocytes [20], showing their great potential application in the clinic.

Adult Human Cells

Adult cells are the most accessible source of cells for reprogramming if there is a need for the iPSC to have a certain haplotype or to create a cell line from a diseased individual. The reprogramming efficiency varies wildly depending on the cell type that is used as a source, the transfection method, and the culture conditions (Table **1**). In most cases, however, it is advisable to choose the cell source based on accessibility rather than efficiency of the reprogramming process, and then try various methods and gene combinations to increase efficiency if need be. This decision, of course, will depend on the exact experiment that will be performed once the iPSC line is established.

Table 1. Most common cell sources and their average reprogramming efficiency. ADSCs: adipose-derived stem cells. PBMCs: peripheral blood mononuclear cells.

Donor age	Cell type	Efficiency	Reference
Newborn	Fibroblasts	0.01–1%	Yu *et al.*, 2009 [10]
Newborn	HUVECs	2.5–3%	Panopoulos *et al.*, 2011 [16]
Adult	Fibroblasts	0.002%	Takahashi *et al.*, 2007 [8]
Adult	Keratinocytes	1%	Aasen *et al.*, 2008 [23]
Adult	Melanocytes	0.19%	Utikal *et al.*, 2009 [25]
Adult	Hepatocytes	0.1%	Liu *et al.*, 2010 [26]
Adult	Neural stem cells	0.004%	Kim *et al.*, 2009 [24]
Adult	ADSCs	0.5%	Sugii *et al.*, 2010 [27]
Adult	PBMCs	0.0008–0.001%	Loh *et al.*, 2010 [28]
Adult	T lymphocytes	0.1%	Seki *et al.*, 2010 [29]
Adult	Exfoliated urinary cells	4%	Zhou *et al.*, 2012 [30]

Fibroblasts typically have a reprogramming efficiency of 0.002% [8], although it varies depending on the method used and the age of the actual cells (neonatal cells have shown a better efficiency than more adult cells) [21]. Given that frequently, adult ***dermal fibroblasts*** are available, and adult skin has typically been exposed to UV radiation, it is likely that these cells already carry mutations and chromosomal aberrations prior to the reprogramming process [22]. Interestingly, epidermal ***keratinocytes*** have been shown to form colonies twice as rapidly as fibroblasts [23].

Neural stem cells, on the other hand, display no mutations but achieve a much lower efficiency, with less than 0.004% of reprogramming events [24]. On the other hand, ***melanocytes*** have shown a very high reprogramming efficiency, reaching as much as 0,19% [25], and ***hepatocytes*** have been reprogrammed with up to 0.1% efficiency [26].

Adipose tissue derived mesenchymal stem cells have reached efficiencies of 0.5% [27]. This cell type is especially interesting because the extraction protocol is relatively easy and poses few risks, with high numbers of cells extracted. The downside is that the procedure must be performed in an operating room, thus elevating the cost of the samples.

One of the easiest cell types to obtain are ***peripheral blood mononuclear cells*** (PBMCs), given they can be extracted from a large enough blood sample using a minimally invasive procedure. Although dermal fibroblasts have a slightly higher reprograming efficiency, a skin biopsy is a much more invasive and risky procedure than blood extraction and it also requires an expansion period in order to obtain the amount of cells needed [31]. Blood cell-derived iPSCs have been extensively studied, with the efficiency of the reprograming compared among the different subpopulations: CD34+ cells showed a reprogramming efficiency of 0.002%, whereas PBMCs showed a very low efficiency, ranging from 0.0008% to 0.001% [28] with the regular OSKM factors. Some other approaches to ensure the safety of reprogramed cells for *in vivo* use have also been performed [29] and have managed to increase efficiency to as much as 0.1% in T lymphocytes.

Finally, exfoliated urinary cells/renal epithelial cells have been found to require less than 2 weeks to produce reprogrammed colonies, with an efficiency of up to 4%, and are perhaps the simplest and least risky option because they are extracted from simple urine samples [30]. The major downside is that, like fibroblasts, they require expansion time in culture.

How to Choose the Cell Source

Cell source should be chosen depending on availability, type of patient and research goals Fig. (**1**). Overall, the most accessible sources of cells for reprogramming are exfoliated urinary cells and PBMCs from adults, HUVECs from neonates, and amniotic fluid cells from fetuses. Among these, exfoliated urinary cells and HUVECs present the best efficiencies; however, PBMCs are a faster and safer option because they do not require previous expansion in culture. In the case of equal availability, PBMCs will produce more rapid results, whereas other cell types will produce more reprogrammed cells; thus, the choice will need to be evaluated for every case.

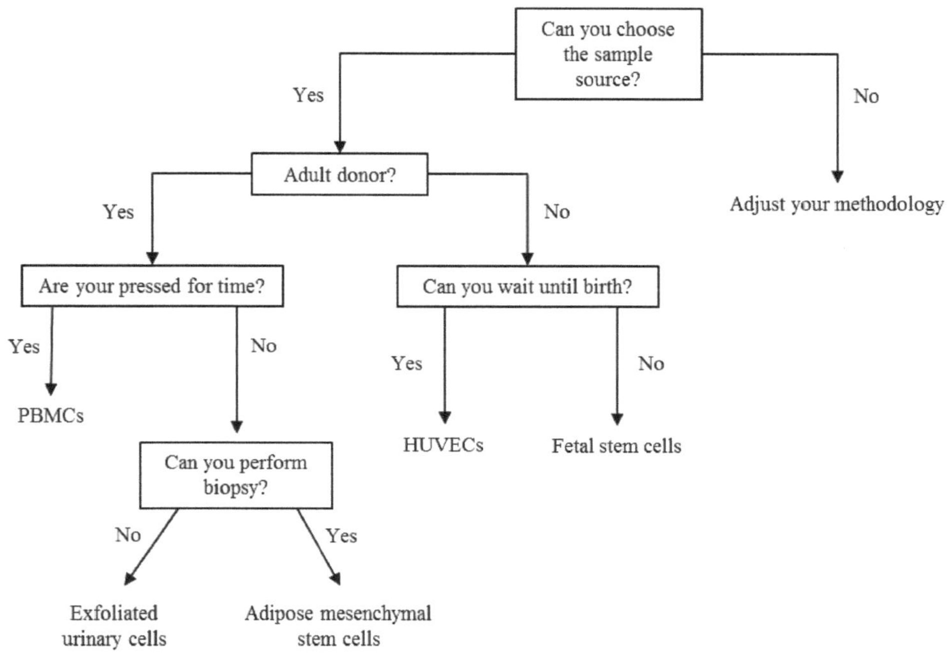

Fig. (1). Workflow for choosing the cell source of origin.

It is also worthy of note that iPSCs retain some epigenetic memory of the cell type from which they come and it is usually easier to differentiate them again into a cell type from the same germ layer as the source type. This effect, however, has been shown to minimize with time and passage events, suggesting that after a sufficiently long time this epigenetic memory can disappear [32]. This disappearance will be very important in the clinical context, when personalized cells will need to be created and used as soon as possible, thus, time is a limiting factor.

METHODS

Human iPSCs can be generated by inducing the expression of reprogramming factors that can be delivered by various methods, depending on the factors, the cell type, and the intended use of the derived cells. In general, the approaches can be classified into virus-mediated and viral-free. Viral-based delivery can be achieved by either genome integrating or nonintegrating viral vectors. Virus-free approaches are nonintegrating and are nucleic acid-dependent or nucleic acid-free.

Viral Methods

Several methods of iPSC derivation use viruses as vectors to deliver the reprogramming genes. These technologies include vectors that integrate into the host genome, such as lentivirus and retrovirus, as well as adenoviruses or Sendai viruses, which do not integrate.

Integrative

This reprogramming approach started with the work of Yamanaka's group, inducing the expression of the four reprogramming factors (OSKM) by **retroviral**-mediated transduction in MEFs or TTFs [7]. After transduction, drug selection with G418 resulted in reprogramming of 0.02% of the MEFs or TTFs at 14–21 days post-transduction. These vectors had strong constitutively active CAG promoters for higher efficiency, integrating their genome into the host chromosome.

Lentivirus is another type of integrating vector, capable of transducing dividing and nondividing cells, unlike retroviruses, which require mitosis for transduction; thus, lentivirus is better for nondividing cells such as neurons, PBMCs, and other terminally differentiated cells with efficiencies ranging from 0.01-1% [33].

These techniques are simple to implement in various cell types and are sufficiently efficient; however, they are followed by many challenges for their therapeutic use. The two main concerns are the suboptimal stoichiometry achieved incorporating four or more individual vectors [34] and the incorporation of the potential risks of random and permanent integration of viruses into the host genome at multiple sites, which can result in random insertional mutations and even in tumorigenesis.

Bacteriophages can also be used to deliver genetic information to human cells. Typically these viruses use an integrase enzyme to bind its phage attachment sites (attP) into bacterial DNA in bacterial attachment sites (attB). AttB-like binding sites have been found in the human genome, mostly in intergenic locations; however, some can also be located in introns. Reprogramming efficiency with this method is 0.0029% [35].

Additionally, integrative methods that use enzymes, such as site-specific zinc finger nucleases, can be designed to generate patient-specific human-induced pluripotent stem cells (hiPSCs) by site-specific insertion of stem cell transcription factor genes flanked by LoxP sites [36]. Similar to nucleases, the combination of CRISPR/Cas9 with iPSCs opens new perspectives for the use of hiPSCs. However, Cas9 nuclease can lead to genetic damage, and off-target DNA breaks because of its imprecise reparation. The precision of this method is expected to improve, but for now the integrity of iPSCs modified using CRISPR/Cas9 must be verified; thus, this approach is not yet ready for clinical use [37].

Non-Integrative Viruses

Many studies have investigated transgene delivery *via* nonintegrating viruses: adenoviral vectors and Sendai virus vectors.

Adenoviruses are one of the most common vectors used for generating iPSCs. Like lentiviruses, they are capable of transducing dividing and nondividing cells, and there is no integration into the genome [38]. However, the reprogramming efficiency of this approach is very low: 0.001%–0.0001% in mouse cells [39], and adenovirus sometimes induces an innate immune response [40]. Although these viral vectors do not integrate into the genome, further work must be performed to find any useful application in this field to increase the reprogramming efficiency, because now in human cells it is 0.0002%.

Sendai Viruses are a respiratory pathogen of rats and mice, with the helpful quality that they are RNA viruses that replicate in the cytoplasm of infected cells. This method has efficiencies as high as 0.1% in blood cells and 1% in human fibroblasts [41]. However, the cost and difficulty is higher than other viral methods, and it takes 10 passages for the virus to be diluted out of cells after infection. On the other hand, the Sendai virus is nonpathogenic to humans, and thus this method is considered an essential tool for reprogramming [42].

Transgene Excision

After viral transduction and reprogramming, it is advisable to excise integrated viral sequences when using generated iPSCs in clinical practice. Two systems enable cells to excise integrated viral sequences and incorporate large DNA fragments.

The Cre-loxP system is a single cassette reprograming vector in which each of the four reprogramming factors was separated by a self-cleaving peptide signal. The loxP site allows the integrated sequence to be excised by the overexpression of Cre recombinase [43]. This vector, called STEMCCA, provides reprogramming efficiencies of 0.1-1% in mouse and human cells [3]. However there is a risk of insertional mutations because some residual vector sequences remain [44].

The main transposable elements used in reprogramming are PiggyBac (PB) and Sleeping Beauty (SB) [45, 46]. PB was identified as an active element in insects. SB originates from fish, where it remains as an inactive element and reawakens by an *in vitro* mutagenesis approach. Both transposon systems have been shown to be suitable for the derivation of iPSCs, integrating large DNA fragments and subsequently excising them from the cell genome. However, the integration profile of PB tends to target genes and their upstream regulatory regions, and the one from SB is near random [47].

Nonviral Methods

For nonviral methods of reprogramming, some nucleic acid-based approaches have been developed, given they are more stable than thawing and refreezing viruses as needed. The virus-free approaches also use recombinant proteins and small reprogramming molecules. To use these approaches, transfection of the factors into the cells needs to be accomplished. There are several methods of introducing nucleic acids into cells, such as chemical and physical transfection. Chemical transfection was the first method used to introduce genes into mammalian cells, using cationic polymers, calcium phosphate, cationic lipids (the most used), or cationic amino acids. All of these positively charged compounds create complexes with nucleic acids that are attracted to the negatively charged cell membrane. Some examples of this method are diethylaminoethyl-dextran, polyethyleneimine, polybrene, calcium phosphate, and the widely used lipofectamine. The transfection efficiency varies depending on the nucleic acid/ chemical ratio, solution pH, and cell type used, all of which are highly efficient and commercially available products. They have consistency and reproducibility without the size limitation of packaged nucleic acid and low toxicity with no mutagenesis. Physical transfection methods are simple in principle and straightforward, are less dependent on cell type, and can be used in single cell-transfection; however, they require special instruments, are more laborious, and are skill dependent. The most frequently used physical method is electroporation (a short electrical pulse disturbs the cell membrane, making holes through which nucleic acids can pass); however, there are also others, such as biolistic particle delivery (using gold particles to conjugate with nucleic acids, which shoot them into the cells with a "gene gun"), direct microinjection (injecting the nucleic acids directly into the cytoplasm, although this causes significant cell death), laser-based transfection (also called optoporation, using a pulse to irradiate the cell membrane and form a pore, which is useful in small cells but requires expensive equipment), ultrasound (sonoporation), and magnetic fields (magnetofection).

Combinations of both chemical and physical methods, such as nucleofection, is the most recommended method for difficult-to-transfect cells or for primary cells, but this requires specific equipment. One of the most commonly used instruments is the Amaxa Nucleofector (Lonza), but there are also others, such as the Neon Transfection System (Thermo Fisher Scientific) [48].

Nucleic Acids

Episomal vectors express two components of the Epstein-Barr virus, OriP and EBNA-1. These elements lead to expression of reprogramming factors for a longer time than standard plasmids. They are capable of permanent extrachromosomal replication once per cycle, without integration [10]. The efficiency using the OSKM factors with this method was lower than that of using retrovirus (0.002% in mouse cells and <0.001% in human cells); however, modifications of the vectors yielded a higher efficiency using OCT4, SOX2, KLF4, L-MYC, LIN-28, and shRNA for p53, up to 0.1% in $CD34^+$ cells [49]. However, using a p53 knockdown for therapeutic use is problematic due to the possibility of unexpected p53-related oncogenesis. Overall, episomal reprogramming is an effective method to generate footprint-free iPSCs, meaning there will be no traces of the vector once the iPSCs are generated, so there would be no dangers when used in the clinic, with the only negative trait being an inability to easily reprogram certain cell types such as fibroblasts, bone marrow mononuclear cells, PBMCs, and lymphoblast B cells.

Minicircles are vectors containing only cDNA and a eukaryotic promoter to express it. They have an advantage with respect to plasmids because they might be less immunogenic; however, the efficiency is lower, at approximately 0.005% [50]. To prevent DNA integration, **RNAs** can be transfected directly into cells as modified mRNA. These mRNAs have a 5' guanine cap to increase their half-life, but they might induce an immune response because not all of them are capped, showing 5'triphospates [51]. This response can be reduced by various methods, but there are other disadvantages, such as short half-lives, requiring various rounds of transfection [52]. To reduce the frequency of electroporation and other transfection methods, a self-replicating positive-strand RNA replicon was developed from the Venezuelan equine encephalitis virus [53]. Successful generation of iPSCs has also been achieved with **microRNAs,** which is more efficient than retroviral transduction. This method also avoids reactivation of transgenes and insertional mutations and is a promising method that could be clinically applied [54].

Nucleic Acid-Free

Reprogramming factors can be introduced as ***recombinant proteins*** for successfully generating iPSCs *via* a protein transduction domain that allows translocation across the membrane, retaining their biological activity [55]. This method prevents genetic modification of the iPSCs, but the efficiency is very low (0.006% in mouse and 0.001% in human cells at approximately 2 months); thus, more research is required for it to be useful in the future. While the risks of random integration and viral reactivation are lowered by many of these nonviral methods, the efficiency is not comparable. ***Small molecules*** can improve efficiency and iPSC quality and do not require a delivery method into the cells. For example, iPSCs can be generated with seven molecules: valproic acid (a wide histone deacetylase inhibitor) CHIR99021 (a GSK3β inhibitor), RepSox (E616452, an ALK5 inhibitor), tranylcypromine (a lysine-specific demethylase 1 inhibitor), forskolin (a cyclic adenosine monophosphate signaling activator), 3-deazaneplanocin A (DZNep, a DNA methyltransferase inhibitor), and TTNPB (a synthetic retinoic acid receptor ligand) [56]. These small molecules trigger multiple pathways and together take over the function of the classical reprogramming factors; thus, they are a promising approach to creating iPSCs, with up to 0.2% reprogramming efficiency at 28 days [57].

Various Gene Combinations

The first combination of reprogramming factors for generating iPSCs is known as Yamanaka factors, a mixture of OCT3/4, SOX2, KLF4, and C-MYC [7]. There are also methods using only three of the factors, excluding C-MYC given it is a known oncogene, but its reprogramming efficiency is reduced compared with the Yamanaka factors. Thomson's reprogramming mix includes OCT4, SOX2, and NANOG, which encode three well-established pluripotency-associated transcription factors, along with a fourth, *Lin28*, which encodes an RNA-binding protein [21]. Another combination is OCT4, SOX2, KLF4, and TBX3, which increases efficiency [58]. Using Zic3 instead of C-MYC enhances reprogramming efficiency, but it also reactivates the endogenous Zic3 protein that is essential for the generation of iPSCs [59]. Additionally, using L-MYC instead of C-MYC increases the efficiency and generation of iPSCs [60]. Another factor capable of replacing C-MYC is GLIS1, which shows more reprogramming efficiency at early stages and reduced tumor formation [61].

Additionally, lineage-specific genes can replace OCT4 and SOX2 in reprogramming [62]. Gata3 and other mesodermal-specifying genes can replace Oct4 as a reprogramming factor [63], and SOX2 was replaced with ZNF521, an early ectodermal lineage marker, although the efficiency was not as high as with the OSKM factors [62]. These data might indicate that OCT4 and SOX2 can function as lineage specifiers and are not only limited to pluripotency genes [63].

Additional Factors/Epigenetic Modulators

Transcription factor p53 is a tumor suppressor that induces cell cycle arrest, senescence, DNA repair, and apoptosis, preventing the initiation of tumor formation. As stated above, inhibiting this protein or functional mutations of the gene enhances nuclear reprogramming [64, 65]. Combined with non-integrating methods, suppression of p53 leads to an improvement in iPSC generation [49]. Small molecules could be a useful tool for inhibiting this pathway in a reversible manner; for example, pifithrin-α transiently blocks this pathway, enhancing reprogramming efficiency without long-term genetic instability and malignancies [66].

More of these small molecules, natural and synthetic, boost the efficiency of iPSC generation by targeting pathways that are involved in stem cell renewal, such as the TGF-β pathway, MEK pathway, tyrosin kinase receptor pathways, and ROCK pathways, and have been used in addition to the classical reprogramming factors:

TGF- β: RepSox [67], SB431542 [68]
MAPK/ERK modulators: PD0325901 [68]
PKA modulators: forskolin [56], IBMX [56]
PI3K modulators: rapamycin [69]
Wnt modulators: CHIR99021 [70]
ROCK modulators: Y27632 [71]
Nuclear receptor modulators: TTNPB [72]

Epigenetic Modulators

During reprogramming, a variety of epigenetic modifications take place. First, chromatin modifications occur, and then are followed by changes in DNA methylation (for two comprehensive reviews see [73, 74]).

The first histone methylation mark is H3K4me2, which is related to rapidly silence the transcription of lineage-specific genes. Later, H3K4me3 increases in promoters of active transcription genes and is associated with DNA hypomethylation. H3K36me2/3 marks are globally increased during reprogramming.

In contrast, repressive chromatin marks such as H3K9me3, are reduced during reprogramming. H3K27me3 follows a similar pattern. H3K79me2 marks are reduced in genes associated with epithelial to mesenchymal transition.

Histone acetylation marks also change during reprogramming. The global acetylation of histones H3 and H4, as well as histone acetylation marks H3K9ac, H3K27ac and H4K5ac are increased during reprogramming.

In the late phase of reprogramming, DNA methylome changes, mostly outside CpG islands. Pluripotency associated genes get demethylated, whereas methylation of somatic genes is extensive. In fact, DNA demethylation enzymes are indispensable for reprogramming [75]. Genome wide 5-hmC levels are increased during reprogramming, associated to TET1 activation.

Accordingly, several epigenetic modulators are added to the culture for increasing reprogramming efficiencies:

DNA methylation modulators: hydrocortisone, 5-AZA [76],

Histone methylation modulators: ascorbic acid [77], tranylcypromine [56, 78], DZNep [72], and vitamin C [79].

Histone acetylation modulators: inhibitors such as valproic acid, trichostatin A [80], sodium butyrate [81], and resveratrol [82].

It is important to mention that, although no modulator is available yet, transcription of Dppa2/4 together with OSKM, has been recently shown to initiate global chromatin decompaction and yield reprogramming efficiencies exceeding 80% in mouse cells [83].

Culture Conditions

Initially, mouse and then human iPSCs were established in ESC media with calf serum, with leukemia inhibitor factor (LIF) and bFGF as pluripotency maintaining factors [84]. In addition to growth factors, hypoxia (5% O_2 tension) enhances expression of pluripotency genes, thus promoting reprogramming [85].

Traditionally, iPSCs are cultured over a ***feeder layer*** of mitotically inactivated MEFs that provide essential cytokines, growth factors, and other extracellular matrix (ECM) compounds to support the pluripotency of the cells [86]. These animal-derived products involve a risk of transferring zoonotic pathogens, unknown viruses, and exogenous antigens to those newly generated iPSCs, making the clinical use of iPSCs unsuitable [87].

Thus, studies have been focusing on ***feeder-free culture systems***. The first step was using human fibroblasts as a feeder layer instead of mouse cells [88]. This approach was not optimal because of the time and effort needed for the preparation of these cells, and the problems involved in selection mass culture of human iPSCs from the human feeders. Without feeder cells, the cultures are simpler and cleaner and there are fewer variables, making it possible to culture iPSC cells in a xeno-free and chemically defined environment.

Various Substrates: MEFs Versus Vitronectin, Matrigel and Others

A viable alternative for the feeder cells is ***Matrigel***, a mixture of factors derived from tumoral mouse cells enriched with laminin, fibronectin, heparan sulphate proteoglycans and collagen IV [89]. Its composition is undefined and animal derived, so there is great lot-to-lot variability, but it is one of the most commonly used coating mixtures for iPSCs derivation. To solve the problems with Matrigel, as iPSCs express integrin receptors, other ECM molecules have been used, for example laminin, fibronectin, and vitronectin [90]. Recombinant ***laminin*** E8 fragments (LM-E8s), which are truncated peptides composed of the C-terminal regions of the α, β, and γ chains of laminin, allow the propagation of iPSCs in cultures with defined and xeno-free media [91]. In addition, the laminin isoform LN521 supports integration-free reprogramming (>0.18% efficiency, comparable to Matrigel coating) [92]. Other proteins, such as fibronectin, vitronectin, and gelatin, are rich in the arginine-glycine-aspartate (RGD) peptide sequences involved in integrin pathways. ***Vitronectin***, a protein present in plasma and tissues, was tested and proven as a feeder-free substrate to successfully generate and maintain iPSCs. However, vitronectin is expensive and a high concentration is required to coat the culture dishes [90]. Gelatin, or type-I collagen, could be used with conditioned MEF media or other types of supplements, such as insulin, transferrin, bFGF, and heparin, to correctly maintain the iPSC culture. Similar to ECM proteins, an E-cadherin and IgG Fc fusion protein has been developed to replace Matrigel, maintaining the pluripotency in iPSCs [93].

Various Media and Supplements

Although native ECM proteins can be used in animal-free culture systems, they are still not ideal because of batch-to-batch differences, cost, and especially their long-term instability (1 month at most) due to their large, temperature, and pH-sensitive conformations. To solve this, ***synthetic peptides, polymers, and other macromolecules*** have been developed [94]. Synthemax is an example of a commercially available material, based on an acrylate polymer, including RGD peptides to promote iPSC culture *via* integrins. More interestingly, reducing batch-to-batch differences and lowering the cost and efforts, culture-treated polystyrene has been treated with oxygen plasma, also maintaining the conditions necessary to culture iPSCs in conditioned media [95]. Other examples of synthetic materials are calcium alginate hydrogels [96], chitosan and alginate scaffolds [97], and poly-D-lysine [98]. All these substrates contribute to significant advances in iPSC feeder-free long-term expansion.

Moreover, 3D microenvironments have recently been preferred for providing native cell-cell and cell-matrix interactions [98], with a better growth factor gradient and more space for expansion without agglomeration of cells. Many ***biomaterials***, such as alginate, agarose, and poly(N-isopropylacrylamide)-co-poly(ethylene glycol) (PNIPAAm-PEG) were compared. PNIPAAm-PEG showed higher efficiency at promoting iPSC expansion, with an approximate 20-fold expansion rate of iPSCs per passage, maintaining purity and pluripotency [99]. Whatever 2D or 3D culture system, it's clear that feeder-free methods had become the basis for clinical-related iPSC labs.

Developing ***animal product-free media*** started with the use of knockout serum replacement (KSR) and mTeSR1 media, both chemically defined and consistent. However, because mTeSR1 and KSR contain animal-derived products, new media has been commercially developed as xeno-free media, such as TeSR2, NutriStem, Essential E8, and StemFit [100].

TeSR1 is composed of a Dulbecco's Modified Eagle's Medium (DMEM)/F12 base with 4-(2-hydroxyethyl)-1-piperazineethanesulfonic acid, supplemented with human serum albumin, vitamins, antioxidants, trace minerals, and specific lipids, and cloned growth factors such as gamma-aminobutyric acid (GABA), pipecolic acid, bFGF, TGF β1, insulin, transferrin, and glutathione [101]. However, there are concerns about excess spontaneous differentiations in culture, requiring a daily media change routine. Later, a xeno-free medium based on the same formulation was released as TeSR2. A reoptimized version of the formulation with the basic components of ESC and iPSC media, called E8, was also developed, improving the efficiency of human iPSC derivation. The final E8 medium consisted of insulin, selenium, transferrin, L-ascorbic acid, FGF2, and TGFβ (or NODAL) in DMEM/F12 with pH adjusted using $NaHCO_3$ [102]. These media are to be used in feeder-free conditions, using the TeSR family with Matrigel and E8 with recombinant vitronectin.

How to Choose the Reprogramming Method

Choosing a reprograming method depends on the cell type being reprogrammed, the efficiency of the reprogramming method, and the goal of the iPSCs generated Fig. (**2**).

If the cells are not for a clinical purpose, a viral infection with STEMCCA will produce iPSCs in most cell types with good efficiency, and the virus could later be excised with Cre-recombinase. However, if there is a downstream application to patients, the genes should preferably not be integrated; thus, the method should be footprint-free. The Sendai virus works well in most cells but requires at least 10 passages to be considered a footprint-free culture. Episomal vectors work well on blood cells, which are easier to obtain than other cell types, and there is no footprint. The other reprogramming methods discussed above have limitations, such as large footprint (viral) or low efficiency and limits on the cell type used (mRNA, PiggyBac, Minicircles, adenovirus, protein).

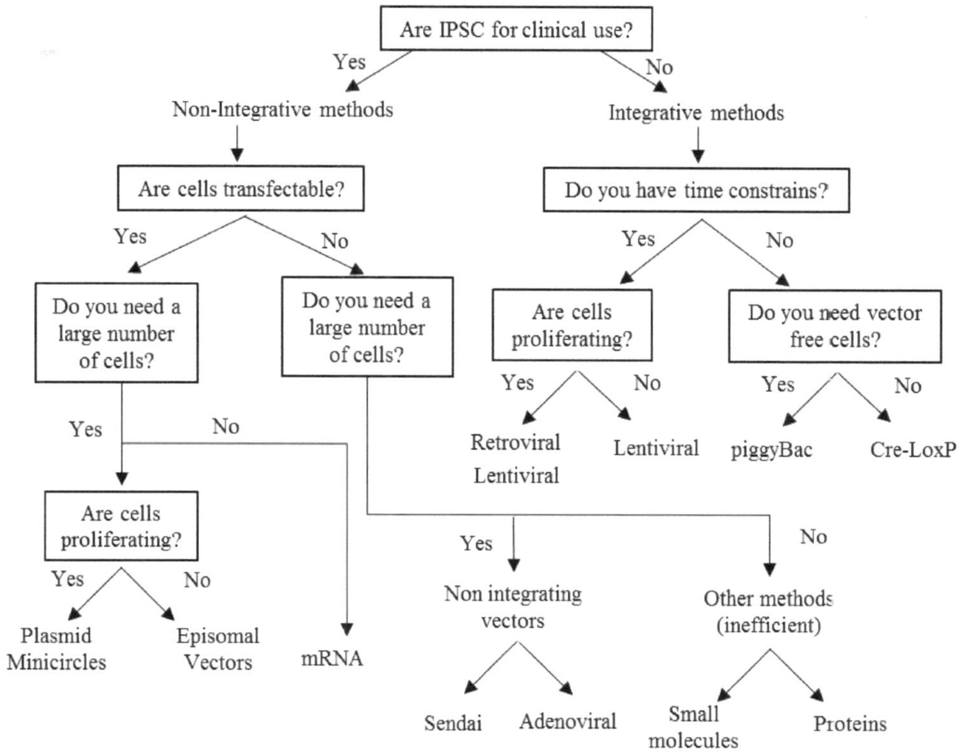

Fig. (2). Workflow for choosing the reprogramming method.

The media used in the culture systems must be the most defined possible and feeder-free compatible, if possible, to facilitate culturing. Most of the commercial tablemedia, such as mTeSR1/2 (STEMCELLA), StemPro (Invitrogen), Nutristem (Stemgent), and PluriSTEM (Invitrogen) contain bovine serum albumin (BSA) with amino acids, hormones, and growth factors. TESR2 contains human serum albumin (HSA), whereas E8 does not have HSA nor BSA, making it a completely defined medium.

Altogether, there is no consensus on culture systems between different laboratories. There are various protocols using a whole range of seeding densities for feeder cells, or protocols not using them at all, preferably relying on xeno-free or synthetic products. Most use Matrigel with TESR2 or vitronectin with E8.

SELECTION

Completely *Versus* Partially Reprogrammed Cells

Various methods could be used for the distinction between completely *versus* partially reprogramed cells and their isolation and selection Fig. (**3**).

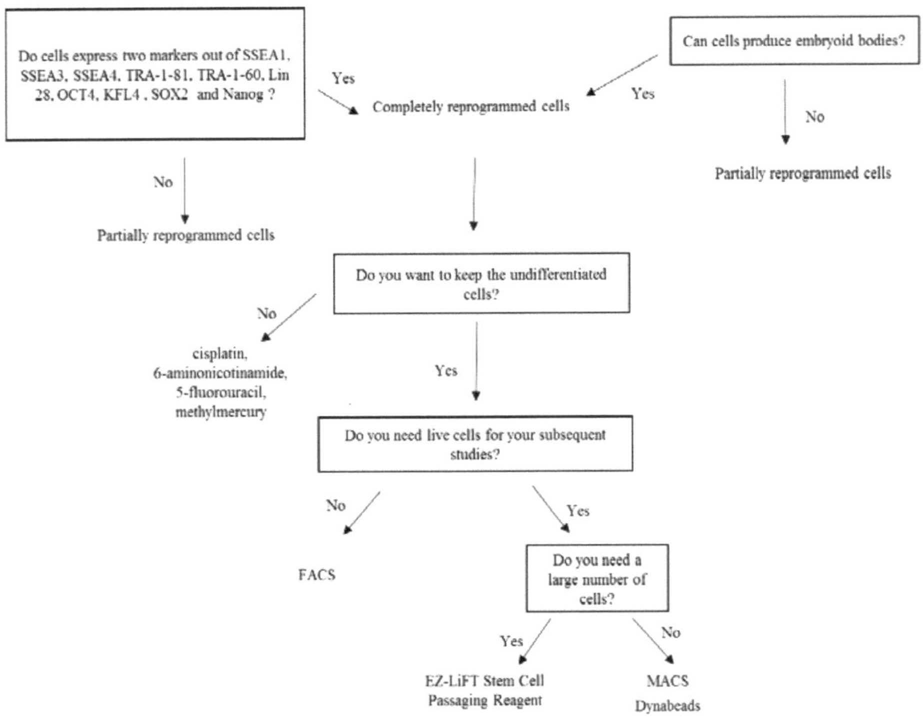

Fig. (3). Workflow for choosing the cell type and isolation method.

One method is based on iPSC morphology. Colonies of iPSCs have clear borders, and they form compact colonies of small cells with a large nucleus/cytoplasm ratio.

Alkaline Phosphatase assay is another common method, which is based on the high phosphatase activity that iPSCs present, allowing a signal even *in vivo* [32].

Pluripotent cells express various proteins that indicate pluripotency that can also be used as markers. These **markers** are stage-specific embryonic antigens (SSEA 1, and SSEA 3 or SSEA 4 for human cells), specific tissue-restricted antigens (TRA-1-81 and TRA-1-60, which is the most reliable marker, with up to 90% prediction success), and pluripotency genes (NANOG, Lin 28, OCT4, KFL4 and

SOX-2, among others). To verify the presence of these markers, techniques can be used based on genomics, such as polymerase chain reaction (PCR) or quantitative PCR, or on proteomics, such as live immunostaining and fluorescence-activated cell sorting [32,103].

In addition to markers, a functional evaluation is important to determining the completion of iPSC reprogramming, that is, **pluripotency**, which is defined as the capability to generate cells of the three embryonic germ layers. Three assays are commonly used for this functional assessment. The easiest is *in vitro* differentiation through embryoid body (EB) formation, in which the cells in suspension form large colonies or bodies; if they are pluripotent cells, they differentiate into ectoderm, mesoderm, and endodermal cells. This differentiation can be verified by immunohistochemistry using specific markers for those germ layers, such as SOX1, Brachyury, and GATA6, respectively [104]. Another assay is the teratoma formation assay, an *in vivo* assay in which the cells are injected, usually subcutaneously, into an immunodeficient mouse, where they will form a teratoma. The teratoma should be histologically comprised of cells of the three embryonic germ layers. The last and more stringent assay is the embryonic chimera formation based on the injection of iPSCs into morulas or blastocysts, assuring that iPSCs can contribute to the development of an animal [105]. Obviously the latter can only be applied to murine iPSCs.

Finally, but no less important, is the ***epigenetic profile*** characterization, to ensure that key genes for pluripotency like Nanog or OCT-4 are demethylated in iPSCs, and that specific genes to the somatic cell type are methylated [32].

Completely reprogrammed cells have characteristics such as indefinite self-renewal capacity, adequate levels of pluripotency proteins Nanog and OCT-4, and they show X chromosome reactivation [106]. On the other hand, **partially reprogrammed cells** are characterized by a lack of Nanog and the marker SSEA1, they do not lose markers associated with the identity of the cell they originated from, and they cannot generate teratomas [107, 108].

However, these partially reprogrammed cells could be of clinical use, given that in many studies they are capable of transdifferentiating into other cell types [107, 109, 110]. For example, human fibroblasts classically reprogrammed into iPSCs for 8 days show a plastic state characterized by the absence of pluripotent markers TRA-1-60 and TRA-1-81. The use of a mesodermal induction medium then promotes the appearance of a CD34+ population at high efficiency. In episomal vectors, the same cells, but reprogrammed with other factors (OCT4, SOX2, KLF4, LMYC, LIN28, and short hairpin RNA against p53) also demonstrated direct CD34+ cell generation [109].

In vivo studies on mice have been performed using partial reprogramming. One study used OSKM inducible by doxycycline, showing that the partial reprogramming of *in vivo* dermal fibroblasts decreased the fibrosis and the scar tissue without risk of tumor production [107]. Also, *in vitro* mouse fibroblasts have been used to produce induced neural stem cells (iNSCs) by the use of a retrovirus encoding SOX2, KLF4 and c-Myc and the use of a lentivirus encoding OCT4, whose activity was controlled by doxycycline. This led to the discovery that the regulated expression of OCT4 was necessary for the induction of iNSCs; however, the prolonged expression of OCT4 would end in the production of iPSCs [110].

Differentiation into Several Cell Types

iPSCs have been differentiated into a large number of cell types, such as neural stem cells, hepatic cells, T lymphocytes, hematopoietic progenitors, photoreceptors, and cardiomyocytes. For this differentiation, the first approaches involved induction of EB formation. Neural differentiation has been achieved by EB differentiation, after which EBs in suspension are attached onto poly-ornithine/laminin coated plates in neural differentiation medium (DMEM/F12 and N2) for 10–12 days and then changed to a neural proliferation medium (neurobasal medium supplemented with B27, L-glutamine, bFGF, and LIF [111]. Another example is the differentiation of hepatocytes. EBs are changed to plates with Matrigel with DMEM/F12 supplemented with activin A, leading to the differentiation toward endoderm lineage. Then, for the next 8 days the cells are grown in DMEM/F12, KORS (knockout serum replacement), nonessential amino acids, dimethyl sulfoxide and hepatocyte growth factor [112]. Photoreceptors can be differentiated from iPSCs by the use of laminin coating and a specific medium containing DMEM, heparin, nonessential amino acids, and N2 supplement. Finally, at 16 days of culture, the media must be changed to a retinal differentiation media containing DMEM/F12 and B27 [113]. *In vivo* studies aimed at obtaining hematopoietic stem cells make use of teratoma formation, by the injection of iPSC into immunodeficient mice together with OP9 stromal cells and hematopoietic cytokines. CD45+ cells appear in mouse peripheral blood at 10 weeks after injection. iPSC-derived hematopoietic stem cells (HSCs) thus have the capacity to migrate through the peripheral blood and home in the host bone marrow, but this method poses some problems in clinical practice because of zoonosis and the remaining undifferentiated iPSCs [114, 115].

Currently, more precisely directed differentiation approaches are available, including specific differentiation stimuli. The differentiation stimuli can be a ***coculture***; for example, iPSCs cultured with the bone marrow stromal cell line OP9 allows to obtain T cells, because O9 expresses the NOTCH ligand delta like

1, crucial for T cell development [116, 117]. The coculture of iPSCs with OP9 cells has also been used to obtain hematopoietic progenitors from iPSCs when stem cell factor and thrombopoietin or hematopoietic cytokines are added to the coculture. After 5 days of coculture, CD34+ cells have been detected and CD45+ cells appear at day 10 [118].

Finally, ***directed differentiation*** of iPSCs has also been achieved for several cell types. For example, cardiomyocytes have been directly derived from iPSCs by the initial activation of BMP and WNT pathways. For that purpose, an initial medium is used containing FGF2 that inhibits neural induction, activin A that boosts cardiac differentiation, BMP4 that allows spontaneous differentiation, and CHIR and insulin that promote the increase of mesodermal markers. After one day, a change of medium is necessary, using a medium that contains transferrin and ascorbate. Finally, the inhibition of the WNT pathway is necessary, using C59 for example, allowing the emergence of mesodermal and cardiac markers [119].

How to Remove Pluripotent Cells In a Differentiated Culture

One of the most important aspects related to the ability to use cells derived from iPSCs is the development of methods that allow the removal of undifferentiated cells in a differentiated culture, given they can cause tumor development due to their high cell division potential [113].

There are many methods for the isolation of differentiated cells. One is the use of cell surface markers and magnetic separation methods such as Dynabeads and MACS, based on the junction of iron spheres with antibodies, allowing isolation with a magnet of cells that have the antigen. For example, iPSC-derived endothelial cells are CD31+; thus, anti-CD31 and magnetic beads can serve for the isolation of the differentiated cells. Another method is fluorescence-activated cell sorting (FACS), which is based on the use of the same antibodies against membrane antigens and secondary antibodies coupled with fluorophores [120, 121]. This technique has been used, for example, for the isolation of midbrain dopaminergic neural progenitors derived from iPSCs based on a surface marker set (CXCR4-CORIN+CD166+) leading to 90% purity [121]. The purification efficiencies using FACS are higher than 95%, but the yield of FACS instruments is low, and this method reduces the cell's viability. Magnetic-activated cell sorting (MACS), on the other hand, has a high throughput (10^9 cells/15 min), and shows less reduction of cell viability. The use of CD34+ cells derived from iPSCs and sorted by MACS for its differentiation to smooth muscle progenitor cells shows lack of tumor formation when injected into mice [122]. The main problem with both methods in many cases is the lack of knowledge of adequate antigen surface markers [123].

An intriguing possibility could be the use of substances that are known to be extremely toxic, particularly for ESCs and embryonic germ cells, such as cisplatin, 6-aminonicotinamide, 5-fluorouracil (our own unpublished observations), methylmercury, chloride, hydroxyurea [124], lithium, hexavalent chromium [125], or retinoic acid [126], and take advantage of this fact to get rid of pluripotent cells in a differentiated culture Fig. (**3**). This possibility needs further research, given these factors have also demonstrated inhibition of pluripotency at lower doses.

Finally, there is a new method that supposedly improves the number of iPSCs in a culture when passaging, the EZ-LiFT Stem Cell Passaging Reagent. This enzyme-free reagent allows the passage of iPSCs to a new well without allowing the passage of the differentiated cells. This method thus could potentially be used to remove the iPSCs and retain only the differentiated cells [127].

How to Prevent Mutations and How to Select for Unmutated Cells

Various mutations have been observed to arise in iPSCs during culturing. Trisomy 8, trisomy 12, and trisomy X [128, 129] are common. The reasons for the mutations are not clear; however, if the mutations confer a proliferation advantage, the mutated cells subsequently have the potential of greater expansion, and thus can predominate in the culture. Therefore, a previous screening for iPSC mutations for their safe use in the clinic is mandatory [128].

One of the best ways to prevent mutation is the use of safe methods for reprograming. First, the use nonintegrative methods, such as recombinant proteins, synthetic mRNA, plasmids, or Sendai viruses, preventing the integration of reprogramming genes, is highly recommended [32, 128, 129]. Culture conditions also affect the rate of iPSC mutation. The culture accommodation to single-cell passaging is a particular example. Specific adaptation to single-cell passaging produce abnormalities like trisomy 12 or 20q11.21 amplification [129]. In addition, the use of CRISPR-CAS9 can cause mutations; thus, this method should be avoided in the clinical setting [128, 129].

Reprogramming in hypoxic conditions could be another way of preventing iPSC mutations [130]. Pluripotent stem cells rely on two distinct mechanisms to preserve their genomic integrity: first, they display reduced levels of reactive oxygen species, increased expression of antioxidants, and higher fidelity of repair mechanisms; and second, if they are unable to repair damaged DNA lesions, stem cells easily undergo senescence or differentiation [131, 132]. Paradoxically, lower mutations in hypoxic reprogramming could be due to the low DNA damage, which will not activate these repair genes, given hypoxia reduces mitochondria number and the likelihood of oxidative stress, which is highly related to genomic instability. Moreover, recent findings suggest that DNA repair plays a fundamental role in maintaining pluripotency [133]; furthermore, DNA repair processes appear to be necessary for cell reprogramming [132]. In fact, cell reprogramming must switch on the powerful systems of genomic stability maintenance required by pluripotent stem cells in order to ensure their ability to generate multiple cell lineages while avoiding mutations. It will be of interest to investigate whether reprogramming under hypoxic conditions favors or prevents the activation of this efficient DNA repair mechanism. In fact, it has been demonstrated that at least half of the protein-coding point mutations detected in 22 analyzed iPSC cell lines were acquired during the reprogramming process, independent of the methods used [134]. In addition, even when the karyotype is normal, microsatellite instability and reduced DNA repair capacities have been encountered [135], highlighting the need to address not only karyotype stability but also DNA repair abilities before iPSC use. The understanding of how pluripotent reprogrammed cells ensures genetic stability is highly relevant for their safe therapeutic application.

Several methods can be used for the selection of unmutated cells, including whole genome analysis. Others that include the analysis of known parts of the genome are fluorescent *in situ* hybridization (FISH) and PCR for the detection of known aberrations. Single nucleotide variations and copy number variants present in iPSCs can be detected with digital droplet PCR. FISH is more efficient than digital droplet PCR, but it is more expensive. The disadvantage of these techniques is the need for previous knowledge of the mutation for the design of primers; other methods, however, such as G band karyotyping, next-generation sequencing and single nucleotide polymorphism, allow for the whole genome analysis. Next generation sequencing would be the preferred technique for the detection of iPSC abnormalities [129].

Clinical Uses: Regenerative Medicine, Disease Modeling, Drug Testing

iPSC-derived cells have important advantages because they can be obtained from the patient and be differentiated into other types of cells, yet they will still be autologous; thus they would not promote an immune response when reinjected into the patient [136]. Many studies are focused on ***regenerative*** methods. For example, cones and rods derived from iPSCs could be reintroduced to the retina of individuals with retinal degeneration [113]; or iPSC-derived cardiovascular cells could be useful for the repair of heart valves, vessels, and ischemic tissues [137].

Other uses could be the ***correction of known mutations*** and transplant in the patients of origin. Several mutations have been corrected with iPSCs: Duchenne muscular dystrophy (DMD) is caused by the dysfunction of the dystrophin gene. The correction of this mutation in fibroblasts from a patient with DMD has been achieved using a human artificial chromosome with the correct dystrophin gene *via* microcell-mediated chromosome transfer. The corrected fibroblast was next reprogrammed successfully to iPSCs using lentiviral infectious and Yamanaka factors [138]. Fanconi anemia (FA) is another example of the generation of corrected iPSCs. Dermal fibroblasts obtained from patients with FA could be successfully reprogrammed to iPSCs after their transduction with lentivirus encoding FANCA or FANCD2. These corrected iPSCs could produce CD34+ hematopoietic progenitors when cocultured with 0P9 stromal cells at a similar frequency as iPSCs derived from healthy individuals. After the purification of CD34+ with MACS, these cells generated a large erythroid and myeloid colony, demonstrating that the FA pathway is functional in these cells [118].

Moreover, iPSC technology allows for the ***identification of mutations***. A clear example is the study of familial dysautonomia. The generation of neural crest cells from the iPSCs of three young patients led to the discovery that this illness presents defects in cellular migration and neurogenesis and a splicing defect in the IKBKAP gene. They tested several drugs with a known effect on this splicing and found that kinetin reduced this defect and modestly improved neurogenesis [139].

Other clinical uses could be in ***immunotherapy***, using adoptive cell transfer (ACT). Today, this method consists of the isolation of anti-tumoral T cells, their *in vitro* expansion, and then cell injection. Thus far, however, these cells are completely differentiated or senescent. iPSC technology allows us to obtain naïve-like T cells that could recognize tumor peptides. These cells have shown expansion capability, and *in vivo* they promote tumor regression [116]. In addition, the differentiation of iPSCs toward immune cells is not only useful for ACT, but also could be useful in many autoimmune diseases [117].

Another use of iPSC-derived cells is ***drug development***, including the study of drug action mechanisms and cytotoxicity [140]. A biotechnology Company, iPierian, focuses on the development of new therapies for neurodegenerative diseases. To this end, this company uses iPSCs from patients with type 1 spinal muscular atrophy (SMA). SMA is an autosomal recessive neuromuscular disease characterized by a mutation that causes decreased levels of survival motor neuron protein (SMN). These iPSCs are differentiated into motor neurons, and screening them against a library of 200,000 compounds shows candidates that promote changes in SMN expression levels [123, 141]. iPSCs derived from a patient with congenital long QT syndrome (LQTS) have been used for generating cardiomyocytes that show the disease phenotype. These cardiomyocytes have been used to study drugs that can improve or aggravate the action potential duration and the arrhythmogenicity. When drugs like potassium channel blockers were tested, the action potential duration was longer and the arrhythmogenicity was increased. Other drugs, nifedipine and pinacidil, removed the spontaneous membrane depolarizations and all arrhythmic activity in the LQTS cardiac-tissue model [142]. These cells could be used to study other compounds that provoke cardiotoxicity [123]. iPSCs are also important in ***modeling disease***. iPSC-derived cardiomyocytes are useful to study diseases like LQTS and the mechanism involved because the cells obtained from patients present the electrophysiological feature that characterizes the disease [119, 143]. Another example is that iPSC-derived cardiomyocytes from patients with leopard syndrome have the disease phenotype allowing for the study of the effects of various drugs and to identify compounds and optimal doses [143]. The study of SMA disease demonstrates that the number of motor neurons derived from patient iPSCs was decreased in culture, as occurs in this illness. This study is linked with the drug discovery of two compounds, valproic acid and tobramycin, which allow the elevation of SMN-rich structures [123]. iPSC-derived cells were also useful for the study of other illnesses, such as Huntington disease (HD). Thanks to the modeling of HD in iPSC, several novel findings have been identified, such as increased activity of caspase 3/7. The correction of the expanded HTT allele in iPSCs derived from patients led to the generation of neural stem cells with changes in E cadherin, TGF-beta and BDNF signaling pathways and to the decreased activity of caspase 3/7 [111, 144]. The generation of 3D organoids for modeling the disease were useful for the study of this illness; for example, the development of neural cortical organoids from iPSCs obtained from families with autism spectrum disorder (ASD) were useful for the study of this illness. An association with the overproduction of GABAergic production and the increased expression of FOXG1 was identified, despite a sequence or structural variation in the FOXG1 locus [136, 145]. Hepatic organoids (HOs) from iPSCs have been generated for the study of Alagille syndrome (ALGS). HO derived from patients show some ALGS characteristics, such as a lack of bile ducts, intrahepatic cholestasis, and fibrosis. At a molecular level, the decreased expression of mRNA for JAG1, NOTCH2, and CK7 was important [146].

Seventy-two ***clinical trials*** involving iPSC technology are currently ongoing (as registered in clinicaltrials.gov) [147]. Forty-two of them are in the recruitment phase. Three of these 42 are for regenerative medicine purposes. One proposes the use of human iPSCs for generating mesoderm cells that could repair degenerated capillaries in diabetic retinopathy. In another study, the capacity of differentiation of iPSCs and the immune response to stem cells for their importance in potential therapies is being evaluated. In a third clinical trial, the effect of transplantation of iPSC-derived hematopoietic stem cells from patients with beta-thalassemia will be studied. Twenty-one of these 42 studies involve modeling disease. Five of them are studying cardiovascular diseases. Another 5 are investigating neurological diseases and nervous system diseases. The rest are studying eye disease, chronic obstructive pulmonary diseases, cerebrovascular diseases, immunological diseases, alpha-1 antitrypsin deficiency, channelopathies, malignant hyperthermia, and craniosynostosis. The last 10 of the 42 studies revolve around drug development and drug toxicity. Two of them are cancer drugs and how they affect the cardiovascular system. Another two are related to the nervous system, and the others are investigating cystic fibrosis transmembrane conductance regulator modulators, the role of a molecule called Prima in ectodermal dysplasia, the use of CYP-001 in the treatment of steroid-resistant acute graft *versus* host disease, antihypertensive drugs, and the generation of tumor-fighting cells from iPSCs. Three of the 42 clinical trials are involved in the creation of an iPSC Bank for modeling diseases, regenerative medicine, and drug development. In another 2 clinical trials the investigators study the molecular mechanism of reprogramming and its differentiation capacity. One clinical trial with the goal of creation of an iPSC bank has been suspended and another with the aim of generating neurons from iPSCs to study genetic variants and neurotoxicity of drugs has been withdrawn.

Four of the 72 clinical trials have an unknown recruitment status and involve the study of autism spectrum disorder, the generation of retinal epithelium with future medicine applications, the modeling of cardiovascular diseases, and the production of haploid stem cells from human haploid germ cells. Five clinical trials are in a "not recruiting" phase. One is for testing a cystic fibrosis transmembrane conductance regulator (CFTR) modulator (Orkambi), another is for the study of the differentiation capacity of iPSCs into skeletal muscle progenitor cells and the biomolecular messages associated, and the last three involve the study of cardiovascular diseases, neurological diseases, and pseudoxanthoma elasticum.

Nine of the 72 clinical trials are active. Four of them are implicated in drug development for nervous system diseases, cardiovascular diseases, pulmonary disease, and cancer, the other five involve modeling eye diseases and cardiovascular diseases, and regenerative medicine for chronic obstructive pulmonary disease. Ten of the 72 clinical trials have been completed. One was to study the clinical phenotype in ASD. Thanks to this neural model, they observed a depletion in Na^+ and fast K^+ voltage-gated current. That observation could be linked to the altered excitability that iPSC-derived neurons show. The genetic study suggests that the defects in ion channels is a phenotype characteristic of cells in patients with ASD [148]. In the clinical trial called "Stem cell study of genetics and drug addiction," the investigators use a dopaminergic (DA) neuron model derived from opioid-dependent individuals with a polymorphism variant through the production of iPSCs from fibroblasts. They show that this polymorphism affects the expression level of the dopamine transporter. The conclusion was that this polymorphism regulates DAT levels in these DA neurons [149]. In another clinical trial, they studied the capacity to produce iPSCs from individuals with Bombay disease, which show a nonsense mutation and a missense mutation in the FUT2 gene. Their results link the differentiation capacity of iPSCs to hematopoietic lineage and gives rise to the possibility of the production of a universal blood group donor by inserting such mutations, because of the derivation of a hematopoietic lineage from iPSC derived from individuals with the failure of expressing ABH antigens [150]. Other clinical trials involved the study of amyotrophic lateral sclerosis (ALS). The use of iPSCs from patients with this disease led to verification that motor neurons express a mutation in the fusion sarcoma gene and have a low frequency of spontaneous potential action, but not in its amplitude. This result correlated with the hypoexcitability described previously in ALS motor neurons. Also, these motors neurons derived from patients with ALS show a progressive defect in axonal transport [151]. The results of the remaining six completed clinical trials have not yet been published, but involve the study of granulomatosis disease, the molecular mechanism in arrhythmias and valvopathies, the study of coronary artery disease (CAD), fibromuscular dysplasia (FMD) and other vascular diseases and cardiac alterations in patients with diabetes mellitus, sickle anemia and histamine intolerance.

Quality Control

Clinical use of iPSC-based transplants requires strict quality control to ensure patient safety. This is essential as iPSCs show genomic aberrations increasing with long term culture, so a set of standards must be agreed upon. In the last few months, several important recommendations have been proposed [152]. The first quality control guidelines white paper [153] has been released while this chapter was reviewed, and although surely will continue to evolve, mandatory critical attributes and recommended test methods have been established:

Identity: Single tandem repeat (SRT) genotyping of both source cells and iPSC

Microbiological sterility: Mycoplasma (by qPCR or culture standard pharmacopeia methods), bacteriology (standard bacterial and fungal sterility test recommended in pharmacopeia) and viral testing (hepatitis B, hepatitis C, HIV and nucleic acid testing, and others as requested specifically in different countries).

Endotoxin free: by standard pharmacopeia methods.

Genetic fidelity and stability: Residual vector (for now by in home testing by qPCR, as episomal vector method is the one recommended and there are no accredited laboratories that offer these services yet). Two different regions common to all plasmids used and sensitivity of at least 1 plasmid copy per 100 cells should be achieved. In addition, karyotype (by G banding) should be considered normal.

Viability: cell viability demonstrated by dye exclusion or flow cytometry.

Cell characterization: At least two markers, one extracellular and other intracellular of the following: SSEA3, SSEA4, TRA-1-60, TRA-1-81, OCT4, SOX2, NANOG should be positive for at least 70% of cells.

Potency: Embryoid body formation assay or directed differentiation of monolayer cultures to produce cells of the three germ layers, ectoderm, mesoderm and endoderm.

CONCLUSIONS AND FUTURE PERSPECTIVES

Since the first reprograming experiments in human fibroblasts in 2007, many other cell types have been tested, and the reprograming protocols adjusted: hepatocytes, keratinocytes, stomach cells, neural stem cells, pancreatic cells, hematopoietic progenitor cells, cord blood cells, and peripheral blood cells have all been successfully reprogrammed.

Many methods to express such reprogramming factors have been developed, from the classical integrative retroviral vectors to the more clinically friendly nonintegrative methods, such as episomal vectors or the transfection of actual proteins.

In addition, several ready-to-use tissue culture well coatings and specific reprogramming media are now commercially available, some of which are animal product-free.

The current and future clinical uses of iPSCs range from disease modeling, mutation identification and correction, drug development, and immunotherapy, to regenerative medicine purposes. The reprogramming and later the differentiation of the iPSCs into any other cell type is a very powerful tool to better understand the biochemical processes and pathophysiology of rare diseases in which certain tissue samples might be difficult or even impossible to obtain. Undoubtedly, the clinical use of patient-derived iPSCs for all these purposes is one of the main highlights in the future of medicine.

Obviously, strict quality control of the iPSC lines and differentiated cells derived from them must be adhered upon, and the first guidelines for clinical grade human iPSC lines have been released.

CONSENT FOR PUBLICATION

Not applicable.

CONFLICT OF INTEREST

The authors confirm that they have no conflict of interest to declare for this publication.

ACKNOWLEDGEMENTS

The authors are most grateful for financing from Foundation Domingo Martínez and Jesús Antolín Garciarena Grants. The authors thank Juliette Siegfried and her team at servingmed.com for the English editing.

REFERENCES

[1] Wilmut I, Beaujean N, De Sousa PA, *et al.* Somatic cell nuclear transfer. Nature 2002; 419(6709): 583-7. http://www.ncbi.nlm.nih.gov/pubmed/12374931 [Internet]. [http://dx.doi.org/10.1038/nature01079]

[2] Kishigami S, Wakayama S, Van Thuan N, *et al.* Production of cloned mice by somatic cellnuclear transfer. Nat Protoc 2006; 1(1): 125-38. http://www.ncbi.nlm.nih.gov/pubmed/17406224 [Internet]. [http://dx.doi.org/10.1038/nprot.2006.21]

[3] Somers A, Jean JC, Sommer CA, *et al.* Generation of transgene-free lung disease-specific human induced pluripotent stem cells using a single excisable lentiviral stem cell cassette. Stem Cells 2010; 28(10): 1728-40. http://www.ncbi.nlm.nih.gov/pmc/ articles/PMC3422663/ [Internet]. [http://dx.doi.org/10.1002/stem.495]

[4] Taranger CK, Noer A, Sørensen AL, *et al.* Induction of Dedifferentiation, Genomewide Transcriptional Programming, and Epigenetic Reprogramming by Extracts of Carcinoma and Embryonic Stem Cells. Mol Biol Cell 2005; 16(12): 5719-35. http://www.molbiolcell.org/doi/10.1091/mbc.e05-06-0572 [Internet]. [http://dx.doi.org/10.1091/mbc.e05-06-0572]

[5] Resnick JL, Bixler LS, Cheng L, Donovan PJ. Long-term proliferation of mouse primordial germ cells in culture. Nature 1992; 359(6395): 550-1. http://www.ncbi.nlm.nih.gov/pubmed/1383830 [Internet]. [http://dx.doi.org/10.1038/359550a0]

[6] Matsui Y, Zsebo K, Hogan BLM. Derivation of pluripotential embryonic stem cells from murine primordial germ cells in culture. Cell 1992; 70(5): 841-7. http://www.ncbi.nlm.nih.gov/pubmed/1381289 [Internet]. [http://dx.doi.org/10.1016/0092-8674(92)90317-6]

[7] Takahashi K, Yamanaka S, Zhang Y, *et al.* Induction of Pluripotent Stem Cells from Mouse Embryonic and Adult Fibroblast Cultures by Defined Factors. Cell [Internet] 2006; 126(4): 663-76. http://linkinghub.elsevier.com/retrieve/pii/S0092867406009767

[8] Takahashi K, Tanabe K, Ohnuki M, *et al.* SUPP/ Induction of pluripotent stem cells from adult human fibroblasts by defined factors. Cell 2007; 131(5): 861-72. http://www.ncbi.nlm.nih.gov/pubmed/18035408 [Internet]. [http://dx.doi.org/10.1016/j.cell.2007.11.019]

[9] Ying QL, Nichols J, Evans EP, Smith AG. Changing potency by spontaneous fusion. Nature 2002; 416(6880): 545-8. http://www.ncbi.nlm.nih.gov/pubmed/11932748 [Internet]. [http://dx.doi.org/10.1038/nature729]

[10] Junying Y, Kejin H, Kim SO, *et al.* Human induced pluripotent stem cells free of vector and transgene sequences. Science (80-) 2009; 324(5928): 797-801.

[11] Lo Sardo V, Ferguson W, Erikson GA, *et al.* Influence of donor age on induced pluripotent stem cells. Nat Biotechnol 2017; 35(1): 69-74. [Internet]. [http://dx.doi.org/10.1038/nbt.3749]

[12] Csobonyeiova M, Krajciova L, Nicodemou A, Polak S, Danisovic L. Induction of pluripotency in long-term cryopreserved human neonatal fibroblasts in feeder-free condition. Cell Tissue Bank 2017; 18(1): 45-52. [http://dx.doi.org/10.1007/s10561-016-9602-5]

[13] Cananzi M, Atala A, De Coppi P. Stem cells derived from amniotic fluid: New potentials in regenerative medicine. Reprod Biomed Online 2009; 18 (Suppl. 1): 17-27. [Internet]. [http://dx.doi.org/10.1016/S1472-6483(10)60111-3]

[14] Abdulrazzak H, Moschidou D, Jones G, Guillot P V. Biological characteristics of stem cells from foetal, cord blood and extraembryonic tissues. J R Soc Interface 2010; 7 (SUPPL. 6).

[15] Guillot PV. Induced pluripotent stem (iPS) cells from human fetal stem cells. Best Pract Res Clin Obstet Gynaecol 2016; 31(June): 112-20.http://linkinghub.elsevier.com/retrieve/pii/S1521693415001510 [Internet]. [http://dx.doi.org/10.1016/j.bpobgyn.2015.08.007]

[16] Panopoulos AD, Ruiz S, Yi F, *et al.* Rapid and highly efficient generation of induced pluripotent stem cells from human umbilical vein endothelial cells. PLoS One 2011; 6(5): 14-20. [http://dx.doi.org/10.1371/journal.pone.0019743]

[17] Abdi M, Zare S, Fathi F, *et al.* Generation of induced pluripotent stem cells from mesenchymal stem cells of human umbilical cord vein. Majallah-i Ilmi-i Danishgah-i Ulum-i Pizishki-i Kurdistan 2014; 19(1): 78-87.

[18] Haase A, Olmer R, Schwanke K, *et al.* Generation of Induced Pluripotent Stem Cells from Human Cord Blood. Cell Stem Cell 2009; 5(4): 434-41. [http://dx.doi.org/10.1016/j.stem.2009.08.021]

[19] Okita K, Yamakawa T, Matsumura Y, *et al.* An efficient nonviral method to generate integration-free human-induced pluripotent stem cells from cord blood and peripheral blood cells. Stem Cells 2013; 31(3): 458-66. [http://dx.doi.org/10.1002/stem.1293]

[20] Haile Y, Nakhaei-Nejad M, Boakye PA, *et al.* Reprogramming of HUVECs into induced pluripotent stem cells (HiPSCs), generation and characterization of HiPSC-derived neurons and astrocytes. PLoS One 2015; 10(3): 1-25. [http://dx.doi.org/10.1371/journal.pone.0119617]

[21] Yu J, Vodyanik MA, Smuga-Otto K, Antosiewicz-Bourget J, Frane JL, Tian S, *et al.* Induced pluripotent stem cell lines derived from human somatic cells. Science (80) [Internet] 2007; 318(5858): 1917-20. Available from: https://www.ncbi.nlm.nih.gov/pubmed/18029452

[22] Singh VK, Kumar N. Mechanism of Induction : Induced Pluripotent Stem Cells (iPSCs) Mechanism of Induction : Induced Pluripotent Stem Cells (iPSCs). J Stem Cells 2016; 2015: 10. http://www.ifctp.org/download/iPSCs/Mechanism of Induction- Induced Pluripotent Stem Cells (iPSCs).pdf [Internet].

[23] Aasen T, Raya A, Barrero MJ, *et al.* Efficient and rapid generation of induced pluripotent stem cells from human keratinocytes. Nat Biotechnol 2008; 26(11): 1276-84. [http://dx.doi.org/10.1038/nbt.1503]

[24] Kim JB, Sebastiano V, Wu G, *et al.* Oct4-Induced Pluripotency in Adult Neural Stem Cells. Cell 2009; 136(3): 411-9.http://www.hubmed.org/display.cgi?uids=19203577 [Internet]. [http://dx.doi.org/10.1016/j.cell.2009.01.023]

[25] Utikal J, Maherali N, Kulalert W, Hochedlinger K. Sox2 is dispensable for the reprogramming of melanocytes and melanoma cells into induced pluripotent stem cells. J Cell Sci 2009; 122(19): 3502-10. http://jcs.biologists.org/cgi/doi/10.1242/jcs.054783 [Internet]. [http://dx.doi.org/10.1242/jcs.054783]

[26] Liu H, Ye Z, Kim Y, Sharkis S, Jang YY. Generation of endoderm-derived human induced pluripotent stem cells from primary hepatocytes. Hepatology 2010; 51(5): 1810-9. [http://dx.doi.org/10.1002/hep.23626]

[27] Sugii S, Kida Y, Kawamura T, *et al.* Human and mouse adipose-derived cells support feeder-independent induction of pluripotent stem cells. Proc Natl Acad Sci USA 2010; 107(8): 3558-63. http://www.pnas.org/cgi/doi/10.1073/pnas.0910172106 [Internet]. [http://dx.doi.org/10.1073/pnas.0910172106]

[28] Loh YH, Hartung O, Li H, *et al.* Reprogramming of T cells from human peripheral blood. Cell Stem Cell 2010; 7(1): 15-9. [http://dx.doi.org/10.1016/j.stem.2010.06.004]

[29] Seki T, Yuasa S, Oda M, *et al.* Generation of induced pluripotent stem cells from human terminally differentiated circulating t cells. Cell Stem Cell 2010; 7(1): 11-3. [Internet]. [http://dx.doi.org/10.1016/j.stem.2010.06.003]

[30] Zhou T, Benda C, Dunzinger S, *et al.* Generation of human induced pluripotent stem cells from urine samples. Nat Protoc 2012; 7(12): 2080-9. [http://dx.doi.org/10.1038/nprot.2012.115]

[31] Zhou Q. Induced Pluripotent Stem Cells: Current Progress and Future Perspectives. Genomics, Proteomics Bioinforma [Internet] 2013; 11(5): 257-8. Available from: http://link.springer.com/10.1631/jzus.B1300215

[32] Brouwer M, Zhou H, Nadif Kasri N. Choices for Induction of Pluripotency: Recent Developments in Human Induced Pluripotent Stem Cell Reprogramming Strategies. Stem Cell Reviews and Reports 2016; 12: 54-72. [http://dx.doi.org/10.1007/s12015-015-9622-8]

[33] Lentz TB, Gray SJ, Samulski RJ. Viral vectors for gene delivery to the central nervous system. Neurobiol Dis [Internet] 2012; 48(2): 179-88. Available from: https://www.ncbi.nlm.nih.gov/pubmed/22001604

[34] Papapetrou EP, Tomishima MJ, Chambers SM, *et al.* Stoichiometric and temporal requirements of Oct4, Sox2, Klf4, and c-Myc expression for efficient human iPSC induction and differentiation. Proc Natl Acad Sci [Internet] 2009; 106(31): 12759-64. Available from: http://www.pnas.org/cgi/doi/10.1073/pnas.0904825106

[35] Ye L, Chang JC, Lin C, *et al.* Generation of induced pluripotent stem cells using site-specific integration with phage integrase. Proc Natl Acad Sci USA 2010; 107(45): 19467-72. http://www.pnas.org/cgi/doi/10.1073/pnas.1012677107 [Internet]. [http://dx.doi.org/10.1073/pnas.1012677107]

[36] Ramalingam S, London V, Kandavelou K, *et al.* Generation and Genetic Engineering of Human Induced Pluripotent Stem Cells Using Designed Zinc Finger Nucleases. Stem Cells Dev 2013; 22(4): 595-610. http://online.liebertpub.com/doi/abs/10.1089/scd.2012.0245 [Internet]. [http://dx.doi.org/10.1089/scd.2012.0245]

[37] Tidball AM, Dang LT, Glenn TW, *et al.* Rapid Generation of Human Genetic Loss-of-Function iPSC Lines by Simultaneous Reprogramming and Gene Editing. Stem Cell Reports 2017; 9(3): 725-31. http://www.ncbi.nlm.nih.gov/pmc/articles/PMC5599229/ [Internet].

[http://dx.doi.org/10.1016/j.stemcr.2017.07.003]

[38] Stadtfeld M, Nagaya M, Utikal J, Weir G, Hochedlinger K. Induced pluripotent stem cells generated without viral integration. Science 2008; 322(5903): 945-9. [http://dx.doi.org/10.1126/science.1162494]

[39] Meng F, Wang X, Gu P, Wang Z, Guo W. Induction of retinal ganglion-like cells from fibroblasts by adenoviral gene delivery. Neuroscience 2013; 250: 381-93. http://europepmc.org/abstract/MED/23856066 [Internet]. [http://dx.doi.org/10.1016/j.neuroscience.2013.07.001]

[40] Petersen GF. Generation of diverse neural cell types through direct conversion. World J Stem Cells [Internet] 2016; 8(2): 32. Available from: http://www.wjgnet.com/1948-0210/full/v8/i2/32.htm

[41] Fusaki N, Ban H, Nishiyama A, Saeki K, Hasegawa M. Efficient induction of transgene-free human pluripotent stem cells using a vector based on Sendai virus, an RNA virus that does not integrate into the host genome. Proc Japan Acad Ser B [Internet] 2009; 85(8): 348-62. Available from: http://www.ncbi.nlm.nih.gov/pmc/articles/PMC3621571

[42] Seki T, Yuasa S, Fukuda K. Generation of induced pluripotent stem cells from a small amount of human peripheral blood using a combination of activated T cells and Sendai virus. Nat Protoc [Internet] 2012; 7(4): 718-28. Available from: https://www.ncbi.nlm.nih.gov/pubmed/22422317

[43] Sommer CA, Stadtfeld M, Murphy GJ, *et al.* Induced Pluripotent Stem Cell Generation Using a Single Lentiviral Stem Cell Cassette. Stem Cells [Internet] 2009; 27(3): 543-9. Available from: http://doi.wiley.com/10.1634/stemcells.2008-1075

[44] Kadari A, Lu M, Li M, *et al.* Excision of viral reprogramming cassettes by Cre protein transduction enables rapid, robust and efficient derivation of transgene-free human induced pluripotent stem cells. Stem Cell Res Ther 2014; 5(2): 47.

[45] Talluri TR, Kumar D, Glage S, *et al.* Non-viral reprogramming of fibroblasts into induced pluripotent stem cells by Sleeping Beauty and piggyBac transposons. Biochem Biophys Res Commun 2014; 450(1): 581-7.

[46] Kaji K, Norrby K, Paca A, *et al.* Virus-free induction of pluripotency and subsequent excision of reprogramming factors. Nature [Internet] 2009; 458(7239): 771-5. Available from: https://www.ncbi.nlm.nih.gov/pubmed/ 19252477

[47] Aronovich EL, McIvor RS, Hackett PB. The Sleeping Beauty transposon system: A non-viral vector for gene therapy. Hum Mol Genet 2011; 20(R1): R14-20. http://www.ncbi.nlm.nih.gov/pmc/articles/PMC3095056/ [Internet]. [http://dx.doi.org/10.1093/hmg/ddr140]

[48] Kim TK, Eberwine JH. Mammalian cell transfection: The present and the future. Anal Bioanal Chem 2010; 397(8): 3173-8. http://www.ncbi.nlm.nih.gov/pmc/articles/PMC2911531/ [Internet]. [http://dx.doi.org/10.1007/s00216-010-3821-6]

[49] Okita K, Matsumura Y, Sato Y, *et al.* A more efficient method to generate integration-free human iPS cells. Nat Methods 2011; 8(5): 409-12.

[50] Narsinh KH, Jia F, Robbins RC, *et al.* Generation of adult human induced pluripotent stem cells using nonviral minicircle DNA vectors. Nat Protoc 2011; 6(1): 78-88. http://www.ncbi.nlm.nih.gov/pmc/articles/PMC3657506/ [Internet]. [http://dx.doi.org/10.1038/nprot.2010.173]

[51] Warren L, Manos PD, Ahfeldt T, *et al.* Highly efficient reprogramming to pluripotency and directed differentiation of human cells with synthetic modified mRNA. Cell Stem Cell 2010; 7(5): 618-30. http://www.ncbi.nlm.nih.gov/pmc/articles/PMC3656821/ [Internet]. [http://dx.doi.org/10.1016/j.stem.2010.08.012]

[52] Warren L, Ni Y, Wang J, Guo X. Feeder-free derivation of human induced pluripotent stem cells with messenger RNA. Sci Rep 2012; 2: 657.

[53] Yoshioka N, Gros E, Li HR, *et al.* Efficient generation of human iPSCs by a synthetic self-replicative RNA. Cell Stem Cell 2013; 13(2): 246-54.

[54] Anokye-Danso F, Trivedi CM, Juhr D, *et al.* Highly efficient miRNA-mediated reprogramming of mouse and human somatic cells to pluripotency. Cell Stem Cell 2011; 8(4): 376-88.

[55] Zhao XY, Li W, Lv Z, *et al.* IPS cells produce viable mice through tetraploid complementation. Nature 2009; 461(7260): 86-90. http://www.ncbi.nlm.nih.gov/pubmed/19672241 [Internet]. [http://dx.doi.org/10.1038/nature08267]

[56] Hou P, Li Y, Zhang X, *et al.* Pluripotent stem cells induced from mouse somatic cells by small-molecule compounds. Science (80-) [Internet] 2013; 341(6146): 651-4. Available from: https://www.ncbi.nlm.nih.gov/pubmed/23868920

[57] Zhang R, Zhang LH, Xie X. IPSCs and small molecules: A reciprocal effort towards better approaches for drug discovery. Acta Pharmacol Sin [Internet] 2013; 34(6): 765-. Available from: https://www.ncbi.nlm.nih.gov/pubmed/23603980

[58] Han J, Yuan P, Yang H, *et al.* Tbx3 improves the germ-line competency of induced pluripotent stem cells. Nature 2010; 463(7284): 1096-0.

[59] Declercq J, Sheshadri P, Verfaillie CM, Kumar A. Zic3 Enhances the Generation of Mouse Induced Pluripotent Stem Cells. Stem Cells Dev [Internet] 2013; 22(14): 2017-5. Available from: http://online.liebertpub.com/doi/abs/10.1089/scd.2012.0651

[60] Nakagawa M, Takizawa N, Narita M, Ichisaka T, Yamanaka S. Promotion of direct reprogramming by transformation-deficient Myc. Proc Natl Acad Sci USA 2010; 107(32): 14152-7. http://www.pnas.org/cgi/doi/10.1073/pnas.1009374107 [Internet]. [http://dx.doi.org/10.1073/pnas.1009374107]

[61] Maekawa M, Yamaguchi K, Nakamura T, *et al.* Direct reprogramming of somatic cells is promoted by maternal transcription factor Glis1. Nature 2011; 474(7350): 225-8.

[62] Montserrat N, Nivet E, Sancho-Martinez I, *et al.* Reprogramming of human fibroblasts to pluripotency with lineage specifiers. Cell Stem Cell 2013; 13(3): 341-50.

[63] Chou BK, Cheng L. And then there were none: No need for pluripotency factors to induce reprogramming. Cell Stem Cell [Internet] 2013; 13(3): 261-. Available from: https://www.ncbi.nlm.nih.gov/pubmed/24012365

[64] Surget S, Khoury MP, Bourdon JC. Uncovering the role of p53 splice variants in human malignancy: A clinical perspective. OncoTargets Ther 2014; 2013(7): 57-67.

[65] Lin T, Lin Y. P53 Switches Off Pluripotency on Differentiation. Stem Cell Res Ther 2017; 8(1): 44. http://www.ncbi.nlm.nih.gov/pmc/articles/PMC5330084/ [Internet]. [http://dx.doi.org/10.1186/s13287-017-0498-1]

[66] Komarov PG, Komarova EA, Kondratov R V, *et al.* A chemical inhibitor of p53 that protects mice from the side effects of cancer therapy. Science (80-) 1999; 285(5434): 1733-7.

[67] Ichida JK, Blanchard J, Lam K, *et al.* A Small-Molecule Inhibitor of Tgf-β Signaling Replaces Sox2 in Reprogramming by Inducing Nanog. Cell Stem Cell [Internet] 2009; 5(5): 491-503. Available from: https://www.ncbi.nlm.nih.gov/pubmed/19818703

[68] Lin T, Ambasudhan R, Yuan X, *et al.* A chemical platform for improved induction of human iPSCs. Nat Methods [Internet] 2009; 6(11): 805-. Available from: https://www.ncbi.nlm.nih.gov/pubmed/19838168

[69] Pan H, Cai N, Li M, Liu GH, Izpisua Belmonte JC. Autophagic control of cell "stemness." EMBO Mol Med [Internet] 2013; 5(3): 327-1. Available from: https://www.ncbi.nlm.nih.gov/pubmed/23495139

[70] Li W, Zhou H, Abujarour R, *et al.* Generation of human-induced pluripotent stem cells in the absence

of exogenous Sox2. Stem Cells [Internet] 2009; 27(12): 2992-3000. Available from: https://www.ncbi.nlm.nih.gov/pubmed/19839055

[71] Lai W-H, Ho JC-Y, Lee Y-K, *et al.* ROCK Inhibition Facilitates the Generation of Human-Induced Pluripotent Stem Cells in a Defined, Feeder-, and Serum-Free System. Cell Reprogramming (Formerly "Cloning Stem Cells") [Internet] 2010; 12(6): 641-53. Available from: http://www.liebertonline.com/doi/abs/10.1089/cell.2010.0051

[72] Liu K, Song Y, Yu H, Zhao T. Understanding the roadmaps to induced pluripotency. Cell Death Dis 2014; 5(5): e1232. http://www.ncbi.nlm.nih.gov/pmc/articles/PMC4047905/ [Internet]. [http://dx.doi.org/10.1038/cddis.2014.205]

[73] Brix J, Zhou Y, Luo Y. The Epigenetic Reprogramming Roadmap in Generation of iPSCs from Somatic Cells. J Genet Genomics 2015; 42(12): 661-70. [http://dx.doi.org/10.1016/j.jgg.2015.10.001]

[74] Papp B, Plath K. Epigenetics of reprogramming to induced pluripotency. Cell [Internet] 2013; 152(6): 1324-43. Available from: http://www.ncbi.nlm.nih.gov/pubmed/23498940 [http://dx.doi.org/10.1016/j.cell.2013.02.043]

[75] Pawlak M, Jaenisch R. De novo DNA methylation by Dnmt3a and Dnmt3b is dispensable for nuclear reprogramming of somatic cells to a pluripotent state. Genes Dev [Internet] 2011; 25(10): 1035-40. Available from: http://genesdev.cshlp.org/cgi/doi/10.1101/gad.2039011

[76] Han J, Sachdev PS, Sidhu KS. A combined epigenetic and non-genetic approach for reprogramming human somatic cells. PLoS One 2010; 5(8): e12297.

[77] Esteban MA, Wang T, Qin B, *et al.* Vitamin C Enhances the Generation of Mouse and Human Induced Pluripotent Stem Cells. Cell Stem Cell [Internet] 2010; 6(1): 71-9. Available from: https://www.ncbi.nlm.nih.gov/pubmed/20036631

[78] Xu X, Duan S, Yi F, *et al.* Mitochondrial regulation in pluripotent stem cells. Cell Metab [Internet] 2013; 18(3): 325-. Available from: https://www.ncbi.nlm.nih.gov/pubmed/23850316 [http://dx.doi.org/10.1016/j.cmet.2013.06.005]

[79] Wang T, Chen K, Zeng X, *et al.* The histone demethylases Jhdm1a/1b enhance somatic cell reprogramming in a vitamin-C-dependent manner. Cell Stem Cell 2011; 9(6): 575-87. http://www.ncbi.nlm.nih.gov/pubmed/22100412 [Internet]. [http://dx.doi.org/10.1016/j.stem.2011.10.005]

[80] Huangfu D, Maehr R, Guo W, *et al.* Induction of pluripotent stem cells by defined factors is greatly improved by small-molecule compounds. Nat Biotechnol 2008; 26(7): 795-7. http://www.ncbi.nlm.nih.gov/pubmed/18568017 [Internet]. [http://dx.doi.org/10.1038/nbt1418]

[81] Liang G, Taranova O, Xia K, Zhang Y. Butyrate promotes induced pluripotent stem cell generation. J Biol Chem 2010; 285(33): 25516-1.

[82] Zhang M, Li K, Xie M, Ding S. Chemical Approaches to Controlling Cell Fate. Princ Dev Genet Second Ed [Internet] 2014; 28: 59-76. Available from: https://www.ncbi.nlm.nih.gov/pubmed/25461450

[83] Hernandez C, Wang Z, Ramazanov B, *et al.* Dppa2/4 Facilitate Epigenetic Remodeling during Reprogramming to Pluripotency. Cell Stem Cell [Internet] 2018; 23(3): 396-411.e8. Available from: https://linkinghub.elsevier.com/retrieve/pii/S1934590918303862

[84] Hirai H, Firpo M, Kikyo N. Establishment of LIF-dependent human iPS cells closely related to basic FGF-dependent authentic iPS cells. PLoS One 2012; 7(6): e39022.

[85] Yoshida Y, Takahashi K, Okita K, Ichisaka T, Yamanaka S. Hypoxia Enhances the Generation of Induced Pluripotent Stem Cells. Cell Stem Cell 2009; 5(3): 237-41.

[86] Hongisto H, Vuoristo S, Mikhailova A, *et al.* Laminin-511 expression is associated with the

functionality of feeder cells in human embryonic stem cell culture. Stem Cell Res 2012; 8(1): 97-108.

[87] Ährlund-Richter L, De Luca M, Marshak DR, *et al.* Isolation and Production of Cells Suitable for Human Therapy: Challenges Ahead. Cell Stem Cell 2009; 4(1): 20-6.

[88] Takahashi K, Narita M, Yokura M, Ichisaka T, Yamanaka S. Human induced pluripotent stem cells on autologous feeders. PLoS One [Internet] 2009; 4(12): e8067. Available from: https://www.ncbi.nlm.nih.gov/pubmed/19956543

[89] Hughes CS, Postovit LM, Lajoie GA. Matrigel: a complex protein mixture required for optimal growth of cell culture. Proteomics 2010; 10(9): 1886-90.

[90] Kalaskar DM, Downes JE, Murray P, Edgar DH, Williams RL. Characterization of the interface between adsorbed fibronectin and human embryonic stem cells. J R Soc Interface [Internet] 2013; 10(83): 20130139. Available from: https://www.ncbi.nlm.nih.gov/pubmed/23554347

[91] Miyazaki T, Futaki S, Suemori H, *et al.* Laminin E8 fragments support efficient adhesion and expansion of dissociated human pluripotent stem cells. Nat Commun 2012; 3: 1236. [Internet]. [http://dx.doi.org/10.1038/ncomms2231]

[92] Rodin S, Antonsson L, Niaudet C, *et al.* Clonal culturing of human embryonic stem cells on laminin-521/E-cadherin matrix in defined and xeno-free environment. Nat Commun 2014; 5: 3195. [http://dx.doi.org/10.1038/ncomms4195]

[93] Nagaoka M, Si-Tayeb K, Akaike T, Duncan SA. Culture of human pluripotent stem cells using completely defined conditions on a recombinant E-cadherin substratum. BMC Dev Biol 2010; 10(1): 60. [Internet]. [http://dx.doi.org/10.1186/1471-213X-10-60]

[94] Wong S, Shim MS, Kwon YJ. Synthetically designed peptide-based biomaterials with stimuli-responsive and membrane-active properties for biomedical applications. J Mater Chem B Mater Biol Med 2014; 2: 595-615. [http://dx.doi.org/10.1039/C3TB21344G]

[95] Mahlstedt MM, Anderson D, Sharp JS, *et al.* Maintenance of pluripotency in human embryonic stem cells cultured on a synthetic substrate in conditioned medium. Biotechnol Bioeng 2010; 105(1): 130-40. [http://dx.doi.org/10.1002/bit.22520]

[96] Siti-Ismail N, Bishop AE, Polak JM, Mantalaris A. The benefit of human embryonic stem cell encapsulation for prolonged feeder-free maintenance. Biomaterials 2008; 29(29): 3946-52. [http://dx.doi.org/10.1016/j.biomaterials.2008.04.027]

[97] Li Z, Leung M, Hopper R, Ellenbogen R, Zhang M. Feeder-free self-renewal of human embryonic stem cells in 3D porous natural polymer scaffolds. Biomaterials 2010; 31(3): 404-12. [http://dx.doi.org/10.1016/j.biomaterials.2009.09.070]

[98] Carlson AL, Florek CA, Kim JJ, *et al.* Microfibrous substrate geometry as a critical trigger for organization, self-renewal, and differentiation of human embryonic stem cells within synthetic 3-dimensional microenvironments. FASEB J 2012; 26(8): 3240-51. http://www.fasebj.org/cgi/doi/10.1096/fj.11-192732 [Internet]. [http://dx.doi.org/10.1096/fj.11-192732]

[99] Lei Y, Schaffer DV. A fully defined and scalable 3D culture system for human pluripotent stem cell expansion and differentiation. Proc Natl Acad Sci USA 2013; 110(52): E5039-48. http://www.pnas.org/cgi/doi/10.1073/pnas.1309408110 [Internet]. [http://dx.doi.org/10.1073/pnas.1309408110]

[100] Meng G, Liu S, Rancourt DE. Synergistic Effect of Medium, Matrix, and Exogenous Factors on the Adhesion and Growth of Human Pluripotent Stem Cells Under Defined, Xeno-Free Conditions. Stem Cells Dev [Internet] 2012; 21(11): 2036-48. Available from: http://online.liebertpub.com/doi/abs/10.1089/scd.2011.0489

[101] Vemuri MC, Schimmel T, Colls P, Munne S, Cohen J. Derivation of human embryonic stem cells in xeno-free conditions. Methods Mol Biol 2007; 407: 185. [Internet]. [http://dx.doi.org/10.1007/978-1-59745-536-7_1]

[102] Chen G, Gulbranson DR, Hou Z, *et al.* Chemically defined conditions for human iPSC derivation and culture. Nat Methods 2011; 8(5): 424-9.http://www.ncbi.nlm.nih.gov/pmc/articles/PMC3084903/ [Internet]. [http://dx.doi.org/10.1038/nmeth.1593]

[103] Bacakova L, Zarubova J, Travnickova M, *et al.* Stem cells: their source, potency and use in regenerative therapies with focus on adipose-derived stem cells – a review. Biotechnol Adv 2018; 36: 1111-26. [http://dx.doi.org/10.1016/j.biotechadv.2018.03.011]

[104] Poh Y-C, Chen J, Hong Y, *et al.* Generation of organized germ layers from a single mouse embryonic stem cell. Nat Commun 2014; 5: 4000. [http://dx.doi.org/10.1038/ncomms5000]

[105] Chhabra A. Derivation of Human Induced Pluripotent Stem Cell (iPSC) Lines and Mechanism of Pluripotency: Historical Perspective and Recent Advances. Stem Cell Reviews and Reports 2017; 13: 757-73. [http://dx.doi.org/10.1007/s12015-017-9766-9]

[106] Maherali N, Hochedlinger K. Guidelines and Techniques for the Generation of Induced Pluripotent Stem Cells. Cell Stem Cell 2008; 3: 595-605. [http://dx.doi.org/10.1016/j.stem.2008.11.008]

[107] Doeser MC, Schöler HR, Wu G. Reduction of Fibrosis and Scar Formation by Partial Reprogramming *In Vivo.* Stem Cells 2018; 36(8): 1216-25. [http://dx.doi.org/10.1002/stem.2842]

[108] Rosa M. Marión MAB. 41 Long Live Partial Reprogramming. Circ Res 2017; 120(9): 1381-3. [http://dx.doi.org/10.1161/CIRCRESAHA.117.310594]

[109] Kurian L, Sancho-Martinez I, Nivet E, *et al.* Conversion of human fibroblasts to angioblast-like progenitor cells. Nat Methods 2013; 10(1): 77-83. [http://dx.doi.org/10.1038/nmeth.2255]

[110] Thier M, Wörsdörfer P, Lakes YB, *et al.* Direct conversion of fibroblasts into stably expandable neural stem cells. Cell Stem Cell 2012; 10(4): 473-9. [http://dx.doi.org/10.1016/j.stem.2012.03.003]

[111] An MC, Zhang N, Scott G, *et al.* Genetic correction of huntington's disease phenotypes in induced pluripotent stem cells. Cell Stem Cell 2012; 11(2): 253-63. [http://dx.doi.org/10.1016/j.stem.2012.04.026]

[112] Ghodsizadeh A, Taei A, Totonchi M, *et al.* Generation of Liver Disease-Specific Induced Pluripotent Stem Cells Along with Efficient Differentiation to Functional Hepatocyte-Like Cells. Stem Cell Rev Reports 2010; 6(4): 622-32. [http://dx.doi.org/10.1007/s12015-010-9189-3]

[113] Lakowski J, Welby E, Budinger D, *et al.* Isolation of Human Photoreceptor Precursors *via* a Cell Surface Marker Panel from Stem Cell-Derived Retinal Organoids and Fetal Retinae. Stem Cells 2018; 36(5): 709-22. [http://dx.doi.org/10.1002/stem.2775]

[114] Suzuki N, Yamazaki S, Yamaguchi T, *et al.* Generation of engraftable hematopoietic stem cells from induced pluripotent stem cells by way of teratoma formation. Mol Ther 2013; 21(7): 1424-31. [http://dx.doi.org/10.1038/mt.2013.71]

[115] Lee J, Dykstra B, Sackstein R, Rossi DJ. Progress and obstacles towards generating hematopoietic stem cells from pluripotent stem cells. Curr Opin Hematol 2015; 22: 317-23.

[116] Vizcardo R, Klemen ND, Islam SMR, *et al.* Generation of Tumor Antigen-Specific iPSC-Derived Thymic Emigrants Using a 3D Thymic Culture System. Cell Reports 2018; 22(12): 3175-90. [http://dx.doi.org/10.1016/j.celrep.2018.02.087]

[117] Schmitt TM, de Pooter RF, Gronski MA, *et al.* Induction of T cell development and establishment of T cell competence from embryonic stem cells differentiated *in vitro*. Nat Immunol 2004; 5(4): 410-7. [http://dx.doi.org/10.1038/ni1055]

[118] Raya Á, Rodríguez-Piz I, Guenechea G, *et al.* Disease-corrected haematopoietic progenitors from Fanconi anaemia induced pluripotent stem cells. Nature 2009; 460: 53-9. [http://dx.doi.org/10.1038/nature08129]

[119] Peischard S, Piccini I, Strutz-Seebohm N, Greber B, Seebohm G. From iPSC towards cardiac tissue—a road under construction. Pflugers Arch 2017; 469: 1233-43. [http://dx.doi.org/10.1007/s00424-017-2003-1]

[120] Welby E, Lakowski J, Di Foggia V, *et al.* Isolation and Comparative Transcriptome Analysis of Human Fetal and iPSC-Derived Cone Photoreceptor Cells. Stem Cell Reports 2017; 9(6): 1898-915. [http://dx.doi.org/10.1016/j.stemcr.2017.10.018]

[121] Paik EJ, O'Neil AL, Ng SY, Sun C, Rubin LL. Using intracellular markers to identify a novel set of surface markers for live cell purification from a heterogeneous hIPSC culture. Sci Rep 2018; 8(1) [http://dx.doi.org/10.1038/s41598-018-19291-4]

[122] Li Y, Green M, Wen Y, *et al.* Efficacy and safety of immuno-magnetically sorted smooth muscle progenitor cells derived from human-induced pluripotent stem cells for restoring urethral sphincter function. Stem Cells Transl Med 2017; 6(4): 1158-67. [http://dx.doi.org/10.1002/sctm.16-0160]

[123] Grskovic M, Javaherian A, Strulovici B, Daley GQ. Induced pluripotent stem cells â€" opportunities for disease modelling and drug discovery. Nat Rev Drug Discov 2011; 10(12): 915-29.

[124] Peters AK, Steemans M, Hansen E, *et al.* Evaluation of the embryotoxic potency of compounds in a newly revised high throughput embryonic stem cell test. Toxicol Sci 2008; 105(2): 342-50. [http://dx.doi.org/10.1093/toxsci/kfn126]

[125] Stummann TC, Hareng L, Bremer S. Embryotoxicity hazard assessment of cadmium and arsenic compounds using embryonic stem cells. Toxicology 2008; 252(1–3): 118-22. [http://dx.doi.org/10.1016/j.tox.2008.08.001]

[126] Festag M, Sehner C, Steinberg P, Viertel B. An *in vitro* embryotoxicity assay based on the disturbance of the differentiation of murine embryonic stem cells into endothelial cells. I: Establishment of the differentiation protocol. Toxicol Vitr 2007; 21(8): 1619-30. [http://dx.doi.org/10.1016/j.tiv.2007.06.018]

[127] http://www.merckmillipore.com/ [homepage on the internet] EZ-LiFT™ Stem Cell Passaging Reagent. Merck KGaA , 2018 [[updated 2018. cited 19 July 2018].]; Available from: http://www.merckmillipore.com/ES/es/product/EZ-LiFT-Stem-Cell-Passaging-Reagent,MM_NF-SCM139-100ML

[128] Lin K, Xiao AZ. Quality control towards the application of induced pluripotent stem cells. Curr Opin Genet Dev 2017; 46: 164-9. [http://dx.doi.org/10.1016/j.gde.2017.07.006]

[129] Assou S, Bouckenheimer J, De Vos J. Concise Review: Assessing the Genome Integrity of Human Induced Pluripotent Stem Cells: What Quality Control Metrics? Stem Cells 2018; 36(6): 814-21. [http://dx.doi.org/10.1002/stem.2797]

[130] De Miguel MP, Alcaina Y, Sainz de la Maza D. Primordial Germ Cell Reprogramming.In IntechOpen. 2018; pp. 43-62. [http://dx.doi.org/10.5772/intechopen.69965]

[131] Facucho-Oliveira JM, John StJC. The relationship between pluripotency and mitochondrial DNA proliferation during early embryo development and embryonic Stem Cell differentiation. Stem Cell Rev Reports [Internet] 2009; 5(2): 140-58. Available from: http://www.ncbi.nlm.nih.gov/pubmed/19521804

[132] Rocha CRR, Lerner LK, Okamoto OK, Marchetto MC, Menck CFM. The role of DNA repair in the pluripotency and differentiation of human stem cells. Mutat Res Rev Mutat Res 2013; 752(1): 25-35. http://www.ncbi.nlm.nih.gov/pubmed/23010441 [Internet]. [http://dx.doi.org/10.1016/j.mrrev.2012.09.001]

[133] Nagaria P, Robert C, Rassool FV. DNA double-strand break response in stem cells: Mechanisms to maintain genomic integrity. Biochim Biophys Acta, Gen Subj 2013; 1830(2): 2345-53. http://www.ncbi.nlm.nih.gov/pubmed/22995214 [Internet]. [http://dx.doi.org/10.1016/j.bbagen.2012.09.001]

[134] Gore A, Li Z, Fung HL, *et al.* Somatic coding mutations in human induced pluripotent stem cells. Nature 2011; 471(7336): 63-7. http://www.ncbi.nlm.nih.gov/pubmed/21368825 [Internet]. [http://dx.doi.org/10.1038/nature09805]

[135] Luo LZ, Gopalakrishna-Pillai S, Nay SL, *et al.* DNA repair in human pluripotent stem cells is distinct from that in non-pluripotent human cells. PLoS One 2012; 7(3): e30541. http://www.ncbi.nlm.nih.gov/pubmed/22412831 [Internet]. [http://dx.doi.org/10.1371/journal.pone.0030541]

[136] Ho BX, Pek NMQ, Soh BS. Disease modeling using 3D organoids derived from human induced pluripotent stem cells. Int J Mol Sci 2018; 19(4.) [http://dx.doi.org/10.3390/ijms19040936]

[137] Singh VK, Kalsan M, Kumar N, Saini A, Chandra R. Induced pluripotent stem cells: applications in regenerative medicine, disease modeling, and drug discovery. Front Cell Dev Biol 2015; 3(2): 1-18. http://journal.frontiersin.org/Article/10.3389/fcell.2015.00002/abstract [Internet].

[138] Kazuki Y, Hiratsuka M, Takiguchi M, *et al.* Complete genetic correction of iPS cells from duchenne muscular dystrophy. Mol Ther 2010; 18(2): 386-93. [http://dx.doi.org/10.1038/mt.2009.274]

[139] Lee G, Papapetrou EP, Kim H, *et al.* Modelling pathogenesis and treatment of familial dysautonomia using patient-specific iPSCs. Nature 2009; 461(7262): 402-6. [http://dx.doi.org/10.1038/nature08320]

[140] Biosciences Axol. Induced Pluripotent Stem Cells : recapitulating disease to facilitate drug discovery. 2015; pp. 1-8. Available from: https://www.axolbio.com/blog/induced-pluripotent-stem-ce-ls-recapitulating-disease-to-facilitate-drug-discovery [cited 19 July 2018]

[141] https://newsbmscom/ [homepage on the internet] Bristol-Myers Squibb Acquires iPierian, Inc | BMS Newsroom [updated 29th April 2014 cited 19 July 2018] 2018. Available from: https://news.bms.com/press-release/partnering-news/bristol-myers-squibb-acquires-ipierian-inc

[142] Itzhaki I, Maizels L, Huber I, *et al.* Modelling the long QT syndrome with induced pluripotent stem cells. Nature 2011; 471(7337): 225-30. [http://dx.doi.org/10.1038/nature09747]

[143] Iglesias-García O, Pelacho B, Prósper F. Induced pluripotent stem cells as a new strategy for cardiac regeneration and disease modeling. J Mol Cell Cardiol 2013; 62: 43-50. [http://dx.doi.org/10.1016/j.yjmcc.2013.04.022]

[144] Connor B. Concise Review: The Use of Stem Cells for Understanding and Treating Huntington's Disease. Stem Cells 2018; 26: 46-60.

[145] Mariani J, Coppola G, Zhang P, *et al.* FOXG1-Dependent Dysregulation of GABA/Glutamate Neuron Differentiation in Autism Spectrum Disorders. Cell 2015; 162(2): 375-90. [http://dx.doi.org/10.1016/j.cell.2015.06.034]

[146] Guan Y, Xu D, Garfin PM, *et al.* Human hepatic organoids for the analysis of human genetic diseases. JCI Insight 2017; 2(17)https://insight.jci.org/articles/view/94954 [Internet]. [http://dx.doi.org/10.1172/jci.insight.94954]

[147] clinicaltrialsgov [homepage on the internet] Search of: ipsc – List results Bethesda (MD): National Library of Medicine (US) , 2000 [[cited 19 July 2018]]; Available from: https://clinicaltrials.gov/ct2/results?cond=ipsc

[148] Liu X, Campanac E, Cheung HH, *et al.* Idiopathic Autism: Cellular and Molecular Phenotypes in Pluripotent Stem Cell-Derived Neurons. Mol Neurobiol 2017; 54(6): 4507-23. [http://dx.doi.org/10.1007/s12035-016-9961-8]

[149] Sheng Y, Filichia E, Shick E, *et al.* Using iPSC-derived human DA neurons from opioid-dependent subjects to study dopamine dynamics. Brain Behav 2016; 6(8) [http://dx.doi.org/10.1002/brb3.491]

[150] Seifinejad A, Taei A, Totonchi M, *et al.* Generation of human induced pluripotent stem cells from a Bombay individual: Moving towards "universal-donor" red blood cells. Biochem Biophys Res Commun 2010; 391(1): 329-34. [http://dx.doi.org/10.1016/j.bbrc.2009.11.058]

[151] Guo W, Naujock M, Fumagalli L, *et al.* HDAC6 inhibition reverses axonal transport defects in motor neurons derived from FUS-ALS patients. Nat Commun 2017; 8(1) [http://dx.doi.org/10.1038/s41467-017-00911-y]

[152] Rohani L, Johnson AA, Naghsh P, *et al.* Molecular Cytogenetics and Quality Control: Clinical Guardians for Pluripotent Stem Cells. Stem Cells Transl Med [Internet] , 2018 [[cited 2018 Oct 8]]; Available from: http://www.ncbi.nlm.nih.gov/pubmed/30218497

[153] Sullivan S, Stacey GN, Akazawa C, *et al.* Quality control guidelines for clinical-grade human induced pluripotent stem cell lines. Regen Med [Internet] , 2018 [[cited 2018 Oct 8]]; Available from: https://www.futuremedicine.com/doi/10.2217/rme-2018-0095

CHAPTER 3

Reprogramming of Adipose-Derived Stem Cells to Neuronal-Lineage Cells is Regulated by Both Cell Signalling and Redox Status

Sajan George, Anine Crous and **Heidi Abrahamse***

Laser Research Centre, Faculty of Health Sciences, University of Johannesburg P.O. Box 17011, Doornfontein, South Africa

Abstract: The generation of specific neuronal-lineages for cellular therapies hold great promise for nervous system disorders. Stem cells can offer regenerative and replacement therapies of the nervous system through cell signalling, which is directed by the addition of cytokines and neuronal growth factors. Adipose-derived Stem Cells (ASCs) are capable of differentiating into neuronal and glial cells through induced cell signalling pathways and altered redox status. In this chapter, we addressed the dynamic changes within ASCs in response to the changes in its *milieu* - a pre-requisite for trans-differentiation *in vitro*. We considered the functional use of ASCs as a regenerative tool in recovering neuronal cells by focusing on ligand expression and their effects on transmembrane receptors. We also discussed various levels of cell signalling capable of modifying epigenetic programming for trans-differentiation processes. Finally, we underlined the fact that harnessing of Reactive Oxygen Species (ROS) and ROS-mediated cellular signalling is a secret recipe for successful differentiation of stem cells *in vitro*.

Keywords: Adipose stem cells, Cell signalling, Differentiation, Epigenetics, Growth factors, Glia, Laser, Light, Neurons, Photobiomodulation, Reactive Oxygen Species.

INTRODUCTION

The most common types of neurological disorders include *spinal cord injuries* caused by accidents, and *stroke* caused by cerebral haemorrhages. These pathologies result from neuronal malfunction or cellular damage that triggers a cascade of signalling events leading to paralysis and cognitive dysfunction. None of the methodologies devised for neuronal differentiation is 'wholly successful'. There is also a need to cogitate on the mechanisms leading to neuroglial transfor-

* **Corresponding author Heidi Abrahamse:** Laser Research Centre, Faculty of Health Sciences, University of Johannesburg P.O. Box 17011, Doornfontein, South Africa; Tel/Fax: +11 559 6550; E-mail: habrahamse@uj.ac.za

Atta-ur-Rahman and Shazia Anjum (Eds.)

mation. Currently there is no research indicating an effective biological or chemical differentiation inducer for Adipose-derived Stem Cells (ASCs), nor the mechanism behind driving ASCs to neurons or glia. Our understanding about various ligands driving the differentiation of ASCs to neurons is very scarce. Researchers have adopted different differentiation approaches or mix-matched them based on literature, where their attempts ended up in 'neuron-like' bodies. These undifferentiated ASCs are electrophysiologically inactive and not useful for targeted therapy. This chapter attempted to connect neuroglial differentiation with relevant and recent literature pertaining to cellular changes, whilst focussing on cell signalling which is often controlled by the redox status.

NERVOUS SYSTEM

Development and Distinction of Neuronal-Lineage Cells

Pre-natal development of the central nervous system (CNS) is characterized by the formation of neural plates that give rise to the neural tubes. A highly integrated and spatially distinct neural stem or progenitor population of neuroepithelial progenitor cells (NEPs) exists within these structures. Initially, NEPs undergo symmetric division to expand, but later switch to asymmetric division for giving rise to lineage-restricted progenitors. Neural Precursor Cells (NPCs) is a mixed population of Neural Stem Cells (NSCs) and neural progenitor cells, which give rise to neurons and supporting cells [1]. The last stage of CNS development is marked by dendrite branching, axonal elongation as well as formation of synaptic junctions to facilitate impulse transmission. In the developing nervous system, NSCs undergo asymmetric division to produce glial-restricted progenitors, which generate astrocytes, oligodendrocytes and Schwann cells [2]. Finally, NSCs are also the source of specific- neurotransmitter release cholinergic, dopaminergic and GABAergic neurons. Table **1** indicates the functional as well as anatomical significance of different glial cell types configuring the human CNS.

Anatomy of Central and Peripheral Nervous Systems

The nervous system is broadly divided into two principal parts—CNS and the peripheral nervous system (PNS). The CNS consists of the brain and the spinal cord, while the PNS consists of sensory and motor nerves that interconnect with the periphery. The latter act as a communication system between the periphery of the body and the spinal cord. The CNS decipher sensory input from the PNS and translate it into an appropriate motor response. The nervous system is also classified according to its control over the body functions. The somatic nervous system (SNS) is responsible for all the sensory or cognitive activities followed by the movement of skeletal and ocular muscles. The autonomic nervous system

(ANS) controls all involuntary activities concerning cardiac and smooth muscles as well as endocrine glands.

Table 1. Glial cell types supporting the central and peripheral nervous systems.

Cell types	Function
Neuroglia (glia)	helper cells of neurons for anatomical and functional support
Schwann Cells	glial cells that produce myelin sheaths around axons of the PNS
Astrocytes	glial cells that function in the BBB in the piamater of the CNS
Oligodendrocytes	glial cells that produce myelin sheaths on axons of the CNS (white matter)
Satellite Cells	glial cells that are helper cells to sensory and autonomic nerves of the PNS
Ependymal Cells	glial cells that are part of the choroid plexus and produce CSF
Microglia	glial cells that are phagocytic immune cells that find & destroy invaders

NOTE: Schwann Cells, Satellite Cells, Oligodendrocytes, Microglia, Astrocytes, and Ependymal Cells are the six types of neuroglia (glia) cells that support neuronal growth and activity. *Abbreviations: CNS, Central Nervous System; PNS, Peripheral Nervous System; CSF, Cerebrospinal Fluid.*

The nervous tissue consists of neuronal and glial cells, which are the functional and supportive cells of both the CNS and PNS, respectively. Morphologically, neurons have a soma or cell body upon which several extensions emerge and interconnect with the neighbouring cell. These processes, referred as dendrites are the structures that receive sensory input from the surroundings. Each neuron will have at least one of these processes as a prominent extension, referred to as the axon. They are the functional units of neurons that are responsible for the transmission of impulses to neighbouring cells or neurons. It is important to note that a bundle of axons in the CNS is mentioned as tract, while the same in the PNS is a nerve. Similarly, a localized collection of neuron cell bodies is referred to as a nucleus in the CNS and as a ganglion in the PNS.

Diseases and Disorders of the Nervous System

Neurodegeneration is a condition characterized by progressive deterioration of neuronal morphology and functional characteristics due to ischaemia or haemorrhage, deposition of intracellular toxic proteins or accumulation of genetic mutations [3]. Alzheimer's disease and amyotrophic lateral sclerosis is characterized by neurovascular dysfunction resulting in the impairment of the blood–brain barrier function. This evidently result in a lack of nutrition to cholinergic nerve cells and the impairment of clearing neurotoxic molecules resulting in neurodegeneration [3]. Another well-studied condition, Parkinson's disease, is caused by the degeneration of dopaminergic neurons resulting in a gradual dysfunction of the motor neuronal system. Any hypofunction of N-methyl-D-aspartic acid-type glutamate (NMDA) receptors at the cerebral cortical

GABAergic interneurons lead to functional abnormalities causing schizophrenia. The generation of specific neuronal-lineage cells might open doors for cellular therapies and modelling of these disease conditions.

Spinal cord injury is a traumatic condition caused by a sudden loss of sensory, motor, and autonomic function resulting in neurological dysfunction and lifelong disability. Pathophysiology of spinal cord injury includes haemorrhage, oedema, necrosis, fluid and electrolyte imbalance, oxidative damage and cavitation [4]. Altogether, this will alter the gap junction coupling between astrocytes causing formation of a glial scar, which interrupts the impulse transmission. A stroke is caused by occlusion of small vessels or haemorrhage due to rupture of vessels resulting in an inflammatory response and neuronal death. This cellular damage triggers a cascade of signalling events leading to paralysis and cognitive dysfunction [3]. Multiple sclerosis is a T-cell mediated disease of the CNS, causing reduction in the number of oligodendrocytes and myelinated axons. Stem cells offer regeneration and replacement therapies by re-myelinating damaged axons, production of neurotropic factors as well as anti-inflammatory cytokines, which altogether are important for nerve repair.

STEM CELLS

Advantages of Using Mesenchymal Stem Cells (MSCs)

Embryonic cells as well as induced pluripotent stem cells that are generated from adult stem cells by epigenetic re-programming, carry the greatest potential for application in regenerative medicine and disease modelling. Unfortunately, the use of embryonic stem cells is strictly controlled and the use of induced pluripotent stem cells make the host genome highly prone to integration of retroviruses, transgenic genes and aberrant reprogramming, all causing a concern for teratoma formation. Stem cells of mesenchymal origin are distributed all around the body, particularly in Wharton jelly of the umbilical cord, adipose tissue, synovium, and the dental pulp of deciduous baby teeth, which can differentiate into specific lineages. They promote axonal growth by regeneration as well as functional plasticity when transplanted to the site of the CNS injury. Further, they synthesize a variety of neurotrophic cytokines such as Brain-derived Neurotropic Factor (BDNF), Nerve Growth Factor (NGF) and vascular endothelial growth factor (VEGF) that stimulate nerve growth and proliferation for repair processes [5, 6].

Adult stem cells have greater viability and are capable of differentiating into different cell and tissue lineages [7]. They are frequently isolated from bone marrow and grown *in vitro* under optimal cell culture conditions. In comparison to highly mitotic embryonic cells, Mesenchymal Stem Cells (MSCs) show better genomic stability. This is seen when inducing differentiation *in vitro* there is less of a chance of neoplasm formation and mosaicism. They also have the ability to secrete multiple bioactive molecules capable of stimulating recovery of injured cells and inhibiting inflammation. MSCs have shown stable cell growth and proliferation *in vitro* for harvest and administration to patients [7, 8]. Moreover, by addition of specific cell growth factors, these cells can differentiate into osteogenic, adipogenic, myogenic and chondrogenic lineages [9 - 11].

Although, MSCs have been proven to promote neurite outgrowth over inhibitory molecules of the glial scar, they only have limited effects at the lesion foci of the brain or spinal cord [12]. Upon transplantation, MSCs may differentiate into oligodendrocytes resulting in re-myelination of damaged or demyelinated axons and neuronal cells. This helps in the building up of action potential for nerve conduction or astrocytes providing scaffolding and nutrition for the injured area. However, the early stage differentiation of MSCs to astrocytes result in the production of growth inhibitory molecules around the lesion foci, impairing the overall nerve regeneration and restoration process [13, 14]. Thus, a comprehensive plan is required for the differentiation of MSCs to neuroglia that can be implemented in regenerative medicine.

ASCs for Replacement Therapy

Adult stem cells are easily attainable from bone marrow and fat depots of the body with minimal surgical procedures. It contains heterogeneous stromal cell populations capable of proliferation and differentiation into both mesodermal and non-mesodermal lineages. Human ASCs have multilineage differentiation potential and act as a source of stem cells with potential clinical application for autologous therapy in neuronal disorders and disease conditions. In comparison to the use of bone marrow stem cells, autologous therapy using ASCs offer a greater yield in terms of stem cells and is free from severe complications. Adult stem cells from the fat depots, which is the energy source of the body, constitute as much as 1% of human adipose stem cells, which is approximately 1000-fold more than stem cells lodging in the bone marrow [15]. By comparison, they yield as much as 5,000 fibroblast colony-forming units (CFU-F) per gram of adipose tissue as compared to bone marrow stem cells, which consists only of 100–1,000 CFU-F per millilitre [16]. Besides this, the ability of proliferation in culture is far superior for ASCs as compared to the bone marrow stem cells [17].

ASCs from adipose tissue are classified as 1) bone marrow 2) brown 3) white 4) mammary and 5) ocular. Bone marrow adipose tissue is the source of energy and the seat of cytokine production. Brown adipose cells that cover the major organs of juveniles possess large numbers of mitochondria and release heat *via* oxidation of fatty acids. White adipose tissue is located in the subcutaneous and visceral regions of adults to provide insulation as well as energy stores. They have a better proliferation and differentiation capacity than the abdominal fat derived stem cells [18]. Brown adipose tissue display reduced plasticity and lesser numbers of ASCs as compared to the white fat tissue [19]. Finally, mammary and ocular fat support physiological functions associated with the secretion of milk or the function of eyes, respectively.

Adult stem cells originating from the various fat depots in the human body can readily be isolated and cultured, therefore, representing a reliable and ethical source of stem cells. They are often isolated from abdominoplasty or lipo-aspirates and grown *in vitro* under optimal cell culture conditions. For isolation of MSCs, adipose tissue is subjected to enzymatic digestion, which yields a mixture of stromal and vascular cells referred to as the stromal-vascular fraction (SVF). Upon plating these cells in tissue culture dishes, ASCs will filter out from the SVF, adhere to surfaces and proliferate (Fig. **1**). ASCs obtained from subcutaneous tissues as well as superficial abdominal regions are resistant to apoptosis and much more likely to be differentiated [20]. In fact, the greatest advantages of using ASCs are its 1) minimally-invasive steps for isolation 2) ability to propagate and differentiate *in vitro* 3) minimal risk of tissue histo-incompatibility and rejection 4) ability to cross the blood-brain barrier and reach lesions 5) omission of ethical and social concerns. Thus, the harvest of ASCs with less invasive procedures from our fat depots and less intensive protocols for differentiation gives preference for adopting stem cell therapy in clinics.

ASCs for Neuroglial Trans-differentiation

Several researchers have successfully differentiated ASCs to neural cells using hormonal and chemical inducers. Human ASCs can differentiate into bone, cartilage, fat, muscle, and neurons, making them a possible source for future applications in clinical settings [11]. The cellular mechanisms involved in the process of trans-differentiation is intriguing and is initiated as early as day one lasting for at least a week or even up to two weeks. Ideally, this process of differentiation should be as short as possible to accomplish a higher percentage of differentiated neuronal and glial cells with the least anomalies due to aging of cells and growth medium.

The processes of proliferation and lineage-specific differentiation is regulated by a complex interplay involving stem cell transcription factors, a wide variety of cellular kinases, growth factors, and receptors. A series of events occur in ASCs due to the autocrine and exogenous cytokines, added growth factors and probably external factors such as temperature, light *etc.* These factors can all interact with each other and can ultimately result in genetic reprogramming of the differentiating ASCs into neuronal-lineage cells. In embryos, the expansion of the CNS is marked by neurogenesis followed by gliogenesis of NSCs with compartmentalization by radial glia, acting as a substrate for migration. To achieve successful differentiation of ASCs into neurons we need to adjust the cell culture conditions to simulate what proceeds in embryonic neurogenesis.

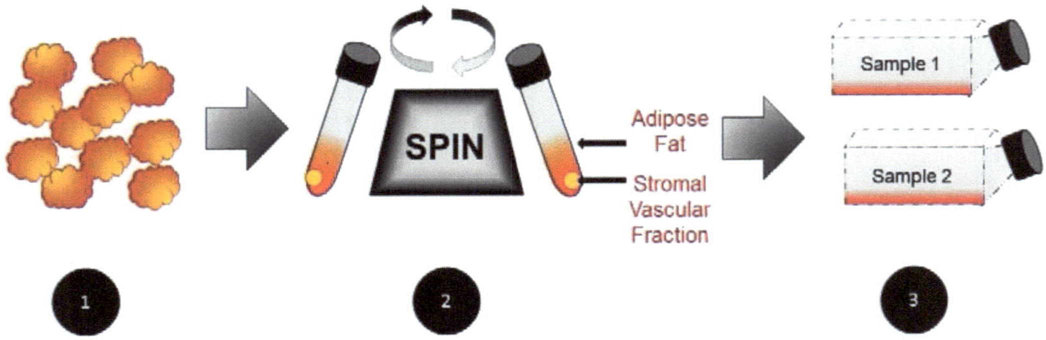

Fig. (1). Schematic harvest of adipose stem cells from fat depots: Following adipose tissue harvest, fat pads were minced thoroughly and incubated with collagenase (step 1), centrifuged to pellet the stromal vascular fraction (step 2) and plated for enrichment of adipose stem cells (step 3).

Trans-differentiation is a complex process where changes in cell signalling machinery redirects cellular biochemical and biophysical mechanisms that alter the functional properties and morphology of a cell. It is a process by which the multipotent cells are changed to another germ layer *in vitro*. Not surprisingly, true differentiation of ASCs can only be achieved by an in-depth understanding of the intriguing pathways of cellular signalling mechanisms. A comparative analysis has found that ASCs accelerates nerve repair and is more efficient in differentiating into neural and vascular endothelial cell types [21]. Microarray analysis has identified that these differentiating ASCs express higher numbers of genes involved in immunity and inflammation [22] and are more likely to differentiate towards myogenic lineages [23]. Revisiting and reanalysing several of these pathways and processes enable researchers to devise novel methodologies for the efficient generation of neuron-lineage cells for therapeutic application.

CELL SIGNALLING

Cell Signalling for Neuronal Differentiation

In mammals, NEPs are seen along the sub ventricular zone of the lateral ventricles and sub granular zone of the hippocampal dendate gyrus. They transform into radial glia, acting as a substrate for migrating NPCs and giving rise to astroglial cells in embryos. Radial glia is also a source of other glial cells, while being able to differentiate into neurons for driving the CNS development. NSCs lodged in the sub ventricular zone of the lateral ventricles and the hippocampus of a growing foetus can differentiate into neurons and glia [24].

Bone morphogenetic proteins (BMPs) is a family sub-divided into several subfamilies. BMP-2/4, BMP-5/6/7/8, growth and differentiation factor (GDF)-5/6/7, and BMP-9/10 [25]. All these BMPs utilize TGFβ receptors to regulate growth and differentiation through phosphorylation of the intracellular *smad* proteins. During development, the levels of noggin (a ligand of BMP receptor) increases, which binds to endogenous BMPs and subsequently prevent activation of the BMP pathway [26]. Since, BMPs block the default neural differentiation pathways; it has to be suppressed using an antagonist such as noggin or chordin for effective neural induction *in vitro* [27].

Growth factors such as BDNF, acidic Fibroblast Growth Factor (FGF) increases the intracellular level of cyclic Adenosine Monophosphate (cAMP), a secondary cell signalling component. The net result is the phosphorylation of transcription factors leading to neuronal differentiation, integrin assembly, neurite outgrowth and neural plasticity [28, 29]. An increase in the intracellular cAMP concentration also result in the activation of signalling by the Shh pathway, critical for normal development of organs and structures in a newly developing embryo. A similar situation occurs when IBMX, indomethacin and insulin are administered for the neuronal differentiation of ASCs *in vitro*. Here, nonspecific inhibition of phosphodiesterase by IBMX increases cAMP levels, and inhibition of COX by indomethacin lead to the arrest of cells in the G2/M phase causing neurogenic differentiation [29]. Signalling by Shh may cause metamorphosis of embryonal cells into many different types of tissues and organs, although, any defect in the activity or function of Shh caused by inhibition or mutations may result in cancer and other maladies.

Pathways, such as notch and smad signalling, and wingless (Wnt) glycoproteins, also control the physiological activities contributing to neuroglia differentiation. *Notch* is a single spanning transmembrane protein that regulates the spatial patterning, timing and outcomes of many different cell fates during embryonic development. Adjacent cells use *notch* for communication while marking their boundaries and laying down specific tissue patterns [25]. *Notch* signalling occurs by lateral inhibition, whereby NPCs upregulate *notch* ligands leading to an increase in transcriptional targets, thus, down-regulating tissue-specific transcription factors leading to control of signals to progenitors during neurogenesis [30]. Wnt, consists of seven-pass transmembrane receptors on the target cell surface that bind to *Frizzled* or the LDL-related protein complex. Wnt regulates neural patterning, axonal growth as well as synaptogenesis during embryonic development. Wnt proteins phosphorylate and deactivate GSK-3β leading to the accumulation of β-catenin, which is an important mediator of canonical Wnt signalling pathways.

Retinoic acid (RA),a metabolite of vitamin A, signals through retinoid X receptors (RXR), which aid in neural patterning and glial cell formation. The primary function of RA is to regulate the cell cycle in order to halt proliferation, and the loss of RA signalling is associated with dedifferentiation and the development of cancer. While several biological and chemical inducers of cell growth promote proliferation largely by inhibiting differentiation, RA is a differentiation-inducing molecule that inhibits cell proliferation [31]. Together, basic FGF and RA downregulate Wnt as well as *notch* signalling in proliferating ASCs, followed by differentiation to a neuronal phenotype [30]. Therefore, these signals control the switch for balancing stem cell proliferation and differentiation, which is fundamental to vertebrate neurogenesis.

Differentiation of Neurotransmitter Release Neurons

Neurotrophins, especially NGF acts through the p^{75NTR} and Trk receptors for the differentiation of early neuronal progenitors and maintenance of cholinergic basal forebrain neurons [32, 33]. Higher concentrations of neurotrophin-3/4 as well as BDNF are required for the neuronal physiological activities including neurotransmitter-release from cholinergic nerve endings [34]. Differentiation of freshly isolated bone marrow-derived MSCs into cholinergic neurons is more achievable by pre-induction with β-mercaptoethanol (BME) followed by NGF [35]. *In vitro* investigations have proved that the PI3K signalling pathway, but not MAPK signalling, is instrumental for the NGF-induced morphological changes and neurotransmitter-release activities of cholinergic neurons [36, 37]. Obviously, more investigations are necessary for unravelling the intricacies of various pathways involved in the differentiation of neurotransmitter-releasing neurons.

Differentiation of specific neurotransmitter-release neurons is also achievable by using RA along with other cell growth inducers. In MSCs, neurotrophins along with RA stimulate vertebrate neurogenesis as well as synaptic induction through the canonical/β-catenin and neuronal differentiation to cholinergic and dopaminergic neurons through the non-canonical/JNK pathways [38]. The carboxylic acid form of RA in combination with Shh and FGF8 (encoded by *fgf8* gene) has synergistic effects on the differentiation of sensory neurons in the PNS. Basic FGF as well as FGF8 differentiate MSCs to functional dopaminergic neurons by binding to Shh through receptor *Patched1* [39, 40]. Besides, the neurotrophic activity of basic FGF and Nurr1 is responsible for the development and survival of dopaminergic neurons *in vitro* [41]. Kondo *et al.* (2005), has already shown that a combination of two signals, Shh and RA may suffice the differentiation of MSCs to glutamatergic sensory neurons [42].

Epigenetic Changes in Neuronal Differentiation

Epigenetic mechanisms consist of DNA methylation, histone posttranscriptional modifications, chromatin organization, and the action of noncoding regulatory RNA. Epigenetics modulate the efficiency of gene transcription by altering chromatin structure causing variations in the accessibility of transcription factors and RNA polymerases. Epigenetic modifications are also critical for regulation of gene expression during the cell cycle, development and differentiation. Any misregulation or dysfunction of these mechanisms could lead to neurological diseases.

Neuronal differentiation is controlled by DNA demethylation as well as methylation in a time-depended manner. Several of these processes have been identified in the development, differentiation, maintenance and self-renewal of neuronal and glial cells. DNA methylases, both DNMT3a and DNMT3b are expressed in early NPCs, which decrease during the progression of neural development. This result in the upregulation of the neuronal genes such as *Dlx2*, *Neurog2* and *Sp8* causing a switch from neurogenesis to gliogenesis. During development, such changes in the DNA methylation pattern are primarily responsible for the differentiation of NPCs to astrocytes [43, 44].

Histone acetylation and deacetylation modify chromatin structures by altering the interactions of histone tails with the DNA helix. Histone acetylation occurs through inhibition of HDAC, which is affected by valproic acid promoting neuronal differentiation of NSCs [45]. NPCs contain a specialized chromatin-remodelling complex, which is the BAF complex that undergoes a variety of subunit changes in an Adenosine triphosphate (ATP)-dependent manner. Additionally, the incorporation of noncoding regulatory RNA alters the composition of subunits in the BAF complex altering the gene expression pattern leading to differentiation of NPCs to neurons [46].

Components of cell signalling pathways also play a role in neuronal differentiation by interfering with the binding of transcriptional regulators to their targets. The bipartite transcription factor formed by a combination of β-catenin and the Wnt effector, TCF7L2 represses Olig2 in the oligodendrocyte precursors. During differentiation, HDAC 1 or 2 competitively inhibits β-catenin to prevent binding to TCF7L2, and the expression of oligodendrocyte-specific genes. Similarly, other signalling pathways can also be affected by moderating nucleosome density during transcription [47].

Undoubtedly, microRNAs play a critical role in maintenance, differentiation and lineage commitment of stem cells. In embryonic stem cells, microRNA-124 targets Sox9 for the formation and maintenance of NSCs, while microRNA-9 controls their proliferation and differentiation [45]. PDGF-BB activates not only JAK-STAT but also the AKT, ERK1/2 and PDGF receptor-β signalling pathways by an ROS dependent mechanism, which upregulates miR-210 expression [48]. To a certain extent, this explains the higher proliferation rate of ASCs after treatment with PDGF-BB, as compared to PDGF-AA or other growth factors. However, there has been contradicting reports about the cellular effects of miR-210 as well as HIF-1α, which are the master controller of cellular response to hypoxia [48, 49]. Further, the expression levels of βIII tubulin are negatively regulated by the micro-RNA segment miR-133b [50]. MicroRNA profiling will help to identify the specific microRNA signature pattern during stem cell developmental stages, and then, microRNAs could be useful biomarkers of neuronal and glial differentiation processes.

CELL MARKERS

Temporal Changes During Differentiation of Neurons

Temporal changes in the configurations of surface proteins help identify the stage of differentiation of neuronal progenitors into neuronal lineage cells. Nestin, Sox1 and Sox2 are some markers on NPCs that disappear at the termination of the differentiation process. Upon transformation, NSCs express increasing levels of early neuroectodermal markers related to the pro-neural genes such as neuroD1, neurogenin-2, MSl1, OTX1 and nestin. NSCs are identified by the presence of markers such as GFAP, CD15, Musashi, EGFR, FABP7 (BLBP) and GLAST [51] Doublecortin is expressed in the very early stages of differentiation while cells are proliferating but still committed to a neuronal fate. This is followed by expression of neuroD1 at the later stages of proliferation towards the onset of terminal neuronal differentiation. A brief account of the changes to surface markers of these differentiating cells is shown in Fig. (**2**).

The generally accepted surface markers on neuronal lineage cells are vimentin, nestin and GFAP. Vimentin is an intermediate filament protein that is responsible for maintenance of structural integrity and cell shape in many cell types of the neural crest. Nestin is downregulated and replaced by GFAP in glial cells and in neuro filaments within neurons [52]. The light/medium/heavy (L/M/H) chains of neuro filament subunits support the structure of neuronal branches and assist in the intracellular transport in axons and dendrites. The expression of these proteins varies in coordination with the developmental patterns of neuronal lineage cells in the process of differentiating from ASCs. Expression of neuro filament L is followed by neuro filament M during the initial neurite outgrowth phase, and later these proteins become part of the mature cytoskeleton of all neurons. However, neuro filament H is more abundant in larger neurons, such as motor neurons [53].

A uniform distribution of MAP2 is observed in the developing neurons and its accumulation on the dendritic tree (as well as its extensions) is coincidental with neuronal maturation [54]. A cytoskeletal dimer, βIII tubulin is important for neurite elongation and has a vital role in neurite outgrowth and neuronal differentiation [55]. The neural marker MAP2 causes changes in the neuronal morphology by polymerization and stabilization of microtubules, while βIII tubulin performs a specific role in organelle transport and synaptic plasticity [56, 57]. Thus, the presence of βIII tubulin and MAP2 are indicative of the formation of immature neurons. The three isoforms of MAP2 are called a, b and c, of which MAP2c appears in axons during early development, and is replaced by MAP2a during maturation. On the other hand, MAP2b is found to be expressed throughout neuron growth and play a role in determining and stabilizing their dendrites [58].

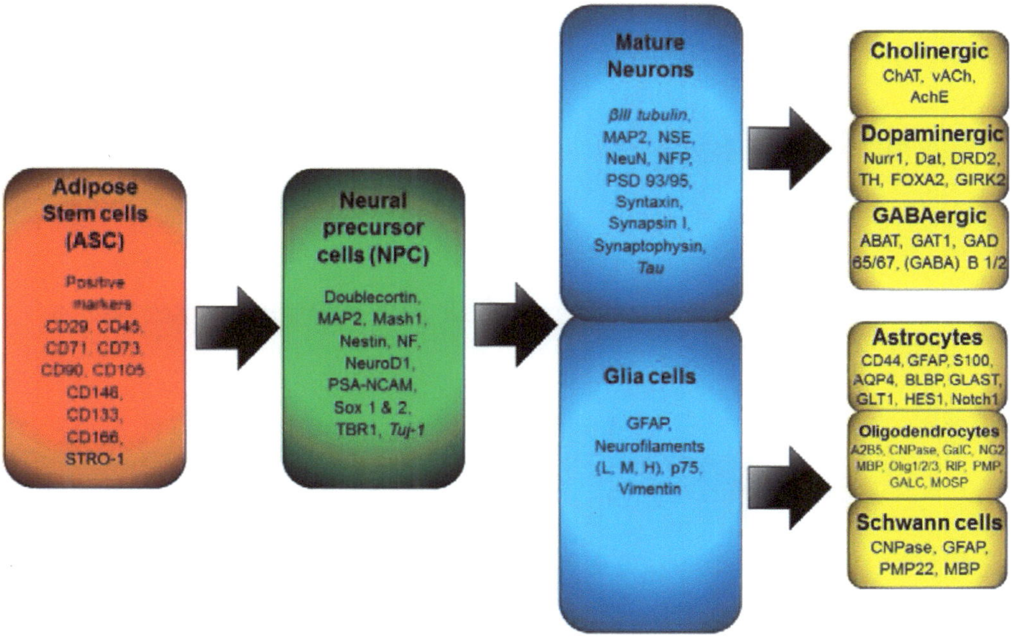

Fig. (2). Cell markers during the neuroglial differentiation of stem cells: Stage-wise changes in cell surface of ASCs during differentiation to neuroglia is diagrammatically represented from left to right. Immature and mature neurons as well as neurotransmitter-releasing neurons and glia are characterized by distinct set of markers. However, individual cell types vary in their expression across the neuronal lineage cells. βIII-tubulin and MAP2 are the widely used for the detection of neuronal differentiation, while GFAP and MBP are identified for glia.

The final stage of differentiation of neurons is characterized by neurite extension, integration of synaptogenesis and exchange of neurotransmitters. These newly emerging dendrites are studded with the cytoplasmic protein PSD95, which plays a role in anchoring receptors and ion channels to the synaptic membranes [59, 60]. Synapsin I has a pivotal role in synapse formation as well as elongation of axons and branching [61]. Synaptophysin, a presynaptic marker is another good indicator of the formation of new nerve terminals and points of synapse formation [62]. Another synaptic vesicle-related protein, syntaxin functions in the fusion of synaptic vesicles with the synaptic membrane of neurons [63]. Overall, changes in cell signalling and paracrine factors regulate the cell surface decoration of membrane proteins.

Markers on Neurotransmitter Release Neurons

Specific identification of neurons according to their pattern of release of neurotransmitters is yet another approach. Cholinergic neurons are identified by their immunoreactivity against acetylcholine, acetylcholinesterase, ChAT, choline transporters and vesicular acetylcholine transporters. It should be noted that, along with Trk receptors, $p75^{NTR}$ mediates the effects of the neurotrophins, and is expressed by ChAT-positive cells in the cholinergic basal forebrain neurons [36]. Mature motor neurons release these markers and stain well for the identification of differentiated cholinergic neurons [64]. Following differentiation, dopaminergic neurons express markers such as. DAT, Foxa2, Girk2 and Nurr-1. Other markers such as, Aldh1α1, *Engrailed1*, LMX1β, Pax5, PITX3, LMX1α and VMAT2 are often affected by culture conditions and changes in the levels of gene expression during the differentiation process [65].

GABA is a mature neuronal marker as well as a neurotransmitter released by GABAergic interneurons. The enzyme for GABA synthesis, Glutamic acid decarboxylase (GAD) exists in two isoforms, GAD65 and GAD67, and may serve as a mature marker of GABAergic interneurons. For neurotransmission, GAD65 is localized at the nerve terminals, while GAD67 synthesizes GABA at synapses [66]. GABAergic neurons express GABA type-A and -B receptors as well as the ionotropic P2X receptors for ATP. Stimulation of GABA-A agonists increases cell growth whereas, stimulation of GABA-B reduces proliferation of ASCs [67]. Serotonergic neurons release serotonin, which is a monoamine neurotransmitter derived from tryptophan. Markers in these cells are serotonin, SERT and tryptophan hydroxylase. The markers found on excitatory glutamatergic neurons are glutaminase, glutamate transporter, glutamine synthetase, vesicular glutamate transporters 1, 2 and 3.

Changes in Surface Markers of Differentiating Glia

A plethora of other surface markers will show changes in their level of expression during differentiation of ASCs to neuroglial cells. In differentiating ASCs, forskolin increases GFAP expression and decreases the expression of nestin and Neurofilament-70 creating an astrocyte-like phenotype [68]. However, the intermediate filament proteins S100β and GFAP are highly conspicuous on these star-shaped cells during various stages of their differentiation. Often, markers such as Aldh1L1, AldoC, ASCGB1, Aquaporin 4, FABP7/BLBP, FGFR3, GLAST and GLT1 are expressed on mature astrocytes [69]. Schwann cells are the chief glia cells in the PNS and in mammals; they possess a natural regenerative capacity. Their activity includes generation of nerve-related extracellular matrix, regulation of neuromuscular synaptic junctions, and secretion of neurotropic factors for tropism [70]. Several markers are expressed on Schwann cells, among which CNP, GFAP, MBP and PMP22 are of diagnostic importance.

In combination with the basic helix-loop-helix proteins, neurogenins and Mash1, Olig2 (Oligodendrocyte transcription factor 2) drives the specialization of early precursors of oligodendrocytes [71]. Later oligodendrocytes become positive for A2B5, CD3, GalC, Nestin, NG2, Olig1, Olig2, Sox9, PDGFRα and PSA-NCAM [72]. The enzyme CNPase is expressed in immature oligodendrocytes and plays a role in tubulin polymerization as well as the outgrowth of their cilia-like processes [73]. When grown in "glia-defined medium" without basic FGF, Epidermal Growth Factor (EGF), and Platelet-derived growth factor (PDGF) these cells develop a network of cytoplasmic extensions and start expressing myelin-specific proteins MAG, MBP, MOG and PLP indicating their maturation [72]. Expression of NSE increases during oligodendrocyte differentiation, but then, become repressed once these cells become fully mature [74]. In short, cell signals are capable of regulating transcription factors, which is determinant for the fate of glia and expression of surface markers.

REDOX STATUS

Cells maintain a subtle balance between Reactive Oxygen Species (ROS) generation and utilization, which turns out to be essential for homeostasis (Fig. **3 A&B**). Depending on its concentrations, high levels of ROS induce apoptosis and necrosis, while lower levels cause changes in cellular growth and differentiation by modulating signalling pathways, cellular enzymes and gene expression levels. For instance, ROS in response to stimulation of external ligands on the members of NADPH (Nicotinamide adenine dinucleotide phosphate) oxidase (NOX) family modulate signals for NPCs survival and proliferation [75, 76]. High levels of ROS, particularly superoxides are found in migratory neurons implying a role of

ROS in the regulation of Guanosine di/tri phosphatases (GTPases) responsible for the cytoskeletal rearrangements and dendritic morphogenesis [77, 78]. ROS can dissociate GDP bound Ras, which results in the Ras-mediated recruitment of cytoplasmic Raf to plasma membrane for phosphorylation of MAP kinases as well. Similarly, an increase in ROS enhances phosphorylation of the neurotrophin receptor TrkA as well as the activation of Rac/Rho GTPases during differentiation [79, 80]. Binding of neurotrophins stimulate PI3K and phosphorylate PIP2 to PIP3. In fact, ROS down regulates the synthesis of PIP3 by inactivating PTEN, which activates signalling by Akt and mTOR for neurogenesis [81, 82]. ROS regulates IκB kinase (IKK) complex activity through activation of various kinases MEKK1 as well as phosphorylation of IKKα by NIK in the NF-kB signalling pathway [83]. In summary, ROS interfere with signalling pathways, which have direct effects on cell morphology, growth and multiplication. While, an overall reduced state predispose cells towards proliferation, induction of oxidation by ROS seems to be an onset of differentiation.

The key signalling molecule in ROS is hydrogen peroxide, which oxidizes sulfhydryl groups of cysteine residues leading to the initiation of protein kinase A (PKA)-induced phosphorylation of calcium channels as well as Ras/Raf and MAPK. Hydrogen peroxide lowers glutathione content of cells affecting the redox status, which alters the balance between self-renewal and differentiation [84]. Downregulation of reduced glutathione is accompanied by a rapid increase in ascorbic acid that regulates the stem cell redox status during differentiation [85]. Intracellular calcium has a direct effect on the pro- and antioxidant enzymes like NOX and glutathione reductase, which maintains redox status (Fig. **3 C&D**). So also, calcium ions directly affect the levels of dismutase, calcium-binding proteins and calmodulin for the down regulation of ROS for homeostasis. Similarly, ROS has a direct effect on the regulatory and catalytic domains of PKC encompassing cysteine-rich regions. It has also been found that the increased levels of ROS deactivate PKC while a controlled or lower dose has stimulatory effect. Hydrogen peroxide is also responsible for the phosphorylation and activation of PLCγ causing production of inositol triphosphate and diacyl glycerol leading to an increase in the intracellular levels of calcium [86]. Further, the activation of PKC by DAG leads to the phosphorylation and recruitment of PKD, which influence the binding of PKCδ and PKD. Mechanistically, this occurs by ROS-mediated phosphorylation of tyrosine 93 and 463 residues and the activation of PKD *via* Sarcoma and tyrosine kinases, respectively [83]. Thus, the induction of ROS by a ligand-receptor interaction or stress due to physical factors may result in the stemness, proliferation and, likely differentiation of ASCs.

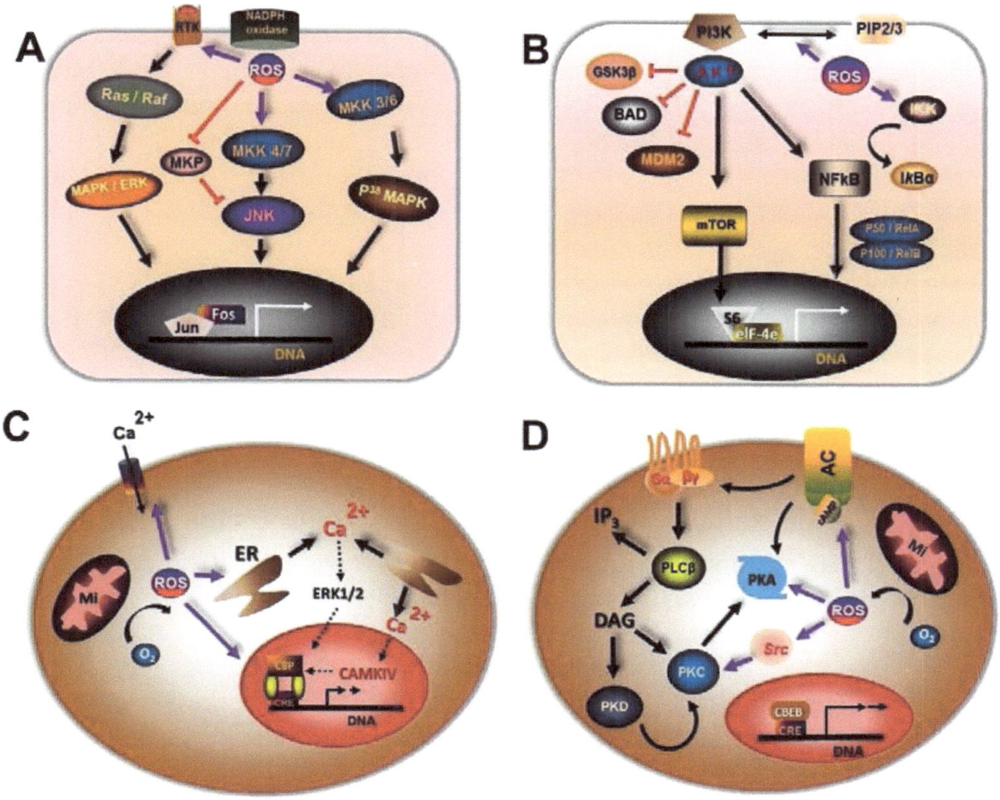

Fig. (3). Generation of ROS and ROS-mediated cellular signalling: Cell surface Nicotinamide Adenine Dinucleotide Phosphate (NADPH) oxidase synthesize Reactive Oxygen Species (ROS) that affect the phosphorylation and activation of Janus Kinases (JNK) and P^{38} Mitogen Activated Protein Kinases (MAPK) pathways (A). ROS interfere with the Phosphatidylinositol 3-Kinase/v-*akt* murine thymoma viral oncogene homolog 1 (PI3K/Akt) and mechanistic Target of Rapamycin (mTOR) signalling. Net result of the Nuclear Factor *kappa*-B cells (NF*k*B) signalling pathways through the Inhibitor of *kappa* B kinase (IKK) and Inhibitor of *kappa* B subunit α (I*k*Bα) is the nuclear translocation of RelA and RelB proteins, respectively (B). Ligands or physical factors activates mitochondria for the production of Reactive Oxygen Species (ROS), which release calcium (Ca^{2+}) and activate transcriptional pathways (C). Any changes in cell redox status stimulates Adenylate Cyclase (AC) enzyme for the synthesis of cAMP and Protein Kinases that has implications on cell survival, proliferation and differentiation (D). *Abbreviations: BAD, Bcl-2 antagonist of cell death; cAMP, cyclic Adenosine monophosphate; CaMK, Calcium/calmodulin kinase; DAG, Diacyl Glycerol; ER, Endoplasmic Reticulum; Gαβγ, G-protein αβγ subunits; GSK3β, Glycogen Synthase Kinase 3β; IP3, Inositol 1,4,5 triphosphate; Mi, Mitochondria; MKK, MAP Kinases; MDM2, Mouse double minute 2 homolog;* O_2*, Oxygen; PLCβ, Phospholipase C subunit β; PK A/C/D, Protein Kinases A/C/D; RTK, Receptor Tyrosine Kinase; Src, Sarcoma kinase.*

PHOTOBIOMODULATION

The visible to Near Infrared region (NIR) of the electromagnetic spectrum is comprised of photons at different wavelengths that can transfer energy to living cells by a process referred as photobiomodulation (PBM). The light energy within the cells can then follow one of three pathways according to the effects seen after application of PBM. The first and most common pathway followed is internal conversion, where the electron goes from a higher to a lower excited state. Here no photons are emitted, also called radiation-less de-excitation. The molecular spin state for internal conversion remains the same and the energy of the electronically excited state is given off to vibrational modes of the molecule, in other words, the excitation energy is transformed into heat. Fluorescence is the second pathway, where molecular absorption of the photon received activates another photon with a longer wavelength. Here again, the energy of the absorbed and emitted photon is converted to molecular vibrations or heat. The third energy transfer pathway is photochemistry whereby light is absorbed by a chromophore so that the chromophores singlet state excitation results in triplet state excitation. Here ground state molecular oxygen, which is a triplet, undergoes a reaction where it becomes singlet oxygen. Another reaction the chromophores can undergo is its triplet state releasing an electron to oxygen, which then forms superoxide. This mechanism is important as the electron transport chain involved in the mitochondrial respiratory chain where the ROS and ATP are generated [87]. Thus, the physical forces like light or laser affects biological systems mostly through production of 'free radicals' and 'bioenergy'. ASCs adopt proliferative or differentiation phases depending on the effect of these free radicals on signalling and transcriptional pathways (Fig. **4**). Our speculations about the effect of light or lasers for neuronal differentiation is based on these changes in cellular redox status [88].

Quantum mechanical theory states that the energy of electromagnetic waves depends on the intensity of photons and their wavelength. These photons are either absorbed by intracellular chromophores or scattered throughout the cells. Light or laser induced changes of the cellular biological processes by activation of the mitochondrial respiratory chain has been shown in stem cells as well as other cell types [89, 90]. The use of Low-Intensity Laser Irradiation (LILI) has reported an increase in cellular functions and chemical production leading to increased proliferation and viability [90, 91]. These effects of PBM are seen without cytotoxicity and act as a useful attribute in tissue engineering and regenerative medicine [92, 93]. In addition to this biostimulatory effect, LILI has also been seen to illicit different responses depending on the power of irradiation, whereby molecular and cellular processes are inhibited (bioinhibition) at different laser intensities [94]. Effects of differing wavelengths and fluences of LILI have been investigated to show an indirect correlation between these parameters and treated cells, as higher wavelengths and fluences seems to inhibit viability and proliferation whereas the opposite was found to be true for lower wavelengths and fluences [95]. Often, these effects on biological activities involving LILI is a therapeutic modality using photons from the visible red to NIR region of the spectrum [96]. The application of visible light on human and animal cells and tissues using visible and NIR radiation has become popular as different process outcomes are being discovered. Of most interest was LILI, for its bio-stimulatory effect, but, where or when larger doses were used, its positive action changes into inhibition [97]. Table **2** details the parameters used for LILI on human stem cells of mesenchymal origin with or without growth factors and biomodulatory effect. Understanding these biostimulatory effects of light or lasers along with various growth factors on ASCs is a befitting option for engendering neurons and glial cells for replacement therapies.

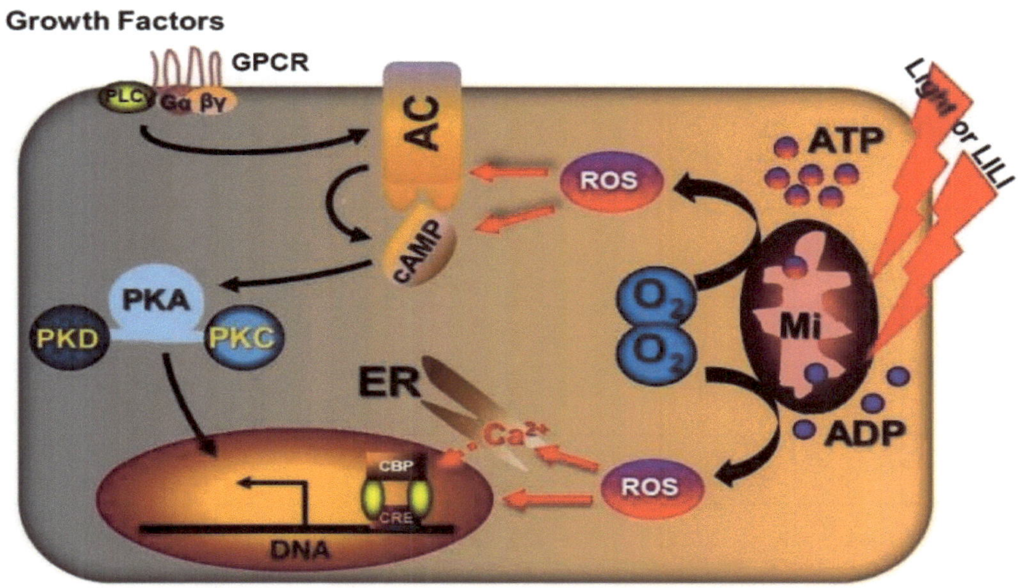

Fig. (4). Mechanism of photobiomodulation: Light or lasers are capable of stimulating respiratory chain reactions in mitochondria (Mi) of stem cells, which increases the ratio of Adenosine Tri/diphosphate (ATP/ADP). Any changes in the microenvironment of a living cell *viz.* growth factors, light or laser, cause Mi to generate Reactive Oxygen Species (ROS) that trigger adenylate cyclase (AC) to release cAMP. This activates protein kinases (PKA/C/D) and cellular signalling pathways affecting gene expression and stem cell growth. ROS is also capable of inducing transcription factors on DNA for lineage-specific differentiation of stem cells. Further, ROS generated from light or lasers are capable of stimulating endoplasmic reticulum (ER) to release Calcium. Any direct photodisruption or mechanical force by laser-generated shockwave to ER results in an increase in intracytoplasmic Calcium. This results in the translocation of Calcium to nucleus by formation of Calcium-calmodulin complex and changes to stem cell morphology and functions. *Abbreviations:* Ca^{2+}, *Calcium; cAMP, cyclic Adenosine Monophosphate; GPCR, G-protein Coupled Receptors; LILI, Low-Intensity Laser Irradiation;* O_2, *Oxygen.*

PRESPECTIVES

New medical strategies and treatment modalities are required to assist the nervous system in repairing the diseases and disorders. Stem cell therapy is a plausible, potential candidate capable of repairing damaged nerves. Adipose tissue is mesenchymal in origin and is often regarded as an easily accessible source for obtaining stem cells with ease and minimal surgical trauma. ASCs are isolated from these harvested tissues, propagated *in vitro*, and have the potential to differentiate into neuronal cell types. However, ASCs in long-term culture often exhibit changes in their growth kinetics, surface markers, cell cycle distribution, and β-galactosidase activity. The morphology of ASCs in culture can show onset of cell senescence pathways beyond passages 10, which will have implications their use in differentiation protocols [109]. These limitations render even a well-defined differentiation protocol towards neuronal lineage unsuccessful.

Table 2. Summary of parameters used for laser irradiation on human stem cells.

Wavelength (nm)	**LASER type**	**CELL type**	**Output power (mW)**	**Density (mW/cm^2)**	**Fluence (J/cm^2) [Seconds (s)]**	**Growth Factor(s)**	**Biological Effect**	**Ref.**
620	Diode	BMSCs	1000	6.67	1, 2, 4 [150 s, 300 s, 600 s]	Osteogenic supplements	Biostimulation (Osteogenic)	[98]
635	Diode	BMSCs	60	6.61	0.5, 1, 2, 5 [75 s, 151 s, 303 s, 757 s]	5-azacytidine	Biostimulation (Myogenic)	[99]
636	Diode	ASCs	50.2	5.5	5 [413 s]	EGF	Biostimulation (Proliferation)	[94]
636	Diode	ASCs	78	8.59	5 [586 s]	Retinoic Acid	Biostimulation (Myogenic)	[100]
647	Diode	MSCs	8.98 to 9.89	9.29	0.093, 0.279, 0.836 [10 s, 30 s, 90 s]	Osteogenic differentiation media	Biostimulation (Osteogenic)	[101]
660	Diode	ASC	50	5.51	3 [600 s]	EGF and FGF	Biostimulation (Neurogenic)	[102]
660	Ga-Al-As	MSCs	50	10	1, 2, 4 [100 s, 200 s, 400 s]	Osteogenic differentiation media	Biostimulation (Osteogenic)	[103]
660	Ga-Al-As	dental	70	15.17	1, 2, 4 [66 s, 132 s, 264 s]	Osteogenic differentiation media	Biostimulation (Osteogenic)	[104]
670	Diode	SaOS-2	400	44.06	1, 2 [23 s, 46 s]	DMEM / Ham's F12	Biostimulation (Osteogenic)	[105]

(Table 2) cont.....

Wavelength (nm)	LASER type	CELL type	Output power (mW)	Density (mW/cm^2)	Fluence (J/cm^2) [Seconds (s)]	Growth Factor(s)	Biological Effect	Ref.
680	Diode	ASCs	35.6	3.92	5, 10, 15 [1276 s, 2551 s, 3887 s]	DMEM / Ham's F12	Biostimulation (proliferation)	[95]
810	Ga-Al-As	BMSCs	50	167	3, 6 [18 s, 36 s]	Differentiation medium	Biostimulation (Neuro/Osteo)	[106]
				167	2, 4 [12 s, 24 s]			
830	Diode	ASCs	35	3.85	5, 10, 15 [1299 s, 2597 s, 3896 s]	DMEM / Ham's F12	Bioinhibition	[95]
830	Ga-Al-As	ASCs	760	83.7	0.05 [0.6 s]	DMEM with 10% fetal bovine serum	Biostimulation (Proliferation)	[107]
1064	Nd: YAG	MSCs	1500	18.75	0.84 each time up to 2.52	Osteogenic differentiation media	Biostimulation (Osteogenic)	[108]
1064	Nd: YAG	MSCs	2250	28.1	1.27 each time up to 3.81	Osteogenic scaffholds	Biostimulation (Osteogenic)	[108]

Abbreviations: ASCs, Adipose-derived Stem Cells;BMSCs, bone marrow stem cells; DMEM, Dulbecco's Modified Eagle's Medium; FGF, Fibroblast Growth Factor; EGF, Epidermal Growth Factor; MSCs, mesenchymal stem cells.

Growth factors are peptides synthesized in the living systems to promote cellular proliferation as well as differentiation. Several of these growth factors either alone, or in combination, are vital to tissue morphogenesis, angiogenesis or wound healing. ASCs induce repair and regeneration of diseased tissues *in vivo* by secretion of cytokines and growth factors. Similar strategy can be adopted for differentiating ASCs to neuroglia *in vitro* for replacement therapies. A growing body of literature suggests that LILI stimulates proliferation and possibly differentiates them to other cell types. The key issue here is to differentiate ASCs into fully functioning neuronal cells in standard laboratory conditions. Determination of the correct cocktail of growth factors together with light or laser has the potential to fulfil this task and make neurons and glia available for replacement therapies.

Several tests need to be cleared by newly differentiated ASCs before being regarded as identical and functional cells of neuronal lineage. These include (i) morphology similar to neurons or neuronal progenitors (ii) viability and proliferative properties of neurons (iii) markers specific to immature and mature neurons (iv) expression of neurotransmitter receptors and (v) electrophysiological properties for impulse transmission. Cell staining procedures will help to identify and sort newly formed neurons in culture. The commonly identified immature markers on neurons *viz.* Doublecortin, nestin and neuroD indicate the onset of differentiation. Further, the presence of β-tubulin III (Tuj-I) and microtubule associated protein 2 (MAP2) are indicative of the formation of mature neurons [110, 111]. At the later stage of differentiation, we expect these mature neurons extend neurites, integrate by synapsis, express electrical activity and exchange neurotransmitters.

CONCLUSION

Stem cells remain credible therapeutic candidates for the repair and regeneration of damaged nerves, and represent an ideal source of precursors of neural lineage cells. Adipose tissue is a highly promising source of stem cells, because it is easy to obtain from subcutaneous sites, from where, ASCs are isolated by enzyme-based procedures. Compelling evidence from pre-clinical studies advocates the use of ASCs in the treatment of neurological conditions associated with cerebral stroke or ischemic brain injury, spinal cord injury, Alzheimer's disease, Parkinson's disease and CNS autoimmune disorders such as multiple sclerosis. Furthermore, differentiated ASCs could be a solution for replacement therapies in patients with critical diseases as well as life-threatening injuries of the CNS. A comprehensive method using biological, chemical and physical inducers of stem cell growth and differentiation is advantageous in terms of time and yield of differentiating neuroglia for clinical use. PBM can be used as an effective treatment for the desired biostimulatory effects on ASCs when used for culture and differentiation. Presumably, use of growth factors along with LILI for the differentiation of ASCs to neuronal cell types offer a promising treatment modality towards nervous system diseases and disorders.

CONSENT FOR PUBLICATION

Not applicable.

CONFLICT OF INTEREST

The authors confirm that they have no conflict of interest to declare for this publication.

ACKNOWLEDGEMENTS

This work is based on the research supported by the South African Research

Chairs Initiative of the Department of Science and Technology and National Research Foundation of South Africa (Grant no. 98337).

REFERENCES

[1] Temple S. The development of neural stem cells. Nature 2001; 414(6859): 112-7. [http://dx.doi.org/10.1038/35102174] [PMID: 11689956]

[2] Gage FH. Mammalian neural stem cells. Science 2000; 287(5457): 1433-8. [http://dx.doi.org/10.1126/science.287.5457.1433] [PMID: 10688783]

[3] Hsu YC, Chen SL, Wang DY, Chiu IM. Stem cell-based therapy in neural repair. Biomed J 2013; 36(3): 98-105. [http://dx.doi.org/10.4103/2319-4170.113226] [PMID: 23806879]

[4] Mothe AJ, Tator CH. Advances in stem cell therapy for spinal cord injury. J Clin Invest 2012; 122(11): 3824-34. [http://dx.doi.org/10.1172/JCI64124] [PMID: 23114605]

[5] Barry FP, Murphy JM. Mesenchymal stem cells: clinical applications and biological characterization. Int J Biochem Cell Biol 2004; 36(4): 568-84. [http://dx.doi.org/10.1016/j.biocel.2003.11.001] [PMID: 15010324]

[6] Pountos I, Corscadden D, Emery P, Giannoudis PV. Mesenchymal stem cell tissue engineering: techniques for isolation, expansion and application. Injury 2007; 38 (Suppl. 4): S23-33. [http://dx.doi.org/10.1016/S0020-1383(08)70006-8] [PMID: 18224734]

[7] Zuk PA, Zhu M, Ashjian P, *et al.* Human adipose tissue is a source of multipotent stem cells. Mol Biol Cell 2002; 13(12): 4279-95. [http://dx.doi.org/10.1091/mbc.e02-02-0105] [PMID: 12475952]

[8] Ogawa R. The importance of adipose-derived stem cells and vascularized tissue regeneration in the field of tissue transplantation. Curr Stem Cell Res Ther 2006; 1(1): 13-20. [http://dx.doi.org/10.2174/157488806775269043] [PMID: 18220849]

[9] Mvula B, Mathope T, Moore T, Abrahamse H. The effect of low level laser irradiation on adult human adipose derived stem cells. Lasers Med Sci 2008; 23(3): 277-82. [http://dx.doi.org/10.1007/s10103-007-0479-1] [PMID: 17713825]

[10] Rodríguez LV, Alfonso Z, Zhang R, Leung J, Wu B, Ignarro LJ. Clonogenic multipotent stem cells in human adipose tissue differentiate into functional smooth muscle cells. Proc Natl Acad Sci USA 2006; 103(32): 12167-72. [http://dx.doi.org/10.1073/pnas.0604850103] [PMID: 16880387]

[11] Zuk PA, Zhu M, Mizuno H, *et al.* Multilineage cells from human adipose tissue: implications for cell-based therapies. Tissue Eng 2001; 7(2): 211-28. [http://dx.doi.org/10.1089/107632701300062859] [PMID: 11304456]

[12] Ponemone V, Choudhury K, Harris K. Stem cell treatment for the spinal cord injury - A concise review. Indian Journal of Neurotrauma 2014; 30-38.

[13] Cao QL, Onifer SM, Whittemore SR. Labeling stem cells *in vitro* for identification of their differentiated phenotypes after grafting into the CNS. Methods Mol Biol 2002; 198: 307-18. [http://dx.doi.org/10.1385/1-59259-186-8:307] [PMID: 11951634]

[14] Setoguchi T, Nakashima K, Takizawa T, *et al.* Treatment of spinal cord injury by transplantation of fetal neural precursor cells engineered to express BMP inhibitor. Exp Neurol 2004; 189(1): 33-44. [http://dx.doi.org/10.1016/j.expneurol.2003.12.007] [PMID: 15296834]

[15] Fraser JK, Schreiber R, Strem B, *et al.* Plasticity of human adipose stem cells toward endothelial cells and cardiomyocytes. Nat Clin Pract Cardiovasc Med 2006; 3 (Suppl. 1): S33-7.

[http://dx.doi.org/10.1038/ncpcardio0444] [PMID: 16501628]

[16] Sakaguchi Y, Sekiya I, Yagishita K, Muneta T. Comparison of human stem cells derived from various mesenchymal tissues: superiority of synovium as a cell source. Arthritis Rheum 2005; 52(8): 2521-9. [http://dx.doi.org/10.1002/art.21212] [PMID: 16052568]

[17] Puissant B, Barreau C, Bourin P, *et al.* Immunomodulatory effect of human adipose tissue-derived adult stem cells: comparison with bone marrow mesenchymal stem cells. Br J Haematol 2005; 129(1): 118-29. [http://dx.doi.org/10.1111/j.1365-2141.2005.05409.x] [PMID: 15801964]

[18] Macotela Y, Emanuelli B, Mori MA, *et al.* Intrinsic differences in adipocyte precursor cells from different white fat depots. Diabetes 2012; 61(7): 1691-9. [http://dx.doi.org/10.2337/db11-1753] [PMID: 22596050]

[19] Prunet-Marcassus B, Cousin B, Caton D, André M, Pénicaud L, Casteilla L. From heterogeneity to plasticity in adipose tissues: site-specific differences. Exp Cell Res 2006; 312(6): 727-36. [http://dx.doi.org/10.1016/j.yexcr.2005.11.021] [PMID: 16386732]

[20] Ong WK, Tan CS, Chan KL, *et al.* Identification of specific cell-surface markers of adipose-derived stem cells from subcutaneous and visceral fat depots. Stem Cell Reports 2014; 2(2): 171-9. [http://dx.doi.org/10.1016/j.stemcr.2014.01.002] [PMID: 24527391]

[21] Ikegame Y, Yamashita K, Hayashi S, *et al.* Comparison of mesenchymal stem cells from adipose tissue and bone marrow for ischemic stroke therapy. Cytotherapy 2011; 13(6): 675-85. [http://dx.doi.org/10.3109/14653249.2010.549122] [PMID: 21231804]

[22] Jansen BJ, Gilissen C, Roelofs H, *et al.* Functional differences between mesenchymal stem cell populations are reflected by their transcriptome. Stem Cells Dev 2010; 19(4): 481-90. [http://dx.doi.org/10.1089/scd.2009.0288] [PMID: 19788395]

[23] Monaco E, Bionaz M, Rodriguez-Zas S, Hurley WL, Wheeler MB. Transcriptomics comparison between porcine adipose and bone marrow mesenchymal stem cells during *in vitro* osteogenic and adipogenic differentiation. PLoS One 2012; 7(3)e32481 [http://dx.doi.org/10.1371/journal.pone.0032481] [PMID: 22412878]

[24] Doetsch F. The glial identity of neural stem cells. Nat Neurosci 2003; 6(11): 1127-34. [http://dx.doi.org/10.1038/nn1144] [PMID: 14583753]

[25] Basson MA. Signaling in cell differentiation and morphogenesis. Cold Spring Harb Perspect Biol 2012; 4(6)a008151 [http://dx.doi.org/10.1101/cshperspect.a008151] [PMID: 22570373]

[26] Cardozo A, Ielpi M, Gómez D, Argibay P. Differential expression of Shh and BMP signaling in the potential conversion of human adipose tissue stem cells into neuron-like cells *in vitro.* Gene Expr 2010; 14(6): 307-19. [http://dx.doi.org/10.3727/105221610X12717040569866] [PMID: 20635573]

[27] Ying QL, Nichols J, Chambers I, Smith A. BMP induction of Id proteins suppresses differentiation and sustains embryonic stem cell self-renewal in collaboration with STAT3. Cell 2003; 115(3): 281-92. [http://dx.doi.org/10.1016/S0092-8674(03)00847-X] [PMID: 14636556]

[28] Stachowiak EK, Fang X, Myers J, Dunham S, Stachowiak MK. cAMP-induced differentiation of human neuronal progenitor cells is mediated by nuclear fibroblast growth factor receptor-1 (FGFR1). J Neurochem 2003; 84(6): 1296-312. [http://dx.doi.org/10.1046/j.1471-4159.2003.01624.x] [PMID: 12614330]

[29] Kompisch KM, Lange C, Steinemann D, *et al.* Neurogenic transdifferentiation of human adipose-derived stem cells? A critical protocol reevaluation with special emphasis on cell proliferation and cell cycle alterations. Histochem Cell Biol 2010; 134(5): 453-68. [http://dx.doi.org/10.1007/s00418-010-0740-8] [PMID: 20945072]

[30] Cardozo AJ, Gómez DE, Argibay PF. Transcriptional characterization of Wnt and Notch signaling pathways in neuronal differentiation of human adipose tissue-derived stem cells. J Mol Neurosci 2011; 44(3): 186-94.
[http://dx.doi.org/10.1007/s12031-011-9503-9] [PMID: 21360053]

[31] Janesick A, Wu SC, Blumberg B. Retinoic acid signaling and neuronal differentiation. Cell Mol Life Sci 2015; 72(8): 1559-76.
[http://dx.doi.org/10.1007/s00018-014-1815-9] [PMID: 25558812]

[32] Yeo TT, Chua-Couzens J, Butcher LL, *et al.* Absence of p75NTR causes increased basal forebrain cholinergic neuron size, choline acetyltransferase activity, and target innervation. J Neurosci 1997; 17(20): 7594-605.
[http://dx.doi.org/10.1523/JNEUROSCI.17-20-07594.1997] [PMID: 9315882]

[33] Auld DS, Mennicken F, Quirion R. Nerve growth factor rapidly induces prolonged acetylcholine release from cultured basal forebrain neurons: differentiation between neuromodulatory and neurotrophic influences. J Neurosci 2001; 21(10): 3375-82.
[http://dx.doi.org/10.1523/JNEUROSCI.21-10-03375.2001] [PMID: 11331367]

[34] Auld DS, Mennicken F, Day JC, Quirion R. Neurotrophins differentially enhance acetylcholine release, acetylcholine content and choline acetyltransferase activity in basal forebrain neurons. J Neurochem 2001; 77(1): 253-62.
[http://dx.doi.org/10.1046/j.1471-4159.2001.t01-1-00234.x] [PMID: 11279281]

[35] Naghdi M, Tiraihi T, Mesbah-Namin SA, Arabkheradmand J. Induction of bone marrow stromal cells into cholinergic-like cells by nerve growth factor. Iran Biomed J 2009; 13(2): 117-23.
[PMID: 19471552]

[36] Madziar B, Lopez-Coviella I, Zemelko V, Berse B. Regulation of cholinergic gene expression by nerve growth factor depends on the phosphatidylinositol-3′-kinase pathway. J Neurochem 2005; 92(4): 767-79.
[http://dx.doi.org/10.1111/j.1471-4159.2004.02908.x] [PMID: 15686478]

[37] Markus A, Zhong J, Snider WD. Raf and akt mediate distinct aspects of sensory axon growth. Neuron 2002; 35(1): 65-76.
[http://dx.doi.org/10.1016/S0896-6273(02)00752-3] [PMID: 12123609]

[38] Tsai HL, Deng WP, Lai WF, *et al.* Wnts enhance neurotrophin-induced neuronal differentiation in adult bone-marrow-derived mesenchymal stem cells *via* canonical and noncanonical signaling pathways. PLoS One 2014; 9(8)e104937
[http://dx.doi.org/10.1371/journal.pone.0104937] [PMID: 25170755]

[39] Nandy SB, Mohanty S, Singh M, Behari M, Airan B. Fibroblast Growth Factor-2 alone as an efficient inducer for differentiation of human bone marrow mesenchymal stem cells into dopaminergic neurons. J Biomed Sci 2014 Sep; 24;21:83,014-0083-1.

[40] Trzaska KA, Rameshwar P. Dopaminergic neuronal differentiation protocol for human mesenchymal stem cells. Methods Mol Biol 2011; 698: 295-303.
[http://dx.doi.org/10.1007/978-1-60761-999-4_22] [PMID: 21431527]

[41] Ratzka A, Baron O, Stachowiak MK, Grothe C. Fibroblast growth factor 2 regulates dopaminergic neuron development *in vivo*. J Neurochem 2012; 122(1): 94-105.
[http://dx.doi.org/10.1111/j.1471-4159.2012.07768.x] [PMID: 22537018]

[42] Kondo T, Johnson SA, Yoder MC, Romand R, Hashino E. Sonic hedgehog and retinoic acid synergistically promote sensory fate specification from bone marrow-derived pluripotent stem cells. Proc Natl Acad Sci USA 2005; 102(13): 4789-94.
[http://dx.doi.org/10.1073/pnas.0408239102] [PMID: 15778294]

[43] Fan G, Martinowich K, Chin MH, *et al.* DNA methylation controls the timing of astrogliogenesis through regulation of JAK-STAT signaling. Development 2005; 132(15): 3345-56.

[http://dx.doi.org/10.1242/dev.01912] [PMID: 16014513]

[44] Hu XL, Wang Y, Shen Q. Epigenetic control on cell fate choice in neural stem cells. Protein Cell 2012; 3(4): 278-90.
[http://dx.doi.org/10.1007/s13238-012-2916-6] [PMID: 22549586]

[45] Maulik N, Karagiannis T. Implications for Epigenetics and Health. Molecular Mechanisms and Physiology of Disease 2014; 14(505): 43.

[46] Lessard J, Wu JI, Ranish JA, *et al.* An essential switch in subunit composition of a chromatin remodeling complex during neural development. Neuron 2007; 55(2): 201-15.
[http://dx.doi.org/10.1016/j.neuron.2007.06.019] [PMID: 17640523]

[47] Arzate-Mejía RG, Valle-García D, Recillas-Targa F. Signaling epigenetics: novel insights on cell signaling and epigenetic regulation. IUBMB Life 2011; 63(10): 881-95.
[http://dx.doi.org/10.1002/iub.557] [PMID: 21901818]

[48] Kim JH, Park SG, Song SY, Kim JK, Sung JH. Reactive oxygen species-responsive miR-210 regulates proliferation and migration of adipose-derived stem cells *via* PTPN2. Cell Death Dis 2013; 4e588
[http://dx.doi.org/10.1038/cddis.2013.117] [PMID: 23579275]

[49] Corn PG. Hypoxic regulation of miR-210: shrinking targets expand HIF-1's influence. Cancer Biol Ther 2008; 7(2): 265-7.
[http://dx.doi.org/10.4161/cbt.7.2.5745] [PMID: 18347426]

[50] Ning H, Lin G, Fandel T, Banie L, Lue TF, Lin CS. Insulin growth factor signaling mediates neuron-like differentiation of adipose-tissue-derived stem cells. Differentiation 2008; 76(5): 488-94.
[http://dx.doi.org/10.1111/j.1432-0436.2007.00240.x] [PMID: 18021255]

[51] Vivian Lee. Sharon Louis, Brent Reynolds. Neural Stem Cells. Stem Cell Technologies 2015; p. 29019.

[52] Michalczyk K, Ziman M. Nestin structure and predicted function in cellular cytoskeletal organisation. Histol Histopathol 2005; 20(2): 665-71.
[PMID: 15736068]

[53] Lariviere RC, Julien JP. Functions of intermediate filaments in neuronal development and disease. J Neurobiol 2004; 58(1): 131-48.
[http://dx.doi.org/10.1002/neu.10270] [PMID: 14598376]

[54] González-Billault C, Engelke M, Jiménez-Mateos EM, Wandosell F, Cáceres A, Avila J. Participation of structural microtubule-associated proteins (MAPs) in the development of neuronal polarity. J Neurosci Res 2002; 67(6): 713-9.
[http://dx.doi.org/10.1002/jnr.10161] [PMID: 11891784]

[55] Katsetos CD, Herman MM, Mörk SJ. Class III beta-tubulin in human development and cancer. Cell Motil Cytoskeleton 2003; 55(2): 77-96.
[http://dx.doi.org/10.1002/cm.10116] [PMID: 12740870]

[56] Lee MK, Tuttle JB, Rebhun LI, Cleveland DW, Frankfurter A. The expression and posttranslational modification of a neuron-specific beta-tubulin isotype during chick embryogenesis. Cell Motil Cytoskeleton 1990; 17(2): 118-32.
[http://dx.doi.org/10.1002/cm.970170207] [PMID: 2257630]

[57] Sánchez C, Díaz-Nido J, Avila J. Phosphorylation of microtubule-associated protein 2 (MAP2) and its relevance for the regulation of the neuronal cytoskeleton function. Prog Neurobiol 2000; 61(2): 133-68.
[http://dx.doi.org/10.1016/S0301-0082(99)00046-5] [PMID: 10704996]

[58] Huber G, Matus A. Differences in the cellular distributions of two microtubule-associated proteins, MAP1 and MAP2, in rat brain. J Neurosci 1984; 4(1): 151-60.
[http://dx.doi.org/10.1523/JNEUROSCI.04-01-00151.1984] [PMID: 6198491]

[59] Hsueh YP, Kim E, Sheng M. Disulfide-linked head-to-head multimerization in the mechanism of ion channel clustering by PSD-95. Neuron 1997; 18(5): 803-14.
[http://dx.doi.org/10.1016/S0896-6273(00)80319-0] [PMID: 9182804]

[60] Prange O, Wong TP, Gerrow K, Wang YT, El-Husseini A. A balance between excitatory and inhibitory synapses is controlled by PSD-95 and neuroligin. Proc Natl Acad Sci USA 2004; 101(38): 13915-20.
[http://dx.doi.org/10.1073/pnas.0405939101] [PMID: 15358863]

[61] Ferreira A, Chin LS, Li L, Lanier LM, Kosik KS, Greengard P. Distinct roles of synapsin I and synapsin II during neuronal development. Mol Med 1998; 4(1): 22-8.
[http://dx.doi.org/10.1007/BF03401726] [PMID: 9513186]

[62] Valtorta F, Pennuto M, Bonanomi D, Benfenati F. Synaptophysin: leading actor or walk-on role in synaptic vesicle exocytosis? BioEssays 2004; 26(4): 445-53.
[http://dx.doi.org/10.1002/bies.20012] [PMID: 15057942]

[63] Zhang R, Maksymowych AB, Simpson LL. Cloning and sequence analysis of a cDNA encoding human syntaxin 1A, a polypeptide essential for exocytosis. Gene 1995; 159(2): 293-4.
[http://dx.doi.org/10.1016/0378-1119(95)00152-V] [PMID: 7622072]

[64] Gil-Bea FJ, García-Alloza M, Domínguez J, Marcos B, Ramírez MJ. Evaluation of cholinergic markers in Alzheimer's disease and in a model of cholinergic deficit. Neurosci Lett 2005; 375(1): 37-41.
[http://dx.doi.org/10.1016/j.neulet.2004.10.062] [PMID: 15664119]

[65] Gonzalez R, Garitaonandia I, Abramihina T, *et al.* Deriving dopaminergic neurons for clinical use. A practical approach. Sci Rep 2013; 3: 1463.
[http://dx.doi.org/10.1038/srep01463] [PMID: 23492920]

[66] Pinal CS, Tobin AJ. Uniqueness and redundancy in GABA production. Perspect Dev Neurobiol 1998; 5(2-3): 109-18.
[PMID: 9777629]

[67] Faroni A, Terenghi G, Reid AJ. Adipose-derived stem cells and nerve regeneration: promises and pitfalls. Int Rev Neurobiol 2013; 108: 121-36.
[http://dx.doi.org/10.1016/B978-0-12-410499-0.00005-8] [PMID: 24083433]

[68] Kim BJ, Kim SS, Kim YI, Paek SH, Lee YD, Suh-Kim H. Forskolin promotes astroglial differentiation of human central neurocytoma cells. Exp Mol Med 2004; 36(1): 52-6.
[http://dx.doi.org/10.1038/emm.2004.7] [PMID: 15031671]

[69] Molofsky AV, Deneen B. Astrocyte development: A Guide for the Perplexed. Glia 2015; 63(8): 1320-9.
[http://dx.doi.org/10.1002/glia.22836] [PMID: 25963996]

[70] Bunge RP. The role of the Schwann cell in trophic support and regeneration. J Neurol 1994; 242(1) (Suppl. 1): S19-21.
[http://dx.doi.org/10.1007/BF00939235] [PMID: 7699403]

[71] Battiste J, Helms AW, Kim EJ, *et al.* Ascl1 defines sequentially generated lineage-restricted neuronal and oligodendrocyte precursor cells in the spinal cord. Development 2007; 134(2): 285-93.
[http://dx.doi.org/10.1242/dev.02727] [PMID: 17166924]

[72] Darbinyan A, Kaminski R, White MK, Darbinian N, Khalili K. Isolation and propagation of primary human and rodent embryonic neural progenitor cells and cortical neurons. Methods Mol Biol 2013; 1078: 45-54.
[http://dx.doi.org/10.1007/978-1-62703-640-5_5] [PMID: 23975820]

[73] Lee J, Gravel M, Zhang R, Thibault P, Braun PE. Process outgrowth in oligodendrocytes is mediated by CNP, a novel microtubule assembly myelin protein. J Cell Biol 2005; 170(4): 661-73.
[http://dx.doi.org/10.1083/jcb.200411047] [PMID: 16103231]

[74] Sensenbrenner M, Lucas M, Deloulme JC. Expression of two neuronal markers, growth-associated protein 43 and neuron-specific enolase, in rat glial cells. J Mol Med (Berl) 1997; 75(9): 653-63. [http://dx.doi.org/10.1007/s001090050149] [PMID: 9351704]

[75] Topchiy E, Panzhinskiy E, Griffin WS, Barger SW, Das M, Zawada WM. Nox4-generated superoxide drives angiotensin II-induced neural stem cell proliferation. Dev Neurosci 2013; 35(4): 293-305. [http://dx.doi.org/10.1159/000350502] [PMID: 23751520]

[76] Kokovay E, Wang Y, Kusek G, *et al.* VCAM1 is essential to maintain the structure of the SVZ niche and acts as an environmental sensor to regulate SVZ lineage progression. Cell Stem Cell 2012; 11(2): 220-30. [http://dx.doi.org/10.1016/j.stem.2012.06.016] [PMID: 22862947]

[77] Puceat M. Role of Rac-GTPase and reactive oxygen species in cardiac differentiation of stem cells. Antioxid Redox Signal 2005; 7(11-12): 1435-9. [http://dx.doi.org/10.1089/ars.2005.7.1435] [PMID: 16356106]

[78] Tsatmali M, Walcott EC, Crossin KL. Newborn neurons acquire high levels of reactive oxygen species and increased mitochondrial proteins upon differentiation from progenitors. Brain Res 2005; 1040(1-2): 137-50. [http://dx.doi.org/10.1016/j.brainres.2005.01.087] [PMID: 15804435]

[79] Suzukawa K, Miura K, Mitsushita J, *et al.* Nerve growth factor-induced neuronal differentiation requires generation of Rac1-regulated reactive oxygen species. J Biol Chem 2000; 275(18): 13175-8. [http://dx.doi.org/10.1074/jbc.275.18.13175] [PMID: 10788420]

[80] Kamata H, Oka S, Shibukawa Y, Kakuta J, Hirata H. Redox regulation of nerve growth factor-induced neuronal differentiation of PC12 cells through modulation of the nerve growth factor receptor, TrkA. Arch Biochem Biophys 2005; 434(1): 16-25. [http://dx.doi.org/10.1016/j.abb.2004.07.036] [PMID: 15629104]

[81] Bórquez DA, Urrutia PJ, Wilson C, van Zundert B, Núñez MT, González-Billault C. Dissecting the role of redox signaling in neuronal development. J Neurochem 2016; 137(4): 506-17. [http://dx.doi.org/10.1111/jnc.13581] [PMID: 26875993]

[82] Leslie NR, Downes CP. PTEN: The down side of PI 3-kinase signalling. Cell Signal 2002; 14(4): 285-95. [http://dx.doi.org/10.1016/S0898-6568(01)00234-0] [PMID: 11858936]

[83] Zhang J, Wang X, Vikash V, Ye Q, Wu D, Liu Y, *et al.* ROS and ROS-Mediated Cellular Signaling. Oxidative Medicine and Cellular Longevity 2016; 2016

[84] Hancock JT, Desikan R, Neill SJ. Role of reactive oxygen species in cell signalling pathways. Biochem Soc Trans 2001; 29(Pt 2): 345-50. [http://dx.doi.org/10.1042/bst0290345] [PMID: 11356180]

[85] Yanes O, Clark J, Wong DM, *et al.* Metabolic oxidation regulates embryonic stem cell differentiation. Nat Chem Biol 2010; 6(6): 411-7. [http://dx.doi.org/10.1038/nchembio.364] [PMID: 20436487]

[86] Banan A, Fields JZ, Zhang Y, Keshavarzian A. Phospholipase C-gamma inhibition prevents EGF protection of intestinal cytoskeleton and barrier against oxidants. Am J Physiol Gastrointest Liver Physiol 2001; 281(2): G412-23. [http://dx.doi.org/10.1152/ajpgi.2001.281.2.G412] [PMID: 11447022]

[87] Hamblin MR, Demidova-Rice TN. Mechanisms of low level light therapy. Proc SPIE 2006; 6140: 614001-12. [http://dx.doi.org/10.1117/12.646294]

[88] George S, Hamblin MR, Abrahamse H. Current and Future Trends in Adipose Stem Cell Differentiation into Neuroglia. Photomed Laser Surg 2018; 36(5): 230-40. [http://dx.doi.org/10.1089/pho.2017.4411] [PMID: 29570423]

[89] Gao X, Xing D. Molecular mechanisms of cell proliferation induced by low power laser irradiation. J Biomed Sci 2009 Jan 12;

[90] Karu TI. Ten lectures on basic science of laser phototherapy. NY, USA: Prima Books 2007.

[91] Abrahamse H, Hamblin MR. Photomedicine and Stem Cells: The Janus face of photodynamic therapy (PDT) to kill cancer stem cells, and photobiomodulation (PBM) to stimulate normal stem cells IOP Concise Physics. CA, USA: Morgan & Claypool Publishers 2017.

[92] AlGhamdi KM, Kumar A, Moussa NA. Low-level laser therapy: a useful technique for enhancing the proliferation of various cultured cells. Lasers Med Sci 2012; 27(1): 237-49. [http://dx.doi.org/10.1007/s10103-011-0885-2] [PMID: 21274733]

[93] de Souza SC, Munin E, Alves LP, Salgado MA, Pacheco MT. Low power laser radiation at 685 nm stimulates stem-cell proliferation rate in Dugesia tigrina during regeneration. J Photochem Photobiol B 2005; 80(3): 203-7. [http://dx.doi.org/10.1016/j.jphotobiol.2005.05.002] [PMID: 15967675]

[94] Mvula B, Moore TJ, Abrahamse H. Effect of low-level laser irradiation and epidermal growth factor on adult human adipose-derived stem cells. Lasers Med Sci 2010; 25(1): 33-9. [http://dx.doi.org/10.1007/s10103-008-0636-1] [PMID: 19172344]

[95] Abrahamse H, Houreld N, Muller S, Ndlovu L. Fluence and wavelength of low intensity laser irradiation affect activity and proliferation of human adipose derived stem cells. Med Tech SA 2010; 24(2): 15-20.

[96] Kneebone WJ, Cnc DIH, Fiama D. Practical applications of low level laser therapy. Pract Pain Manag 2006; 6(8): 34-40.

[97] Vladimirov YA, Osipov AN, Klebanov GI. Photobiological principles of therapeutic applications of laser radiation. Biochemistry (Mosc) 2004; 69(1): 81-90. [http://dx.doi.org/10.1023/B:BIRY.0000016356.93968.7e] [PMID: 14972023]

[98] Peng F, Wu H, Zheng Y, Xu X, Yu J. The effect of noncoherent red light irradiation on proliferation and osteogenic differentiation of bone marrow mesenchymal stem cells. Lasers Med Sci 2012; 27(3): 645-53. [http://dx.doi.org/10.1007/s10103-011-1005-z] [PMID: 22016038]

[99] Hou JF, Zhang H, Yuan X, Li J, Wei YJ, Hu SS. *In vitro* effects of low-level laser irradiation for bone marrow mesenchymal stem cells: proliferation, growth factors secretion and myogenic differentiation. Lasers Surg Med 2008; 40(10): 726-33. [http://dx.doi.org/10.1002/lsm.20709] [PMID: 19065562]

[100] de Villiers JA, Houreld NN, Abrahamse H. Influence of low intensity laser irradiation on isolated human adipose derived stem cells over 72 hours and their differentiation potential into smooth muscle cells using retinoic acid. Stem Cell Rev Rep 2011; 7(4): 869-82. [http://dx.doi.org/10.1007/s12015-011-9244-8] [PMID: 21373882]

[101] Kim HJ, Im GI. Combination of transforming growth factor-beta2 and bone morphogenetic protein 7 enhances chondrogenesis from adipose tissue-derived mesenchymal stem cells. Tissue Eng Part A 2009; 15(7): 1543-51. [http://dx.doi.org/10.1089/ten.tea.2008.0368] [PMID: 19072523]

[102] Shen CC, Yang YC, Chiao MT, Chan SC, Liu BS. Low-level laser stimulation on adipose-tissu--derived stem cell treatments for focal cerebral ischemia in rats. Evid Based Complement Alternat Med 2013; 2013594906 [http://dx.doi.org/10.1155/2013/594906] [PMID: 24363769]

[103] Wu JY, Wang YH, Wang GJ, *et al.* Low-power GaAlAs laser irradiation promotes the proliferation and osteogenic differentiation of stem cells *via* IGF1 and BMP2. PLoS One 2012; 7(9)e44027 [http://dx.doi.org/10.1371/journal.pone.0044027] [PMID: 22962596]

[104] Wu JY, Chen CH, Yeh LY, Yeh ML, Ting CC, Wang YH. Low-power laser irradiation promotes the proliferation and osteogenic differentiation of human periodontal ligament cells *via* cyclic adenosine monophosphate. Int J Oral Sci 2013; 5(2): 85-91.
[http://dx.doi.org/10.1038/ijos.2013.38] [PMID: 23788285]

[105] Stein E, Koehn J, Sutter W, *et al.* Initial effects of low-level laser therapy on growth and differentiation of human osteoblast-like cells. Wien Klin Wochenschr 2008; 120(3-4): 112-7.
[http://dx.doi.org/10.1007/s00508-008-0932-6] [PMID: 18322773]

[106] Soleimani M, Abbasnia E, Fathi M, Sahraei H, Fathi Y, Kaka G. The effects of low-level laser irradiation on differentiation and proliferation of human bone marrow mesenchymal stem cells into neurons and osteoblasts-an *in vitro study*. Lasers Med Sci 2012; 27(2): 423-30.
[http://dx.doi.org/10.1007/s10103-011-0930-1] [PMID: 21597948]

[107] Min KH, Byun JH, Heo CY, Kim EH, Choi HY, Pak CS. Effect of Low-Level Laser Therapy on Human Adipose-Derived Stem Cells: *In Vitro* and *In Vivo* Studies. Aesthetic Plast Surg 2015; 39(5): 778-82.
[http://dx.doi.org/10.1007/s00266-015-0524-6] [PMID: 26183254]

[108] Leonida A, Paiusco A, Rossi G, Carini F, Baldoni M, Caccianiga G. Effects of low-level laser irradiation on proliferation and osteoblastic differentiation of human mesenchymal stem cells seeded on a three-dimensional biomatrix: *In vitro* pilot study. Lasers Med Sci 2013; 28(1): 125-32.
[http://dx.doi.org/10.1007/s10103-012-1067-6] [PMID: 22447402]

[109] Safwani WK, Makpol S, Sathapan S, Chua KH. The impact of long-term *in vitro* expansion on the senescence-associated markers of human adipose-derived stem cells. Appl Biochem Biotechnol 2012; 166(8): 2101-13.
[http://dx.doi.org/10.1007/s12010-012-9637-4] [PMID: 22391697]

[110] Moore TJ, Abrahamse H. Neuronal Differentiation of Adipose Derived Stem Cells: Progress So Far. Int J Photoenergy 2014; 1-8.
[http://dx.doi.org/10.1155/2014/827540]

[111] Cardozo AJ, Gómez DE, Argibay PF. Neurogenic differentiation of human adipose-derived stem cells: relevance of different signaling molecules, transcription factors, and key marker genes. Gene 2012; 511(2): 427-36.
[http://dx.doi.org/10.1016/j.gene.2012.09.038] [PMID: 23000064]

CHAPTER 4

Advances in Skin Regeneration and Reconstruction

Ana-Maria Rosca, Raluca Tutuianu and **Irina Titorencu***

Institute of Cellular Biology and Pathology "Nicolae Simionescu", Bucharest, Romania

Abstract: The skin represents the largest and most accessible organ of the body, and it is subjected to numerous aggressions such as infections, physical wounds and diseases. After a moderate-intensity injury, the intrinsic regenerative mechanisms of the skin lead to restoration of the tissue integrity. However, in some pathological cases such as chronic wounds and extensive burns, the healing capacity of the tissue is overwhelmed. In order to resolve these injuries, innovative therapies based on miRNAs as well as paracrine and trophic activities of stem cells combined with biomaterials are currently being developed. This chapter begins with a description of skin biology, followed by the main stages of wound healing including the key cells and molecules involved. Next, we describe the most studied miRNAs relevant for chronic wounds therapies and the proposed methods of delivery. Regarding cellular therapy, the main adult differentiated cells as well as stem cells available from different sources, are presented. Then, we address the commercially available skin substitutes and also the latest innovative approaches, including 3D bioprinting, for combining biomaterials with the activity of the cells previously described, in order to promote wound repair and regeneration. This chapter concludes with current challenges and future perspectives regarding the use of stem cells for skin regeneration.

Keywords: Biomaterials, Cellular therapy, Chronic wounds, microRNAs, Skin, Skin biology, Stem cells.

INTRODUCTION

The skin is the largest organ of the human body. Situated at the interface between the organism and the exterior environment, it carries out the role of mechanical and waterproof barrier. As a consequence of its position and function, it is often injured as a result of the action of various external factors (mechanical, physical or chemical). Although, in normal physiological conditions, the process of wound healing is extremely effective, resulting in a nearly perfect tissue repair, especially at young ages, there are circumstances in which the repair mechanisms are damaged. The result is either a chronic wound (ulcerative skin damage) or a scar

* **Corresponding author Irina Titorencu:** Institute of Cellular Biology and Pathology "Nicolae Simionescu", Bucharest, Romania; Tel: (+4021)3194518; E-mail: irina.titorencu@icbp.ro

tissue (hypertrophic or keloid). As a consequence of the prolonged requirement of medical care and elevated rate of reoccurrence, chronic wounds represent a significant burden to the healthcare resources all over the world [1]. Furthermore, chronic wounds represent an important problem of the modern society, as they are a significant cause of morbidity, leading to a poor quality of life, disability, pain, depression and social isolation [2]. This type of skin injury is associated with an intensified inflammatory process, which can be a result of mechanical stimuli (pressure ulcers) or a consequence of other severe dysfunctions, among which diabetes is particularly prevalent in our days.

Burns represent another major type of skin injury, which require special treatment and often result in debilitating scars. The healing of a burn depends on the depth of the wound. The deep second-degree and third-degree burns, where the reticular dermis is affected, result in hypertrophic scars and contraction [3]. Moreover, in severe burns, even though they involve a single organ, almost all systems of the body are influenced, so that it becomes a generalized disorder [4]. Furthermore, sepsis and the accompanying invasive infection are still the primary cause for death after the first 24 hours, with a peak at two weeks after the accident.

Last, but not least, the skin can be affected by genetic diseases. Given the fact that the skin comprises various cell types expressing specific molecules with role in maintaining the structural integrity of the skin, even a single mutation in one of these molecules can disrupt the skin structure, leading to cell separation, blistering, and eventually, wounds [5]. For example, epidermolysis bullosa comprises a group of inherited, disfiguring and painful diseases, which determines skin fragility causing blisters in response to even minor trauma [6].

Next, we will begin the with a short description of the skin structure and function, after which the main stages of the wound healing process and the main small non-coding RNAs (microRNAs) will be described. Given the fact that the standard wound care is not effective in severe skin injuries/diseases, we will focus our attention on potential innovative therapeutic approaches such as cellular therapy and the use of available skin substitutes as wells as new strategies to improve them.

SKIN BIOLOGY

The skin has many vital functions, by being an immunologically active sensory and excretory organ and also, by preventing loss of excess water and electrolytes, while having an important role in thermoregulation [7, 8].

The integumentary system is formed by the skin and its derivative structures: sweat glands, sebaceous glands and mammary glands, hair, hair follicles, and nails [9]. The skin is organized in three layers: the stratified cellular epidermis, the underlying dermis, and subcutaneous tissue [10]. The epidermis is a continually renewing stratified epithelium of ectodermal origin, composed of 80% by keratinocytes, which are cells specialized in the synthesis of keratin a fibrous threadlike structural protein with a protective role [2]. These cells are disposed in five distinct layers: stratum basale (basal cell layer), stratum spinosum, stratum granulosum, stratum corneum and stratum lucidum- this last one being present in skin exposed to friction, such as skin on the feet or hands (Fig. **1**). Besides keratinocytes, several other cell populations exist in the epidermis: melanocytes, which have the ability to produce melanin and to donate this pigment to the keratinocytes [11], Langerhans' cells, which have immunological functions, and Merkel cells with tactile functions [10].

The **stratum basale (SB or basal layer**) represents a single layer of mitotically active cells which replaces the cell loss on the outer epidermis [12]. The process of cell kinetics in the epidermis is complex and is represented by the balance between growth with differentiation and cell death [10]. In epidermis, the turnover takes between 40-56 days and plays an important role in maintenance of the skin barrier function [13]. Thus, in the basal layer, there are cells that can persist undifferentiated and, in this way, the self-renewing of the basal cell population is preserved. These cells with unlimited capacity for self-renewal and the ability to generate daughter cells that undergo terminal differentiation are named stem cells. They are situated in small clusters in the basal layer of the epidermis. The basal keratinocytes are characterized by the expression of specific keratins: K5 and K14, and contain gap junctions for cell communication, desmosomes for cell-to-cell attachment, hemidesmosomes for connecting with the basement membrane, and an extracellular matrix [14, 15]. The basal layer synthesizes, secrets, and assembles an extracellular matrix (ECM), which ensure the stability of the connections and communication between the epidermis and the dermis [16]. Immediately above the basal cell layer, keratinocytes become more irregular in shape and form the **stratum spinosum (SS or squamous layer)**. This layer consists of 8 to 10 sheets of keratinocytes which express K1 and K10 and have a reduced potential for cell division. In the middle of this layer, the Langerhans cells are located. The intercellular spaces between squamous cells are bridged by abundant desmosomes which promote mechanical coupling between cells of the epidermis and aid in flexibility, enabling the epidermis to better withstand the effects of friction and abrasion [9] (Fig. **1**).

The **stratum granulosum (SG or granular layer)** is composed of two or three layers of flatten cells containing basophilic keratohyalin granules consisting of profilaggrin, protein that is eventually cleaved into filaggrin peptides [8].

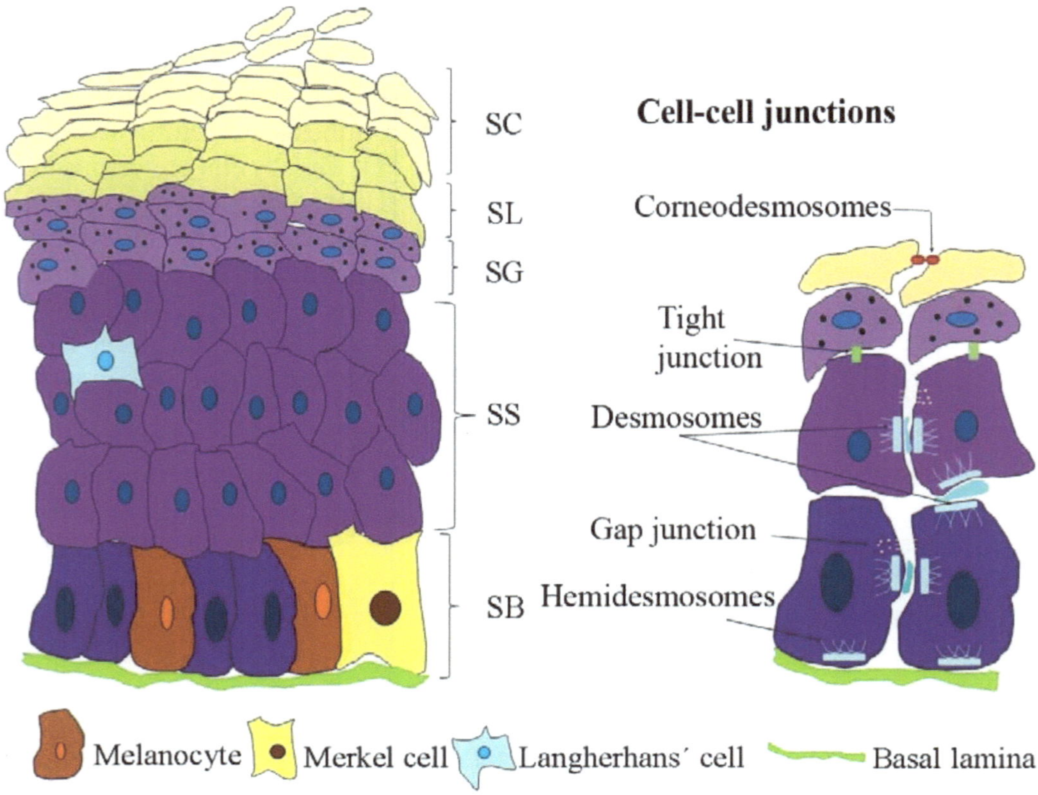

Fig. (1). Schematic representation of epidermis layers.

This structure is considered a transitional (intermediary) link between the metabolically active layers beneath and the non-viable layer above. The passage from the stratum granulosum to the stratum corneum is accompanied by an increase in neutral lipids and sphingolipids [17]. At the transition level between stratum granulosum and stratum corneum, the cells secrete lamellar bodies containing lipids and proteins, into the extracellular space [18, 19]. At this stage, the cells loose the nucleus and also the cytoplasm disappears while the cell membrane is replaced by a cell envelope made of a cross-linked protein (involucrin) with lipids covalently attached to its surface, thus becoming corneocytes or squames that form the **stratum corneum (SC)**. These cells are technically considered to be dead and are characterized by their hexagonal or pentagonal shape. They are disposed in 15-20 layers and function as a barrier between the external environment and the internal medium of the body. In the palmoplantar skin there is an additional zone with nucleated cells which form the **stratum lucidum (SL)**, a transitional zone between the granulosum and corneum layers. The main molecular differentiation markers corresponding to each epidermal layer can be found detailed in Table **1**.

The mechanical resistance of the epidermis is ensured by keratin intermediate filaments which form intercellular attachment plates named desmosomes — a large, continuous intra-epithelial network (Fig. **1**) [20].

The interface between the epidermis and the next layer, the dermis is bounded by the **basement membrane** or **basal lamina**. This thin structure is synthesized by basal keratinocytes and is formed by two zones: lamina lucida composed mainly by adhesive glycoproteins, laminin and fibronectin, and lamina densa formed by type IV collagen [20 - 22]. The basal lamina has two main functions: mechanical support for epidermis and diffusion barrier.

Table 1. Differentiation markers expression pattern.

Layer	**Number of Rows of Cells**	**Key Molecular Differentiation Markers**	**Functions**
Stratum basale	1	Keratins: 5, 14,15 [23], p 63, EGFR, Delta1, IGFR, TβRII, transglutaminase TG 2 [24, 25]	Produces new skin cells [26]
Stratum spinosum	8-10	Keratins: 1,2e,10, AP2, HES1, Notch 1/3 [25]	Synthesis of keratin; make the skin relatively waterproof by releasing a water-repelling glycolipid [27]
Stratum granulosum	3-5	Keratins:2e,9, fillagrin, S100, transglutaminase (TG3) [4] [28, 29]	Formation of the hydrophobic lipid envelope responsible for the skin barrier properties [10]
Stratum corneum	15-20	Loricrin, S100 [28] involucrin	Water proof barrier
Stratum lucidum	3-5	Eleidin	Water barrier [29].

The second layer of skin is the dermis, which has many important roles: provides the structure, strength, pliability, elasticity and flexibility of the skin, regulates body temperature, provides the epidermis with nutrient-saturated blood and also houses the skin annexes. The major cellular constituents of the dermis are mesodermal in origin except for nerves that derive from the neural crest [9]. The predominant cells of the dermis are fibroblasts which synthesize all of the dermal constituents and also modulate the turnover of dermal extracellular matrix components. In the dermis, there are variable numbers of mast cells, lymphocytes, dendritic cells, and vascular endothelial cells (Fig. **2**).

The dermis is divided into two layers: a top *papillary layer* and a denser and thicker layer named *reticular layer* (Fig. **2**). The papillary dermis, as shown in (Fig. **2**) is a relatively thin zone just under the dermal-epidermal junction. It is composed of connective tissues with type I collagen (reticular strands) and elastic strands arranged in a loose mesh-work. It also contains a large number of capillary arches which support the avascular epidermis with vital nutrients and provide a network for thermoregulation [30]. Meissner's bodies in both pear- and capsule-shaped can be found in some dermal papillae.

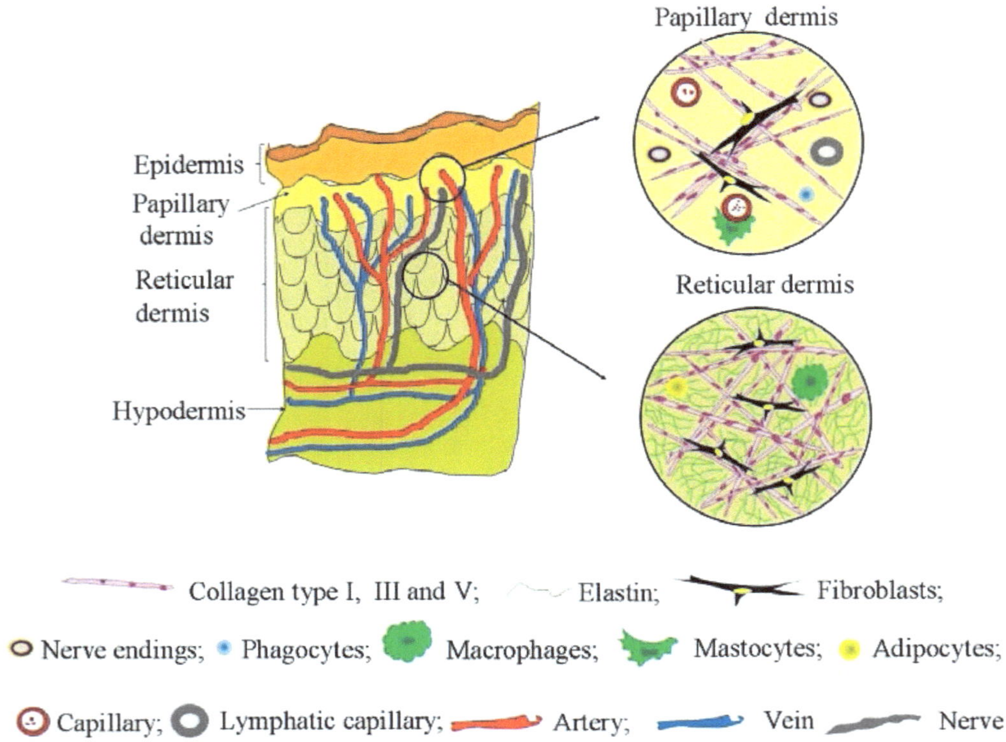

Fig. (2). Schematic representation of the dermis structure.

The border between the papillary layer and the deeper reticular layer of the dermis cannot be distinguished. The reticular layer is much denser than the papillary dermis, more acellular and presents an abundant sensory and sympathetic nerve network. This structure provides the skin with most of its strength, flexibility, and elasticity. The reticular aspect appears due to a tight mesh of collagen fibres. Between the collagen fibrils are elastic fibres, proteoglycans, and glycoproteins. Other components of the skin, such as hair follicles, sweat glands, and sebaceous glands also reside in the reticular dermis (Fig. **3**).

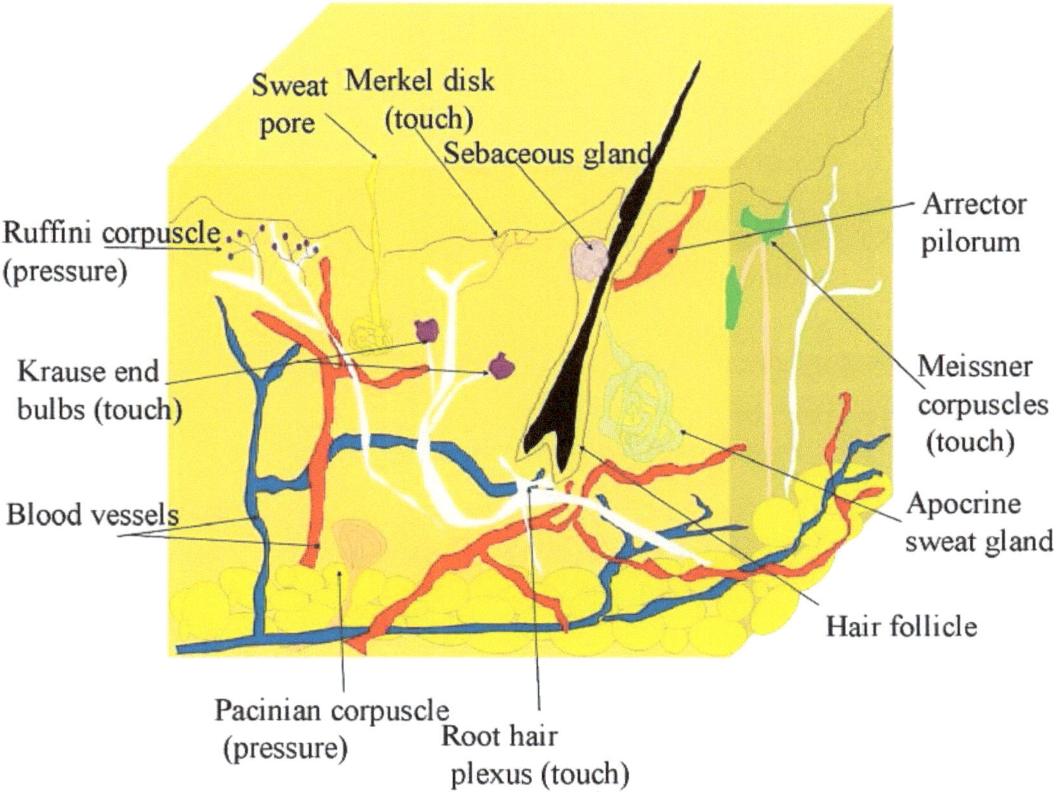

Fig. (3). Skin appendages.

The role of integumentary system is presented in Table **2**.

Table 2. Skin Function.

Cellular Type	**Skin Function**
Keratinocytes, corneocytes	Barrier between the external environment and internal body [31]
Melanocytes	Protection from UV radiation [32]
Sebaceous glands	Absorption and excretion [33]
Subcutaneous layer - adipose tissue	Nutrient and water storage [34]
Langerhans cells	Protection against infection and chemicals [35]

(Table 2) cont.....

Cellular Type	Skin Function
Pacinian corpuscle – vibration Ruffini's corpuscle – joint movements and stretch Meissner corpuscle – tapping and flicker Merkel's disc – vertical dimpling of non-hairy skin Tactile disc - vertical dimpling of hairy skin Hair follicle receptor- speed and direction of movement Nocireceptors – detection of pain	Sensory organ [36]
Small blood vessels presented in the dermis	Thermal regulation [37]
Epidermal basal and suprabasal keratinocytes and dermal fibroblasts	Vitamin D synthesis [38]

THE WOUND HEALING PROCESS

The process of skin wound healing comprises four overlapping phases: haemostasis, inflammation, proliferation and remodelling (Fig. **4**). The outcome of each phase influences the final result which can be a completely healed skin, a scar or a chronic wound.

Haemostasis aims to prevent blood lost upon injury of the vessel walls by constriction and by activation of the platelet adhesion (*via* von Willebrand Factor) and aggregation mechanisms, followed by initiation of the coagulation cascade which results in the deposition and stabilization of a fibrin clot that prevents any further bleeding [39]. The stabilization of the plug depends on the conversion of fibrinogen into fibrin strands *via* thrombin.

This phase stimulates the inflammatory process, by attracting cells like neutrophils, macrophages and T lymphocytes. More exactly platelets release numerous cytokines like TGF-β, IL-1, TNF-α, PDGF and chemokines like CXCL4 (PF4) which stimulate an influx of neutrophils (also known as polymorphonuclear cells) with the main purpose of clearing potential pathogens, and also CXCL12 (SDF-1) which promotes recruitment of progenitor cells and proliferation of local epidermal stem cells [40]. Neutrophils are the first cells of the innate immune system to reach the wound site, reaching a peak in the first 12 hours and their primary role is defending against infections and clearing debris by using proteases (MMPs-2,8,9,25), ROS (reactive oxygen species) and antimicrobial peptides (lysozyme, myeloperoxidase, defensin, azurodin), but they also release multiple cytokines and growth factors that further amplify the inflammatory response [41]. Furthermore, neutrophils also activate local non-immune cells like fibroblasts [42] and endothelial cells- by secreting VEGF [43].

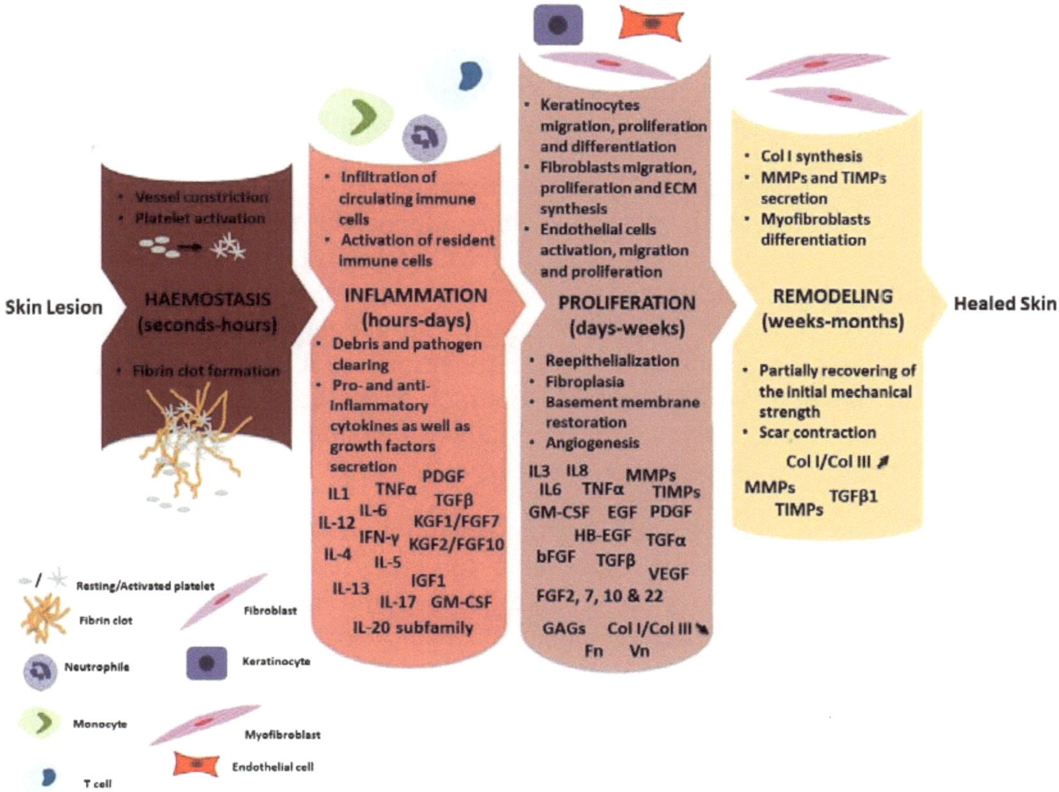

Fig. (4). The four phases of skin wound healing: haemostasis, inflammation, proliferation and remodelling including the main processes together with the cells as well as the major cytokines, growth factors and extracellular matrix components involved. IL-interleukin, TNFα-tumour necrosis factor alpha, PDGF- platelet derived growth factor, TGFβ- transforming growth factor betas, KGF- keratinocyte growth factor, FGF- fibroblast growth factor, IGF1- insulin growth factor 1, GM-CSF- granulocyte-macrophage, MMPs- matrix metalloproteinases, TIMPs-tissue inhibitor of metalloproteinase, EGF- epidermal growth factor, HB-EGF- heparin binding EGF, TGFα- transforming growth factor alpha, VEGF- vascular endothelial growth factor, FGF- fibroblast growth factor, GAGs- glycosaminoglycans, Col I/III- collagen type I/III, Fn-fibronectin, Vn-vitronectin, TGFβ1- transforming growth factor beta 1.

The second type of immune cells to reach the wound is represented by the monocytes which differentiate into macrophages upon migration from the blood stream into the wounded tissue. Besides their immunological role, through the recognition, phagocytosis and destruction of pathogenic microorganisms as well as clearance of cellular debris and apoptotic cells, macrophages also have a trophic role, Willneborg *et al.* demonstrating the importance of a VEGF-A monocyte subtype for early angiogenesis [44]. The skin resident macrophages population mostly found within the skin appendages are activated upon injury *via* DAMPs (damage-associated molecular patterns or alarmins) and PAMPs (pathogen associated molecular patterns). Although out-numbered by the blood infiltrating monocytes which arrive 1-3 days after neutrophils at the wound site, the resident macrophages become activated *via* TLRs (Toll-like receptors) and secrete pro-inflammatory mediators (TNF-α, IL-1β, IL-6) which consolidate the initiation of the immune response by attracting neutrophils and monocytes [45].

In the early stage of inflammation, M1 macrophages (pro-inflammatory) infiltrate the tissue in order to clear bacteria, debris and dead cells, and as the repair progresses macrophages transition to the M2 phenotype (anti-inflammatory) which stimulate the migration and proliferation of fibroblasts, keratinocytes and endothelial cells. However, it must be taken into account that the macrophage population is heterogeneous and different over time, and does not perfectly overlap with the *in vitro* M1 and M2 distinct phenotypes [46]. Also, animal studies revealed that delayed healing diabetic wounds are associated with impaired transition of macrophage phenotype from M1 to M2 [47].

Leucocytes reach the wound site last from the blood vessels *via* diapedesis, with lymphocytes classically being responsible for the adaptive immune response. However, in the major barrier organs like the skin and the mucosae, besides the Langerhans cells (dendritic cells), innate-like lymphocytes reside which can act rapidly in a non-self way, without antigen presentation. More exactly, in normal conditions, most T cells can be found at the dermal-epidermal junction, in the vicinity of blood vessels as well as in the skin appendages. From these levels they can respond to infection and damaged cells by secreting growth factors and cytokines involved in the proliferation and activation of other cells to infiltrate the area [48].

Besides T cells, B cells are also involved in the inflammatory phase of wound healing. The conventional B2 cells oscillate between lymphoid tissue and blood vessels, and are implicated in the adaptive immune response *via* antibodies production. The innate-like B1 cell respond to Toll-like receptor (TLR) activation and can trigger rapid antibody responses independent of T cells in order to exert phagocytosis and anti-microbial effects [49] and also secrete cytokines specific to the innate response such as G-CSF [50]. The proliferative phase begins after maximum one day with keratinocytes migration from the wound periphery with the purpose of restoring the epidermis integrity through the re-epithelialization process and can extent to several weeks. On the other hand, at the dermal level, fibroblasts and endothelial cells are actively involved in the restoration of this layer, through fibroplasia and angiogenesis [51]. This is visually translated in the reddish granulation tissue composed of fibroblasts and the ECM they produce as well as the newly formed blood vessels. The re-epithelialization process involves several stages: migration of the keratinocytes adjacent to the wound, their proliferation and differentiation into a stratified epithelium as well as the reformation of the basement membrane between the epidermal and dermal compartments. Keratinocytes activation at the wound edges is associated with specific keratins: 6, 16 and 17 as the cells acquire a migratory, proliferative phenotype, under the influence of inflammatory cytokines like IL-1 and TNFα. On the other side, TGFβ, a growth factor mainly released by fibroblasts, induces the expression of keratins 5 and 14 specific for the basal epithelial cells [52]. Keratinocytes migration into the bed wound involves an epithelial-mesenchymal transition (EMT), and *in vitro* studies showed that the inflammatory cytokine TNF-α induces EMT features *via* BMP (bone morphogen protein) in keratinocytes from both acute and fibrotic samples [53]. Considered a partial EMT [54], the re-epithelialization process includes a reduction in junctional complexes like desmosomes and adherent junctions, modification of the intermediate filaments-loss of keratin expression and expression of mesenchymal markers like FSP1, vimentin, accompanied by a flattened and migratory morphology. Besides detaching from the substrate and each other, keratinocytes must also make their way through the newly formed fibrin based extracellular matrix. This is accomplished using MMPs like MMP-1, 2, 3, 10, 14, 19 and 28 as well as their inhibitors (TIMPs), a fine balance of the MMPs/TIMPs ratio making the difference between healing and chronic wounds [55, 56]. Cytokines like IL-3, IL-6, IL-8 and GM-CSF, secreted by both activated keratinocytes and fibroblasts are detected at high levels in human wounds fluids and act as mitogen and motogen factors on keratinocytes, stimulating the re-epithelialization process [57]. Multiple growth factors such as HB-EGF, EGF, TGFα, members of the FGF superfamily (FGF2, FGF7/KGF1, FGF10/KGF 2 and FGF 22) and cytokines like IL-1, IL-6 and TNFα act in a synergistic way to ensure the proper migration of keratinocytes [58].

In normal conditions, keratinocytes proliferation and differentiation represent a continuous process and this regenerative capacity of the epidermis mainly relies on the local populations of epidermal stem cells, which reside in specific niches that control their fate: hair follicles, basal layer of the interfollicular epidermis and the base of the sebaceous glands. Mascré *et al*, have shown that, during wound healing, committed progenitor cells play a discrete part in the regeneration process, the major contribution being made by the slow-cycling stem cells from the interfollicular epidermis [59].

The restoration of the basement membrane represents an important step in the wound healing process, as this structure represents the support for the basal keratinocytes and also stabilizes this compartment to the underlying dermis. The major role of the proteins that compose the basement membrane are obvious in diseases like junctional epidermolysis bullosa, in which case, defects in laminin 5 (or 332) – an essential protein associated with type IV collagen, lead to life threatening conditions [60]. Recently, laminin-γ2 was found to be the elected substrate for keratinocytes in partial thickness human wounds, while type IV and VII collagen were deposited after wound closure [61].

Fibroplasia involves the fibroblasts proliferation, migration and production of extracellular matrix components. Around the fourth day after injury, fibroblasts at the wound periphery start to migrate into the fibrin clot where they secrete a collagen-based matrix that also includes proteoglycans and elastin fibres. Some of the chemotactic factors for this process like PDGF and TGF-β are secreted by macrophages, while the hypoxic environment, especially in the wound central area, stimulates fibroblasts proliferation together with the growth factors like EGF and FGF. Also, it has been shown on animal models that intradermal adipocytes are involved in fibroblast recruitment as they are also capable of secreting PDGF [62]. Once they reach the wound bed, fibroblast change their phenotype to a secretory one- involved in matrix synthesis- and later to a contractile phenotype- myofibroblasts, involved in wound contraction. The newly synthesized matrix is characterized by the abundance of type III collagen- also called immature collagen- involved in regulating the maturation of collagen fibres; as the healing progresses, the ratio between type I and type III collagen changes in favour of the first one. A high persistence of collagen III after wound closure is found in impaired healed wounds like keloid [63] and hypertrophic scars [64].

The capillary blood vessels from the dermis initiate the process of angiogenesis upon injury through the activation of endothelial cells, degradation of the basement membrane, cell sprouting and proliferation into the wound area resulting in tubule formation and maturation. Besides the cellular elements, all these stages are dependent on the cytokines present as well as on the components and organization of the newly formed matrix. Inflammatory cells release chemotactic and mitogenic factors for endothelial cells, more exactly platelets and macrophages secrete several growth factors like PDGF, TGF-β, IL-8, bFGF and VEGF [65] while fibroblasts synthesize a matrix rich in glycosaminoglycans, collagens, fibronectin and vitronectin which constitutes the substrate for vessel elongation and tube formation [66, 67].

The remodelling phase begins in the first 2-3 weeks after injury and can extend to up to one year or even more; in the end, the tissue recovering is maximum 80% of its original mechanical strength. The scar dimensions decrease in this stage *via* contraction of the wound and the tensile strength increases *via* the depositions of type I collagen. Wounds for which the re-epithelialization process takes more than 3-4 weeks are prone to develop a visible scar which is usually devoid of skin appendages, more sensitive to ultraviolet radiation and has a diminished resistance to mechanical stress [68].

The initial loose matrix consisting of type III collagen, fibrin, fibronectin and hyaluronic acid which allowed cell migration, proliferation and angiogenesis is substituted mainly by type I collagen- which normally makes up to 80% of the dermis. All through the healing process, the extracellular matrix goes through cycles of synthesis and degradation modulated *via* MMPs and TIMPs. Myofibroblasts coupled *via* integrin mediated focal adhesions with the matrix components induce contractile forces in order to pull the wound margins towards the centre and thus minimize the lesion [69]. The activation of fibroblasts in order to differentiate into myofibroblasts has been linked to the presence of TGFβ1 but also to the stiffness of the substrate [70]. Granato *et al*, showed on *in vitro* spheroids of myofibroblasts that in this 3D system, the cells can become deactivated and re-acquire a contractile phenotype once switch to a 2D culture system, suggesting the importance of the surrounding milieu [71] At the end of a normal healing process, the myofibroblasts disappear *via* apoptosis, while the persistence of this type of cells is linked to hypertrophic and keloid scars [72].

In conclusion, the complex process of wound healing involves 4 overlapping phases: haemostasis, inflammation, proliferation and remodelling, all of which are orchestrated by both skin local resident cells as well as precursors and immune cells attracted to the wounded area, and finely tuned by the cytokines and growth factors released by these cells as well as by the extracellular matrix components which are constantly being modified in order to achieve the restoration of the skin structure and function.

MICRO-RNAS AND WOUND HEALING

MicroRNAs (miRNAs) are small, noncoding RNAs, implicated in the posttranscriptional regulation of gene expression, by interacting with the 3' untranslated region (3' UTR) of target mRNAs, thus inducing its degradation and translational repression. A single miRNA is able to target up to hundreds of genes, while one specific gene can be regulated by several microRNAs. Thus, miRNAs act as central regulators of gene expression and, therefore, these small molecules are implicated in various biological processes [73].

Most of currently known miRNAs are encoded within an intragenic region, primarily in introns, few in exons and their transcription is regulated by the gene promoter. However, miRNAs can be also encoded within intergenic regions, in which case the transcription is regulated by their own promoters [74]. The miRNAs biogenesis begins with the processing of a transcript produced either by RNA polymerase II, or by RNA polymerase III. The canonical miRNA biogenesis involves the generation of a primary miRNA (pri-miRNA) transcript, which is long and contains a stem-loop structure, in which the mature sequence of the miRNA is located. The pri-miRNA is further cleaved by the microprocessor complex, composed by the nuclear RNAse III Drosha and DiGeorge Syndrome Critical Region 8 (DGCR8), resulting the precursor miRNA (pre-miRNA) [75]. Next, the precursor is translocated by the Exportin 5/RanGTP complex into the cytoplasm where the final steps of producing the mature miRNA are accomplished. Thus, the mature miRNA is a result of the pri-miRNA cleavage by the RNase III endonuclease Dicer. Both strands (5p and 3p) can be loaded into the Argonaute (AGO) proteins and form the RNA-induced silencing complex (RISC) [75, 76].

The role of miRNAs in epidermal development and adult skin repair has been well documented and extensive reviews on this topic have recently been published [73, 77, 78]. Therefore, the main miRNAs implicated in the wound healing process and the possibility to use them for the treatment and management of chronic wounds will be discussed below.

MiR-21

Several recent studies have implicated miR-21 as important regulator of multiple aspects of the healing process, such as wound contraction, re-epithelialization and collagen deposition [79]. Thus, it has been shown that miR-21 promotes keratinocyte migration and re-epithelialization during wound healing through TIMP3 and TIAM1, direct targets of miR-21 [80], but also by mediating the TGF-β1-induced migration of keratinocytes *via* targeting phosphatase and tensin homolog (PTEN) [81]. The regulatory role of miR-21 on PTEN was also assessed *in vivo*, on a rat model of skin healing [82] with focus on dendritic cells. The results indicated that miR-21 overexpression increased the ratio and determined the differentiation of dendritic cells, thus significantly improving the wound healing process. Moreover, miR-21 is also enriched in exosomes, which are implicated in intercellular communication. As shown by Hu and collaborators [83], the local transplantation of exosomes derived from umbilical cord blood cells in a mouse wound healing model was followed by accelerated re-epithelialization, reduced scar dimensions, and enhanced angiogenesis, mediated by miR-21-3p through inhibition of PTEN and sprouty homolog 1 (SPRY1). In addition, miR-21 expression was found to be decreased in the wound tissue of diabetic mice and to be important for the fibroblast's migration towards wound area [84].

MiR-31

MiR-31 is an important mediator during the transition from inflammation to re-epithelialization phases of wound healing. It was shown its implication in keratinocytes proliferation and migration by more than one mechanism: by targeting epithelial membrane protein 1 (EMP-1) [85] or by activating the Ras/mitogen-activated protein kinase signalling [86]. Another interesting target of miR-31 is factor-inhibiting hypoxia-inducible factor 1 (FIH-1). Up-regulated levels of FIH-1 in primary human epidermal keratinocytes (HEKs) and human corneal epithelial keratinocytes (HCEKs) impairs differentiation by decreasing Notch signalling. Moreover, down-regulation of FIH-1 enhances keratinocyte differentiation; therefore, by being an endogenous negative regulator of FIH-1 expression, miR-31 is implicated in keratinocyte differentiation [87]. However, Wang and collaborators [88] showed that level of hsa-miR31-5p is significantly higher in hypertrophic scar in comparison to normal skin. Furthermore, hsa-miR31-5p knockdown suppressed the proliferation of hypertrophic scar fibroblasts in hypoxic conditions and inhibited the expression of Collagen I, Collagen III and Fibronectin, suggesting that inhibition of hsa-miR31-5p could be effective for the reduction of hypertrophic scar formation, caused by excessive ECM synthesis and deposition in HSFBs under hypoxia. Altogether, these data indicate a therapeutic potential for miR-31 in skin wound healing.

MiR-203

It has been shown that mir-203 has a significant contribution to skin re-epithelialization. Viticchiè and collaborators showed that, after the injury in mouse epidermis, miR-203 expression is downregulated in the highly proliferating keratinocytes, whereas in the differentiating cells of the skin outside the wound it is robustly expressed [89]. Another recent study showed that IL-17 up-regulated the expression of miR-203, which activated the Janus kinase signalling pathway and promoted VEGF secretion in HaCaT cells, suggesting miR-203 is a positive regulator of IL-17-induced VEGF secretion [90]. Thus, local treatment of anti-miR-203 could accelerate wound closure and reduce scar formation *in vivo*, together with increased re-epithelialization, skin attachment regeneration, and collagen reassignment [91].

MiR-210

miR-210 is a major hypoxia-inducible miR or hypoxamir, whose expression is induced by Hypoxia Inducible Factor-1 alpha (HIF-1alpha) [92] and which is ubiquitously expressed in a large variety of cells, having multiple functions [93], one of which being the modulation of the endothelial cell response to hypoxia, affecting cell survival, migration, and differentiation [94]. Regarding the implication of miR-210 in ischemic wounds, it was demonstrated that HIF-1alpha, which induces miR-210 expression, results in the down-regulation of the cell-cycle regulatory protein E2F3 and by consequence, the re-epithelialization process is impaired [95].

MiR-377

The angiogenic process is regulated by several miRNAs, one of which is miR-377 [174]. Wen and collaborators assessed the implication of miR-377 in mesenchymal stem cell-induced angiogenesis in ischemic hearts and confirmed that it directly targeted VEGF in MSCs, and knockdown of endogenous miR-377 promoted MSC-induced angiogenesis [96]. Moreover, the involvement of miR-377 in the regulation of angiogenesis was assessed in cancer. Thus, MicroRNA-377 suppresses initiation and progression of esophageal cancer by inhibiting CD133 and VEGF [97]. Nevertheless, our group identified miR-377-VE-PTP (vascular endothelial protein tyrosine phosphatase) axis as being involved in the angiogenic effects of endothelial progenitor cells in ischemia-affected myocardium. Thus, it was shown that VE-PTP level is regulated by the anti-angiogenic miR-377, whose expression level is kept low by collagen I [98].

MiRNAs Delivery Approaches for Skin Therapy

The use of miRNAs for the development of new therapeutic targets for wound healing therapy is an appealing idea. Nevertheless, there are limitations regarding the delivery system, which should be designed to prevail over the extracellular and intracellular obstacles in order to ensure the specific up-take of the therapeutic cargo without inducing an immune response [78].

The expression of miRNAs in the target cells can be modulated by using plasmid DNA coding for either a specific microRNA in order to overexpress it, or an antagomir for silencing the expression. The most efficient delivery system of genetic material is represented by viral vectors [79], which can use retroviruses, lentiviruses, and adenoviruses or adeno-associated viruses [99]. All these viral vectors are genetically modified so that they cannot replicate and therefore their safety is increased. Employing viral vectors for the delivery of microRNAs to the target cells has advantages, but has also some drawbacks.

The major advantage of using the viral delivery system is to provide high transfection efficiency and an elevated level of stable expression on microRNAs or antagomirs [100]. Retroviral vectors, which are able to integrate into the host genomic DNA during the mitotic phase of cell cycle, produce a stable and strong expression of microRNAs. In a similar manner, lentiviruses, which also integrate into the host genome and which are able to infect both dividing and non-dividing cells, produce high transfection efficiency and long-term stable expression of miRNAs [101]. The expression of microRNAs can be efficiently induced as well by using adeno-associated viruses, which can also transfect both dividing and nondividing cells [102]. However, despite these advantages and the lack of replication risk, there are several problems associated with the use of a viral system for gene therapy: first, toxins are produced, which can affect cells, second, the viral infection is responsible for inducing inflammatory response and tissue degeneration and third, the inserted genetic material can cause mutations, leading to unforeseeable outcomes [101]. For this reason, nonviral delivery systems, which include polymers of cationic nature, liposomes, and peptides, which have the capacity to package genetic material, and deliver it to the nucleus [79], are preferred.

Several recent studies used cationic polymers to deliver microRNAs *in vitro* in order to develop therapeutic strategies for several injuries/disorders. Song and collaborators synthetized a biodegradable and biocompatible cationic polymer cross-linked with 2,6-pyridinedicarboxaldehyde for transfecting miR-221/222 into rat Schwann cells. The mouse sciatic nerve crush injury model was used to evaluate the efficacy of PDAPEI/miR-221/222 complexes for nerve regeneration *in vivo* [103]. This approach proved to be safe, having low cytotoxicity and an adequate transfection efficiency. Another group used with good results EGFR-targeted cationic polymeric mixed micelles in order to co-deliver Gemcitabine and miR-205 for the treatment of advanced pancreatic cancer [104].

Cationic liposomes are most frequently used in nonviral delivery systems due to their ease of production, high affinity with the cell membrane, nonpathogenic, and nonimmunogenic response [79]. Maroof and collaborators used hydration-o--freeze-dried-matrix formulated liposomes for effective delivery of miR-34b to anaplastic thyroid carcinoma both *in vitro* and *in vivo*, in a xenotransplantation mouse model [105]. Thus, the delivery of microRNAs using cationic liposome could be a functional therapeutic strategy for various diseases, including skin conditions. However, when applied on skin surface, the liposomes remain in the upper layer of the *stratum corneum*, acting as a drug reservoir [106]. Thus, ultra-deformable liposomes (UDLs) that can effectively enter through the skin pores and deliver the loaded drugs into the epidermis and dermis layers have been developed [107]. Two categories of specialized liposomes with augmented skin penetration capability can be described: ethosomes, which include ethanol in their composition as a penetration enhancer and transfersomes, which are made of phospholipids and edge activators (sodium cholate, sodium deoxycholate, Span 60, Span 65, Span 80, Tween 20, Tween 60, Tween 80, and dipotassium glycyrrhizinate) [108]. Such UDLs have been successfully used for the delivery of silencing RNAs across the skin to epidermal cells on mouse models or porcine skin [109 - 111]. Therefore, the use of UDLs for the delivery of other small RNAs could be used.

Interestingly, miRNAs can also be released into extracellular space by a diversity of cell types *via* exosomes into the circulation, thus modulating the cellular activity of target cells, contributing to cell-to-cell communication [74]. Exosomes are an efficient microRNA delivery system, by having a membrane which preserve and deliver functional microRNAs into recipient cells [112]. For example, MSCs-derived exosomes contain various microRNAs involved in both physiological (organism development, epigenetic regulation, immunoregulation) and pathological processes (tumorigenesis, and tumour progression) [113]. Moreover, exosomes producing cells can be genetically modified in order to over-express certain microRNAs with therapeutic properties [114]. Recently, several groups have reported the modification of the exosomal cargo in order to use these nanovesicles as carriers for therapeutic microRNAs in various pathological conditions. For instance, Ma and collaborators investigated the effects of miR-210 over-expressing endothelial progenitor cell-derived exosomes on hypoxia/reoxygenation injured endothelial cells, showing that miR-210 enrichment could heighten the effects of exosomes as such [115].

However, at present, there is still an unsatisfied need for innovative topical agents producing better long-term outcomes with fewer side effects and modulation of the expression and activity of microRNA represents an exciting approach due to the fact that changes in the expression of one microRNA can impact many cellular processes associated with the disease [116].

CELLULAR THERAPY FOR SKIN REGENERATION

Since the epithelium has a remarkable ability of self-renewal, in healthy individuals, the wound healing process is efficient and the normal structure of the skin is quickly re-established. Nevertheless, in chronic wounds, the repair mechanism is impaired either due to the deep and extensive skin damage (pressure ulcers or burns) or to a pathological condition such as diabetes, ischemia, cancer or AIDS [117]. A large part of the population, especially elderly people, is affected by chronic wounds as a result of other systemic conditions (in particular diabetes) and the consequence is a poor quality of life or even disability, increased pain, depression and social isolation. A major problem in the treatment of these patients is the high rate of reoccurrence and the resistance to conventional therapy [118]. Hence, cellular therapy for chronic wound healing is taken into consideration as an alternative cure for this disability. Several cell types, including skin cells (keratinocytes and fibroblasts), but also stem cells, have been tested on animal models or even in clinical studies with promising results. Here, we discuss the use of these cells for skin regeneration.

Resident Skin Stem Cells

The stem cells residing in the skin are able to participate to the renewal of the tegument, either constitutively or after injury. There are several important stem cells reservoirs (niches) in the skin that are involved in the epidermal regeneration. In normal conditions, each niche maintains its respective stem/progenitor cells in a unipotent mode [119]. The most important and well described stem cells niches within the skin are the following: the epidermal niche, the hair follicle bulge and sebaceous gland niche [74] (Fig. **5**).

The first two categories have been used in clinical trials for skin regeneration with impressive results.

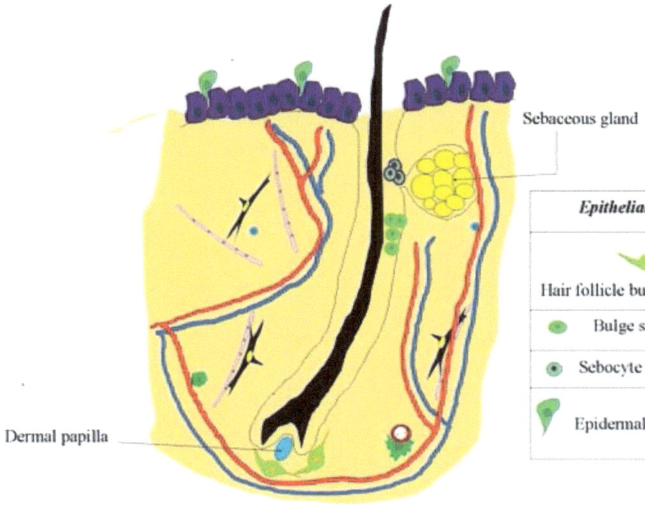

Epithelial stem cell	*Markers*
Hair follicle bulge of stem cell	DNA label retention; TCF3; Sox9; Lhx2; CD34; NFATc1; vitamin D receptor
Bulge stem cell	CD34; K15; Lgr5, Sox9; Tcf3
Sebocyte stem cells	Blimp 1
Epidermal stem cells	β1 integrin; α6 integrin; p63

Fig. (5). Stem cells niches within the skin.

Epidermal Basal Layer Stem Cells

In the basal layer, the stem cells can be found among basal keratinocytes. By using primary cultures of human keratinocytes, Barrandon and Green have established the relationship between stem cell capacity and colony forming ability of primary keratinocytes [120]. Thus, it was observed that single cells generated 3 types of colonies (holoclones, meroclones and paraclones), having different proliferative abilities. The holoclones were described as large and circular colonies, containing small and regularly shaped cells, having the highest proliferative potential. Moreover, holoclones were the only ones having a self-renewal capacity, which made them become synonyms with epidermal stem cells [121, 122]. Due to their high proliferative ability, these cells can be expanded *in vitro* until obtaining a sufficient amount for cell therapy. Moreover, they can be modified in order to repair a genetic defect. In this manner, a recent and impressive study showed that autologous transgenic keratinocyte cultures were able to regenerate an entire and functional epidermis of a child suffering from junctional epidermolysis bullosa [Hirsch 2017]. Prior to the transplant, the cells were transduced in order to express the functional laminin-322, leading to the restoration of the normal formation of junctions [123]. The authors showed that the newly formed epidermis was maintained by a limited number of transgenic epidermal stem cells (holoclones), able to self-renew both *in vitro* and *in vivo*. During the 21 months follow-up, the epidermis was robust and did not develop blisters.

Stem Cells Residing in the Hair Follicle Bulge

Two main subpopulations of stem cells are alleged to exist in hair follicles: a subcategory positioned within the hair germ underneath the bulge, which gives rise to the hair shaft and inner root sheath and the quiescent cluster, which can be found in the bulge region, responsible for generation of the basal outer root sheath keratinocytes [74]. Interestingly, several recent clinical studies using the hair follicle bulge have been reported. For example, Jimenez and collaborators [124] used autologous scalp follicular grafts transplanted into the wound bed of chronic leg ulcers in 10 patients. After 18 weeks follow up, a 27.1% ulcer area reduction in the experimental zone was observed as compared with 6.5% in the control. Moreover, histological analyses showed enhanced epithelialization, neovascularization, and dermal reorganization. Another study aimed to validate the role of hair follicles in venous leg ulcer healing by making an intraindividual comparison of hair follicle-containing *versus* nonhairy punch grafts. The results showed that a 75.15% ulcer area reduction was obtained for the hair follicle group in comparison to with 33.07% in the control group (nonhairy grafts) [125].

Besides the enhancement of wound healing, another interesting application of using hair follicle transplantation is replenishing melanocytes in vitiliginous patches utilizing undifferentiated stem cells of the hair follicles (follicular unit transplantation). A recent clinical study performed on 50 patients with 63 lesions of stable vitiligo showed a reduction in the size of vitiligo patches as well as improvement in the associated leukotrichia. Thus, out of the 63 patches, a good response was obtained in 39 and fair in 16%. Moreover, this medical approach resulted in a quick patient recovery time, very low morbidity, and good colour match [126].

Keratinocytes

Currently, the use of skin grafts is mandatory in the treatment of unresponsive and extensive wounds. In order to spare donor tissue, alternative methods using cultured epidermal cells applied as multi-layered stratified epidermal sheets have been used. However, the results have not been adequate due to poor grafting and fragility of the transplanted tissue [127]. Recently, several studies reported efficient use of autologous keratinocytes suspensions in wound healing therapy. Hartmann *et al* showed that enzymatic and surgical debridement of 19 patients with deep dermal burns of the face and neck, followed by sprayed application of cultured epithelial autograft suspensions resulted in very good cosmetic outcomes in comparison to other currently use methods [128]. Another clinical study showed that the treatment with cultured autologous epidermal cells only was efficient in nine patients with deep second degree burns in the absence of skin grafting. Moreover, combined technique of using split thickness skin grafts covered with cultured epidermal cells was used in 27 patients with 85% success rate. Nevertheless, cultured keratinocytes allowed rapid healing of donor sites and deep second degree burns in extensively burned patients [129]. Another recent paper showed that fibrin added to the keratinocytes suspension which was sprayed onto challenging wounds helped the anchoring of keratinocytes to the wound surface and helped the healing of severe burns [130].

Taken together, these clinical studies indicated keratinocytes as important cellular type needed for the skin regeneration after extensive injuries, when the autologous reserve of pre-keratinocytes and epidermal stem cells is depleted.

Fibroblasts

Fibroblasts are important elements of the dermal layer, critical not only for the normal biology of the skin, but also for an efficient wound healing process. For this reason, attempts have been made in order to use fibroblasts in cellular therapy for various skin injuries, frequently associated with a biomaterial as a support. Thus, Yun and collaborators [131] showed that fibroblasts encapsulated in a poly(ethylene glycol)-b-poly(l-alanine) (PEG-l-PA) thermogel were able to accelerate the wound closure and improve epithelialization and the formation of hair follicles, and sebaceous glands on a full-thickness wound rat model. The use of fibroblasts resulted in similar beneficial results in clinical studies. Venugopal and collaborators used allogeneic cultured fibroblasts in suspension in albumin solution for the treatment of generalized severe recessive dystrophic epidermolysis bullosa. The results indicated that the injection of both allogeneic fibroblasts and suspension solution alone improved wound healing in comparison to noninjected wounds [132]. Another clinical study used a bioengineered skin substitute composed by dermal autologous fibroblasts and hyaluronic acid for the treatment of diabetic foot ulcer [133]. After 12 weeks follow-up period, 84% patients of the treatment group achieved complete wound healing in comparison to 34% patients of the control group and also the time required for complete wound healing was significantly shorter for the treatment group. An additional clinical study addressed to patients with diabetic foot ulcer evaluated the incidence of amputations/bone resections in a randomized controlled trial comparing human fibroblast-derived dermal substitute plus conventional care with conventional care alone [134]. The results showed that significantly fewer amputations/bone resections were needed in patients who received human fibroblast-derived dermal substitute in comparison to control patients. Taken together, the clinical use of fibroblasts for the treatment of skin injuries proved to be a safe and efficient approach.

Mesenchymal Stem Cells

Since their first report by Friedenstein and collaborators [135, 136], mesenchymal stem cells (MSCs) have been seen as main candidates for regenerative therapy. Described as plastic adherent fibroblast-like cells, it is generally agreed that MSCs lack the expression of hematopoietic markers such as CD14, CD11b, CD34, CD45 and HLA-DR and they are positive for other surface markers such as CD90 (Thy-1), CD105 (endoglin), and CD73 (5' nucleotidase). Regarding their plasticity, a minimal tri-lineage differentiation (adipogenic, osteogenic and chondrogenic) potential is required in order to be considered "stem cells" [137]. However, MSCs isolated from various sources (bone marrow, adipose tissue, peripheral blood, cord blood, Wharton's Jelly, dental pulp) have been shown to generate not only adipocytes, osteoblasts and chondrocytes, but also myoblasts, tenocytes and neural cells [138].

The most commonly used MSCs in various *in vitro*, *in vivo* and clinical studies for skin regeneration are represented by bone marrow (BM) and adipose-derived MSCs (ADSCs). Several clinical studies used either BM-MSCs or ADSCs for the treatment of chronic wounds, diabetic and ischemic ulcers, severe radiation burns and hypertrophic scars following burns. The administration route included intra-dermal, intra-muscular or intra-arterial injection, fibrin spray and collagen wound dressing [139 - 149]. The clinical outcomes showed, in most cases, improved or even complete wound healing and reduction of skin graft contracture in the case of scars.

The mechanism by which MSCs of either origin improve the process of wound healing is still under debate. However, although there is some evidence that MSCs could be driven to differentiate towards an epithelial phenotype [150 - 152] and thus they could participate directly to the regenerative process by giving rise to new keratinocytes, the paracrine effect is generally considered to be important supporting the wound healing process [153]. By secreting a rich mixture of growth factors and extracellular vesicles, MSCs promote the reepithelialization, enhance angiogenesis and attenuate scar formation [154]. Thus, MSCs classically secrete pro-angiogenic, chemotactic and remodelling factors such as: platelet-derived growth factor-B (PDGFB), transforming growth factor beta (TGF-β), stromal-derived factor-1 (SDF1), angiopoietin-1 (Ang1), regulator of G-protein signaling-5 (Rgs5), matrix metalloproteinase-9 (MMP-9), epidermal growth factor (EGF), keratinocyte growth factor (KGF), hepatocyte growth factor (HGF), vascular endothelial growth factor-α (VEGF-α), macrophage inflammatory protein (MIP) [155, 156].

An important component of MSCs secretome, which was recently shown to be responsible for the regenerative properties, is represented by the extracellular vesicles, mainly by exosomes. Several recent studies demonstrated that MSCs-released exosomes are able to enhance the skin wound healing process by promoting the proliferation and migration of dermal fibroblasts, keratinocytes migration, improving the angiogenic process and by regulating extracellular matrix remodelling [157 - 159].

All these data indicate that MSCs can contribute to wound repair and they could be a major source for regenerative therapy.

Hematopoietic Stem Cells

Hematopoietic stem cells (HSCs) represent the source for blood cells production that takes place continuously during the entire lifespan of an organism. HSCs are pluripotent, meaning that a single cell is able to generate all the hematopoietic cell types [160]. HSCs can be isolated from the bone marrow, umbilical cord blood, and peripheral blood by using markers such as CD34 and CD133. Commonly, these cells can be easily isolated from the peripheral blood after mobilization into circulation by administrating cytokines such as granulocyte macrophage colony stimulating factor (GM-CSF) [119]. The use of CD34+ cells was shown to be effective for accelerating wound closure [161, 162]. For example, Whiteley and collaborators showed that *in vitro* cultured umbilical cord blood CD34+ cells from frozen units can accelerate wound healing and regenerate the skin on a mouse model of full thickness wound [163]. However, CD34+ blood cells include not only HSCs, but also endothelial progenitor cells, which could promote therapeutic neovascularization [163] and therefore contribute to tissue regeneration. Even though data regarding the use of purified HSCs as a wound repair strategy are scarce, CD34+ cells from bone marrow of blood, which certainly contain a combination of various progenitor cells and have regenerative properties, could be a candidate for therapeutic investigation [162].

Induced Pluripotent Stem Cells

Induced pluripotent stem cells (iPSCs) where generated for the first time in 2006 by Yamanaka and collaborators, by inducing the expression of four reprogramming transcription factors (Oct-3/4, Sox2, c-Myc, and KLF4) into differentiated adult somatic cells such as dermal fibroblasts [164]. Since then, it was shown that these four original inducers could be replaced by other factors such as related transcription factors, miRNAs, small molecules, or lineage specifiers [165]. Although similar to embryonic stem cells regarding their self-renewal capacity and differentiation potential, iPSCs do not give rise to ethical issues, thus offering a non-controversial approach to generate patient-specific stem cell lines and also, they reduce the probability of immune rejection after transplantation. Since their discovery, iPSCs have been exploited in various research areas such as disease modelling, drug development/cytotoxicity studies and, due to their great potential to generate multiple cell types, in regenerative medicine [166].

iPSC can produce epidermal keratinocytes, which are affected both in chronic wounds and in various skin disorders, such as epidermolysis bullosa. It was shown that the resulted keratinocytes were able to produce a stratified epidermis, both *in vitro* (in organotypic three-dimensional skin cultures) and *in vivo* (transplantation in animal models) [167, 168]. Direct reprogramming of somatic cells to obtain keratinocytes from patients suffering of various skin conditions endows the researchers with a perfect approach for studying the disease mechanisms [169]. Moreover, it is achievable to correct the faulty genes in order to obtain fully functional cells for patient specific cellular transplant.

However, the efficiency of gene correction, among other safety concerns, is still a limitation of translation of this therapeutic approach into the clinics. In this context, a recent study made use of a natural occurring process named revertant mosaicism, which involves the spontaneous rectification of a pathogenic gene alteration in a somatic cell [170]. Thus, Umegaki-Arao and collaborators generated iPSCs from revertant keratinocytes of an epidermolysis bullosa patient with heterozygous COL17A1 mutations. The keratinocytes obtained from these iPSC were fully functional and could generate *in vitro* three-dimensional skin equivalents and reconstitute the skin *in vivo* on a mouse model of chronic wounds. In both cases, the cells in the both in the basal layer expressed the initially faulty Col17, which suggested that this approach could be an alternative to *in vitro* genetic modification.

The production of functional keratinocytes from iPSCs was recently assessed in a relevant primate model. Thus, Domingues and collaborators obtained monkey pluripotent stem cells and successfully differentiated them into functional keratinocytes, which were able to reproduce both *in vitro* and *in vivo* a stratified epidermis [171]. Moreover, the generated keratinocytes presented a low level of major histocompatibility complex class I antigens, suggesting that these cells could be a viable approach for autologous or allogeneic epidermal long-term grafting.

A full regeneration of skin, including appendices, was obtained *in vivo* in a mouse model by using progenitor cells obtained from mouse neonatal skin, which was dissociated into single cell suspension and grown as 3D cultures on cell-culture inserts [172]. Thus, Lei *et al.* showed that these epidermal aggregates displayed morphological changes during culture going through six consecutive stages of a self-organization process, leading to the *in vivo* generation of planar hair-bearing skin. However, the adult cell failed to self-organize and formed scarce hairs compared with the robust hair follicle regeneration gotten with new-born mouse cells. However, by using mouse iPSCs, skin organoids, constituting epidermal and dermal layers, spontaneously produced de novo hair follicles in a process that mimics embryonic hair folliculogenesis [173]. Another recent study [174], showed that human iPSC, induced into $CD200^+/ITGA6^+$ epithelial stem cells, seeded on a human acellular amniotic membrane, and transplanted onto skin defects of mice, were able to repair the skin damage and promoted the formation of hair follicles and interfollicular epidermis.

By concluding, all the above-mentioned cell types have been proven efficient, in various extents, for the therapy of chronic wounds or skin disorders caused by faulty gene expression of various cellular or extracellular matrix components. The beneficial effect which was observed was mediated by two mechanisms: (i) a direct regeneration of the skin layers, in the case of skin cells or iPSCs derived keratinocytes transplantation and (ii) generally paracrine effect mediated by the soluble pro-angiogenic and mitogenic factors released by MSCs and HSCs. The cell types with potential application in skin therapy are schematized in Fig. **(6)**.

SKIN SUBSTITUTES

Depending on the surface and depth of the skin wound, its clinical management ranges from topical administration of antibacterial substances and drugs, common wound dressings like hydrogels and hydrocolloids to aggressive surgical debridement and grafts or cell-based dressings. The most challenging skin lesions remain burns as well as chronic wounds, which pose a high burden for both patients and the health care systems, being associate with high morbidity and mortality [175].

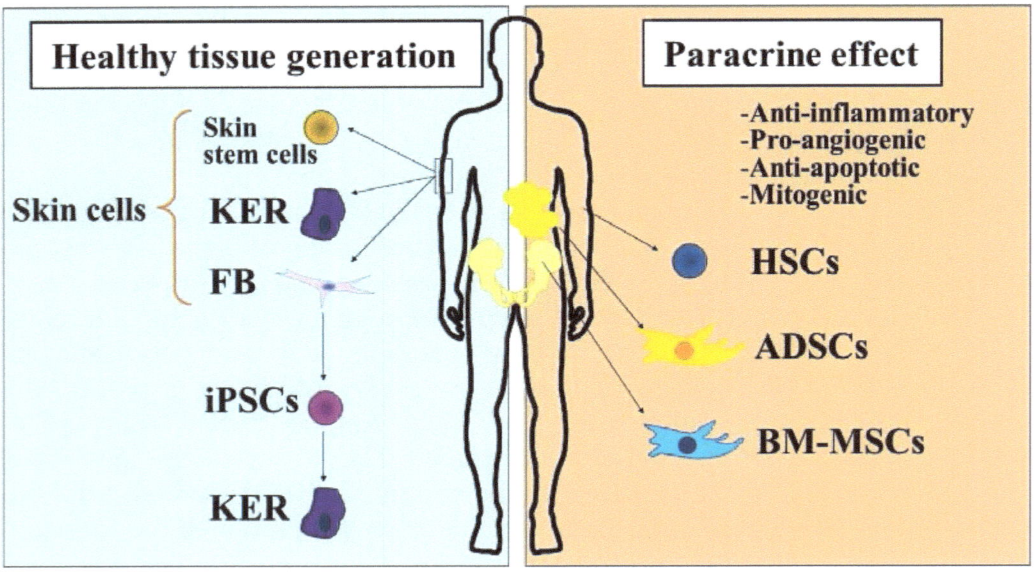

Fig. (6). Cell types for skin regeneration therapy (KER: keratinocytes, FB: dermal fibroblasts, iPSCs: induced pluripotent stem cells, HSCs: haematopoietic stem cells, ADSC: adipose-derived mesenchymal stem cells, BM-MSCs: bone marrow-derived mesenchymal stem cells).

Skin substitutes represent tissue engineered constructs that aim to restore the skin structure and function, being a more expensive alternative than the classical inert dressings made of alginates, foams or hydrofibers [176], but with a good cost-effectiveness ratio. These bioscaffolds also avoid the disadvantages of the full-thickness or split-thickness skin grafts in terms of donor pain, limitation of the harvesting sites and the risk of scarring. The ideal skin substitute should be able to prevent wound infection while lacking immunogenicity, preserve a moist environment, and promote all main phases of wound healing previously discussed in this chapter: attract the cells important for the synthesis of extracellular matrix and neovascularization, while being biodegradable and guiding the correct matrix deposition and also promote migration and proliferation of the patient's cells [177, 178].

Until now, several skin substitutes have made their way to the market and they are either cellular bioengineered scaffolds like Apligraf, Dermagraft and TheraSkin or accelular scaffolds which can be made of natural-Oasis, Integra, DermACELL or composite polymers- Hyalomatrix, and even exclusively synthetic ones: NovoSorb [179].

APLIGRAF (Organogenesis, Inc) is a bilayered allogeneic bioengineered skin substitute intended for the treatment of venous leg and diabetic foot ulcers which do not respond to conventional therapies. It consists of an epidermal layer made of human keratinocytes and a dermal layer of bovine type I collagen with embedded human fibroblasts [180]. A recent study showed that in the case of venous leg ulcers, the application of the bilayered cellular construct induced a shift from a chronic profile to a healing acute one, by modulating the inflammatory cytokines, activation of patient keratinocytes and regulating Wnt/β catenin signalling which is known to be involved in the perturbations of stem cell niches associated with this pathology [181].

TheraSkin (Soluble Systems, LLC, LifeNet Health) is a cryopreserved human skin allograft containing both epidermal and dermal layer, with a high collagen content and supply of growth factors, intended for the treatment of chronic venous leg ulcers. This is actually a split-thickness allograft harvested 24 hours post-mortem and cryopreserved for up to 5 years, and is available in 3 sizes depending on the surface area. In a recent study, which compared TheraSkin and Apligraf for the treatment of patients with venous leg ulcer, the investigators concluded that both grafts presented similar healing rates, but the costs for the TheraSkin cohort was lower [182].

DERMAGRAFT (Organogenesis, Inc) is a human dermal substitute indicated in the treatment of chronic wounds like diabetic foot ulcers, and consists of a bioresorbable scaffold containing human fibroblasts, therefore supporting the synthesis of ECM components and providing cytokines and growth factors in order to stimulate the wound healing process [183]. The dermal fibroblasts are derived from new-born human foreskin tissue and cultured onto a three-dimensional polyglactin mesh which is bioresorbable.

StrataGraft (Stratech Corporation) is a tissue engineered construct which has been designated in 2017 as a regenerative medicine advanced therapy by the FDA, and is under evaluation in a phase 3 clinical trial for treating burns with intact dermal elements. This allogeneic skin substitutes, considered a fully differentiated tissue, consists of a completely stratified epidermis (derived from GMP compliant NIKS cells- near diploid immortalized keratinocyte S) and a dermal layer consisting of a collagen matrix with fibroblasts [184].

Acellular double layer dermal substitutes like Integra or Nevelia are composed of a thick layer of bovine type I collagen and a thin exterior layer made of silicone; they are intended for the treatment of non-infectious burn wounds as well as scar contracture and can be used in combination with epidermal grafts in order to regenerate the skin structure [185]. Inner amnion and outer chorion are mostly used as wound dressing in ophthalmology, but in the last years, commercial human amniotic products have been considered for chronic wounds, in either cryopreserved (Amniofix) or dehydrated/lyophilized form (EpiFix, Integra Biofix, NEOX FLO) [186, 187]. The beneficial effects of these products are based on the cytokines and growth factors from amniotic membranes and umbilical cord tissues that can modulate inflammation and promote wound healing.

DermACELL (LifeNet Health) is a human acellular dermal matrix recommended for chronic wounds [188] and also used in breast reconstruction procedures as it offers soft tissue support. The substitutes are recovered from cadaveric donors and decellularized, resulting a matrix with a DNA content under 3%, while still retaining the major components (collagen, elastin) and growth factors. DermaAcell is provided as a sterile, room temperature ready to use product and compared to other human acellular dermal matrix- like Alloderm, the clinical outcome is similar but the expenses are 15-35% lower [189].

Although the medical market provides several skin substitutes as the ones previously described, their inefficiency in tackling severe skin lesions like burns and chronic wounds supports the need for an interdisciplinary research in this area, with biomaterials still playing a major part in the skin regeneration approach. Among the natural polymers, the most extensively studied remain collagen, gelatine, alginate and chitosan [190]. Due to their low mechanical properties and rapid degradation, composite materials which are built on the synthetic polymers tailored characteristics have emerged as better alternatives.

From the adult multipotent stem cells category, bone-marrow derived mesenchymal stromal/stem cells (BM-MSC) have been the most studied including in applications for skin regeneration, as it has been shown that they are capable of differentiation under proper conditions into epidermal keratinocytes [191]. A proposed therapeutic approach consists in combining genetic modified BM-MSC with bioactive matrices. Commercially available human dehydrated amniotic membrane seeded with BM-MSC which transiently expressed TGFβ3 were used for the healing of a full-thickness wound in rats. The newly formed tissue from the group that received the construct with modified BM-MSC showed more organized collagen bundles with parallel disposition and, overall, a reduced scarring effect [192].

Gene-activated scaffolds have been considered as a better alternative to classical biomaterials as they can provide functional genes for selected growth factors ensuring the *in situ* transfection and therefore expression of important cytokines involved in the skin healing process. The combination of this type of scaffold and BM-MSC has been proven to be a feasible approach for the regeneration of skin with sweat glands on animal model. More exactly, Kolakshyapati *et al.*, have used a bovine collagen and chitosan porous composite scaffold loaded with complexes of Lipofectamine and plasmid DNA-encoding EGF (epidermal growth factor), and seeded with BM-MSC for skin regeneration, including skin appendages. *In vitro,* the construct showed modest results toward sweat gland differentiation, but once transplanted *in vivo* (rodent model), besides the association with a better healing, after 8 weeks the regenerated tissue samples revealed the presence of sweat gland-like structures positive for keratin 8 and 14 [193].

Acellular dermal matrices, since they preserve in various proportion the composition and structure of native extracellular matrix, are intensively used not only in burn and wounds therapies but also in reconstructive plastic surgery. However, the ideal source for patients would be human acellular dermal matrices in order to minimize the inflammatory response. Recently, Qi *et al.* demonstrated on a mouse model, that burn skin waste can be successfully used to obtain denaturized acellular dermal matrix which, seeded with BM-MSC can perform as an autologous substitute that is able to ensure faster healing of mouse full-thickness wounds, facilitating angiogenesis and skin appendages regeneration [194]. Another animal study has shown the potential of a synthetic substitute of electrospun polycaprolactone seeded with adipose derived mesenchymal stem cells (ADSC); the cell seeded construct promoted a better reepithelialization of the full-thickness wounds compared to the bare material, thus suggesting that this type of multipotent cells can favour a better skin regeneration [195].

The potential of human hair follicles stem cells combined with dermal papilla cells (mesenchymal cells including multipotent stem cells) each seeded on the appropriate face of a porcine acellular dermal matrix was evaluated both in terms of *in vitro* skin like architecture and *in vivo* as a xenograft for a full-thickness wound model on nude mice [196]. This study revealed that the presence of dermal papilla cells was associated with a higher neovascularization and structures similar to embryonic hair buds compared to the constructs seeded with dermal fibroblasts.

Most adult stem cells, like bone-marrow and adipose tissue derived, involved an invasive procedure for isolation of different degree. Placental and umbilical derived stem cells however have emerged as more available candidates for cell therapy from this point of view. A study comparing the influence of different human stem cell population isolated from these sources: amnion derived MSCs, blood vessel-derived MSCs- chorionic plate and Wharton's jelly derived MSCs seeded on a commercial dermal substitute (Matriderm) in a murine full-thickness skin wound model, showed that all 3 types of stem cells support, in a similar manner, the healing process, including the neovascularisation [197].

New technologies like 3D printing which allow a better space resolution of the construct components, especially cellular ones, are expected to further improve the performances of skin substitutes by allowing the incorporation of multiple cell types and tailoring of the bioink properties to meet the skin tissue requirements [198]. Most 3D bioprinted skin models involved fully differentiated skin cells: keratinocytes and fibroblasts, as well as melanocytes, and collagen, gelatine, fibrin, alginate or decellularized matrices as basic materials for bioinks.

The major bioprinting techniques are microextrusion, electrodynamic based, drop based (inkjet) and laser based assisted. The constituent material of the bioink should meet the requirements presented in Fig. (7) just as the ones used in classical dermal substitutes while also having adequate rheological properties which allow its printing. Because of the shrinkage of the construct associated with the use of natural polymers like collagen, the bioinks need to be photo- or chemically crosslinked [199]. The printed construct can be either left to mature *in vitro* (static condition or bioreactor) or can be grafted to differentiate *in vivo*. In 2012, a study using an *in situ* bioprinting of a construct consisting of fibrin-collagen matrix with human BM-MSC and ADSC suggested that the beneficial effects observed on the wound healing of nude mice were mainly due to the trophic factors secreted by these stem cells, with ADSC favouring the new blood formation [200]. Due to topological accuracy that can be achieved with 3D bioprinting, the incorporation of skin appendages - a requirement for authentic skin regeneration- can also be achieved as shown on animal models. For example, using an extrusion-based system, Huang *et al.*, developed a bioink composed of gelatine and alginate in which they incorporated epithelial progenitor cells and dermal homogenates as well as EGF. The resulted construct provided an adequate environment to mimic the stem niche needed for sweat gland differentiation as shown by the *in vitro* results as well as the regeneration observed in murine burn wound [201]. Cubo *et al.* developed a 3D bioprinted skin equivalent *via* microextrusion, using fibrin and differentiated cells: human keratinocyte and dermal fibroblast. Their system provided 100 cm^2 of construct in approximately 35 minutes and both the *in vitro* maturated as well as the grafted skin equivalents showed an architecture similar to the native skin [202]. Another skin substitute was composed of 3D fibrillar polycaprolactone and bacterial cellulose matrix obtained *via* electrodynamic inkjet printing and seeded with rat ADSC. The construct was used for the treatment of full-thickness wound model in mice and the neo-tissue was characterized by a skin morphology similar to healthy skin and the presence of appendages [203].

Furthermore, Albana and collaborators demonstrated the feasibility of an *in situ* 3D bioprinting robotic system consisting of a laser scanner for the reconstruction of the wound site (pre-design step) and an inkjet based print head with 8 nozzles which ensures the deposition of each cell type (keratinocytes and dermal fibroblasts) embedded in a collagen-fibrinogen matrix *via* a different nozzle as well as the deposition of thrombin *via* atomization in order for the fibrin to polymerize [204]. The 3D printed autograft on a porcine full-thickness wound model was associated with a better performance in terms of wound closure, reduced wound contraction and re-epithelialization compared to the 3D printed allograft. This proof of concept study opens the avenue for more complex 3D printed structures that could harness the previous experience of classical manually obtained substitutes in terms of the benefits of different cell types like several population of adult stem cells as described above.

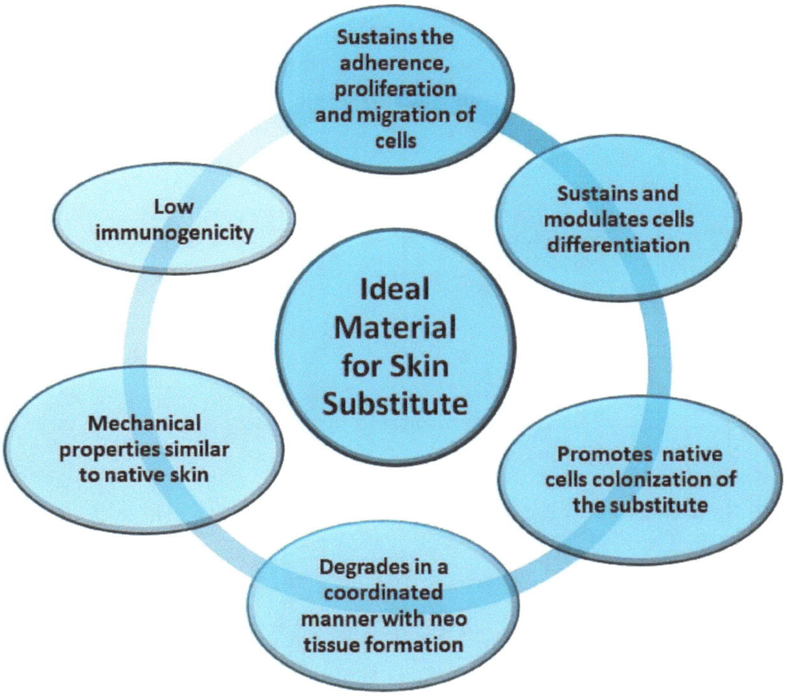

Fig. (7). The main requirements for the ideal material to be used for obtaining skin substitutes.

CONCLUDING REMARKS

The therapy of various skin pathologies/injuries has been an important subject for decades and, even nowadays, despite the advances made in understanding skin biology, the wound healing process and the progressions made in developing innovative therapies, even the most advanced approaches are not enough for chronic or large wounds. However, the advances that were made in the fields of molecular biology, stem cells and biomaterials open new therapeutic opportunities, which could lead to the complete restoration of the skin structure (including epidermal appendages such as hair, sebaceous and sweat glands), skin pigmentation, reduction of scar formation and rectification of congenital diseases.

CONSENT FOR PUBLICATION

Not applicable.

CONFLICT OF INTEREST

The authors confirm that they have no conflict of interest to declare for this publication.

ACKNOWLEDGEMENTS

This work was supported by PN-III-P1-1.2-PCCDI-2017-0749, PN-III-P1-1.1-TE-2016-1592 and the Romanian Academy.

REFERENCES

[1] Hu MS, Borrelli MR, Lorenz HP, Longaker MT, Wan DC. Mesenchymal stromal cells and cutaneous wound healing: a comprehensive review of the background, role, and therapeutic potential. Stem Cells Int 2018; 20186901983
[http://dx.doi.org/10.1155/2018/6901983] [PMID: 29887893]

[2] Järbrink K, Ni G, Sönnergren H, *et al.* Prevalence and incidence of chronic wounds and related complications: a protocol for a systematic review. Syst Rev 2016; 5(1): 152.
[http://dx.doi.org/10.1186/s13643-016-0329-y] [PMID: 27609108]

[3] Goel A. Post burn sequelae and their management.Principles and practice of wound care. 1st ed. 468-515.
[http://dx.doi.org/10.5005/jp/books/11164_34]

[4] Tiwari VK. Burn wound: How it differs from other wounds? Indian J Plast Surg 2012; 45(2): 364-73.
[http://dx.doi.org/10.4103/0970-0358.101319] [PMID: 23162236]

[5] DeStefano GM, Christiano AM. The genetics of human skin disease. Cold Spring Harb Perspect Med 2014; 4(10)a015172
[http://dx.doi.org/10.1101/cshperspect.a015172] [PMID: 25274756]

[6] Hubail AR, Belkharoeva RK, Tepluk NP, Grabovskaya OV. A case of a patient with severe epidermolysis bullosa surviving to adulthood. Int J Gen Med 2018; 11: 413-21.
[http://dx.doi.org/10.2147/IJGM.S180464] [PMID: 30532579]

[7] Archer CB. Functions of the skin.Rook's textbook of dermatology. 8th ed. Oxford, UK: Wiley-Blackwell 2010; pp. 4.1-4.11.
[http://dx.doi.org/10.1002/9781444317633.ch4]

[8] Hwa C, Bauer EA, Cohen DE. Skin biology In: Torello Lotti, Ed. Dermatologic Therapy. Wiley Periodicals Press 2011; pp. 464-70.

[9] Kolarsick PAJ, Kolarsick MA, Goodwin C. Anatomy and physiology of the skin. JDNA 2011; 3(4): 203-13.

[10] McGrath JA, Eady RAJ, Pope FM. Anatomy and organization of human skin.Atlas of normal human skin. New York: Springer 1992; pp. pp.3.1-15.

[11] Cichorek M, Wachulska M, Stasiewicz A, Tymińska A. Skin melanocytes: biology and development. Postepy Dermatol Alergol 2013; 30(1): 30-41.

[http://dx.doi.org/10.5114/pdia.2013.33376] [PMID: 24278043]

[12] Richard W, Hunter J, Savin J, Dahl M. Clinical Dermatology. 4th ed. Sussex, UK: Blackwell Publishers 2008; pp. 10-34.

[13] Maeda K. New method of measurement of epidermal turnover in humans. Cosmetics 2017; 4(47): 1-5. [http://dx.doi.org/10.3390/cosmetics4040047]

[14] Johnson JL, Najor NA, Green KJ. Desmosomes: regulators of cellular signaling and adhesion in epidermal health and disease. Cold Spring Harb Perspect Med 2014; 4(11)a015297 [http://dx.doi.org/10.1101/cshperspect.a015297] [PMID: 25368015]

[15] Walko G, Castañón MJ, Wiche G. Molecular architecture and function of the hemidesmosome. Cell Tissue Res 2015; 360(3): 529-44. [http://dx.doi.org/10.1007/s00441-015-2216-6] [PMID: 26017636]

[16] Blanpain C, Fuchs E. Epidermal stem cells of the skin. Annu Rev Cell Dev Biol 2006; 22: 339-73. [http://dx.doi.org/10.1146/annurev.cellbio.22.010305.104357] [PMID: 16824012]

[17] Piérard GE, Goffin V, Hermanns-Le T, Piérard-Franchimont C. Corneocyte desquamation. Int J Mol Med 2000; 6(2): 217-21. [PMID: 10891569]

[18] Oashi M, Sawada Y, Makita R. Odland body and intercellular substances. Acta Derm Venereol Suppl (Stockh) 1973; 73: 47-54. [PMID: 4360960]

[19] Wickett RR, Visscher MO. Structure and function of the epidermal barrier. Am J Infect Control 2006; 34(10): 98-110. [http://dx.doi.org/10.1016/j.ajic.2006.05.295]

[20] Breitkreutz D, Koxholt I, Thiemann K, Nischt R. Skin basement membrane: the foundation of epidermal integrity--BM functions and diverse roles of bridging molecules nidogen and perlecan. BioMed Res Int 2013; 2013179784 [http://dx.doi.org/10.1155/2013/179784] [PMID: 23586018]

[21] Lin MS, Mascaró JM Jr, Liu Z, España A, Diaz LA. The desmosome and hemidesmosome in cutaneous autoimmunity. Clin Exp Immunol 1997; 107(1) (Suppl. 1): 9-15. [PMID: 9020929]

[22] Masunaga T, Shimizu H, Ishiko A, *et al.* Localization of laminin-5 in the epidermal basement membrane. J Histochem Cytochem 1996; 44(11): 1223-30. [http://dx.doi.org/10.1177/44.11.8918896] [PMID: 8918896]

[23] Narisawa Y, Hashimoto K, Kohda H. Immunohistochemical demonstration of keratin 19 expression in isolated human hair follicles. J Invest Dermatol 1994; 103(2): 191-5. [http://dx.doi.org/10.1111/1523-1747.ep12392730] [PMID: 7518856]

[24] Kypriotou M, Huber M, Hohl D. The human epidermal differentiation complex: cornified envelope precursors, S100 proteins and the 'fused genes' family. Exp Dermatol 2012; 21(9): 643-9. [http://dx.doi.org/10.1111/j.1600-0625.2012.01472.x] [PMID: 22507538]

[25] Fuchs E. Skin stem cells: rising to the surface. J Cell Biol 2008; 180(2): 273-84. [http://dx.doi.org/10.1083/jcb.200708185] [PMID: 18209104]

[26] Humbert P, Ed. Rousselle P, Gentilhomme E, Neveux Y, Markers of Epidermal Proliferation and Differentiation in Springer International Publishing Switzerland. Agache's Measuring the Skin 2017.

[27] OpenStax College (USA) Anatomy and Physiology. Rice University 2013; 171-91.

[28] Eckert RL, Broome AM, Ruse M, Robinson N, Ryan D, Lee K. S100 Proteins in the Epidermis J Invest Dermatol 2004; 123(1): 232-33.

[29] Sumena KB, Lucy KM, Chungath JJ, Kuttinarayanan P, Asho N. Regional histology of stratum

lucidum of epidermis in large white Yorkshire pigs. J Vet Anim Sci (Lahore) 2012; 43: 24-6.

[30] Nilforoushzadeh MA, Ashtiani HRA, Jaffary F, Jahangiri F, Nikkhah N, Mahmoudbeyk M, *et al.* Dermal fibroblast cells: Biology and function in skin regeneration. J Skin Stem Cell 2017; 4(2)e69080 [http://dx.doi.org/10.5812/jssc.69080]

[31] Murphrey MB, Zito PM. Histology, Stratum Corneum.StatPearls TreasureIsland. FL: StatPearls 2018.

[32] Tsatmali M, Ancans J, Thody AJ. Melanocyte function and its control by melanocortin peptides. J Histochem Cytochem 2002; 50(2): 125-33. [http://dx.doi.org/10.1177/002215540205000201] [PMID: 11799132]

[33] Szöllősi AG, Oláh A, Bíró T, Tóth BI. Recent advances in the endocrinology of the sebaceous gland. Dermatoendocrinol 2018; 9(1)e1361576 [http://dx.doi.org/10.1080/19381980.2017.1361576] [PMID: 29484098]

[34] Driskell RR, Jahoda CAB, Chuong C-M, Watt FM, Horsley V. Defining dermal adipose tissue. Exp Dermatol 2014; 23(9): 629-31. [http://dx.doi.org/10.1111/exd.12450] [PMID: 24841073]

[35] West HC, Bennett CL. Redefining the role of Langerhans cells as immune regulators within the skin. Front Immunol 2018; 8: 1941. [http://dx.doi.org/10.3389/fimmu.2017.01941] [PMID: 29379502]

[36] Park B, Kim SJ. Cooling the skin: Understanding a specific cutaneous thermosensation. J Lifestyle Med 2013; 3(2): 91-7. [PMID: 26064844]

[37] Romanovsky AA. Skin temperature: its role in thermoregulation. Acta Physiol (Oxf) 2014; 210(3): 498-507. [http://dx.doi.org/10.1111/apha.12231] [PMID: 24716231]

[38] Mostafa WZ, Hegazy RA. Vitamin D and the skin: Focus on a complex relationship: A review. J Adv Res 2015; 6(6): 793-804. [http://dx.doi.org/10.1016/j.jare.2014.01.011] [PMID: 26644915]

[39] Golebiewska EM, Poole AW. Platelet secretion: From haemostasis to wound healing and beyond. Blood Rev 2015; 29(3): 153-62. [http://dx.doi.org/10.1016/j.blre.2014.10.003] [PMID: 25468720]

[40] Guo R, Chai L, Chen L, *et al.* Stromal cell-derived factor 1 (SDF-1) accelerated skin wound healing by promoting the migration and proliferation of epidermal stem cells. *In Vitro* Cell Dev Biol Anim 2015; 51(6): 578-85. [http://dx.doi.org/10.1007/s11626-014-9862-y] [PMID: 25636237]

[41] Wilgus TA, Roy S, McDaniel JC. Neutrophils and wound repair: positive actions and negative reactions. Adv Wound Care (New Rochelle) 2013; 2(7): 379-88. [http://dx.doi.org/10.1089/wound.2012.0383] [PMID: 24527354]

[42] Su Y, Richmond A. Chemokine regulation of neutrophil infiltration of skin wounds. Adv Wound Care (New Rochelle) 2015; 4(11): 631-40. [http://dx.doi.org/10.1089/wound.2014.0559] [PMID: 26543677]

[43] Gaudry M, Brégerie O, Andrieu V, El Benna J, Pocidalo MA, Hakim J. Intracellular pool of vascular endothelial growth factor in human neutrophils. Blood 1997; 90(10): 4153-61. [http://dx.doi.org/10.1182/blood.V90.10.4153] [PMID: 9354686]

[44] Willenborg S, Lucas T, van Loo G, *et al.* CCR2 recruits an inflammatory macrophage subpopulation critical for angiogenesis in tissue repair. Blood 2012; 120(3): 613-25. [http://dx.doi.org/10.1182/blood-2012-01-403386] [PMID: 22577176]

[45] Krzyszczyk P, Schloss R, Palmer A, Berthiaume F. The role of macrophages in acute and chronic wound healing and interventions to promote pro-wound healing phenotypes. Front Physiol 2018; 9:

419.
[http://dx.doi.org/10.3389/fphys.2018.00419] [PMID: 29765329]

[46] Novak ML, Koh TJ. Phenotypic transitions of macrophages orchestrate tissue repair. Am J Pathol 2013; 183(5): 1352-63.
[http://dx.doi.org/10.1016/j.ajpath.2013.06.034] [PMID: 24091222]

[47] Mirza R, Koh TJ. Dysregulation of monocyte/macrophage phenotype in wounds of diabetic mice. Cytokine 2011; 56(2): 256-64.
[http://dx.doi.org/10.1016/j.cyto.2011.06.016] [PMID: 21803601]

[48] Schaerli P, Britschgi M, Keller M, *et al.* Characterization of human T cells that regulate neutrophilic skin inflammation. J Immunol 2004; 173(3): 2151-8.
[http://dx.doi.org/10.4049/jimmunol.173.3.2151] [PMID: 15265952]

[49] Baumgarth N. The double life of a B-1 cell: self-reactivity selects for protective effector functions. Nat Rev Immunol 2011; 11(1): 34-46.
[http://dx.doi.org/10.1038/nri2901] [PMID: 21151033]

[50] Rauch PJ, Chudnovskiy A, Robbins CS, *et al.* Innate response activator B cells protect against microbial sepsis. Science 2012; 335(6068): 597-601.
[http://dx.doi.org/10.1126/science.1215173] [PMID: 22245738]

[51] Li J, Chen J, Kirsner R. Pathophysiology of acute wound healing. Clin Dermatol 2007; 25(1): 9-18.
[http://dx.doi.org/10.1016/j.clindermatol.2006.09.007] [PMID: 17276196]

[52] Freedberg IM, Tomic-Canic M, Komine M, Blumenberg M. Keratins and the keratinocyte activation cycle. J Invest Dermatol 2001; 116(5): 633-40.
[http://dx.doi.org/10.1046/j.1523-1747.2001.01327.x] [PMID: 11348449]

[53] Yan C, Grimm WA, Garner WL, *et al.* Epithelial to mesenchymal transition in human skin wound healing is induced by tumor necrosis factor-alpha through bone morphogenic protein-2. Am J Pathol 2010; 176(5): 2247-58.
[http://dx.doi.org/10.2353/ajpath.2010.090048] [PMID: 20304956]

[54] Haensel D, Dai X. Epithelial-to-mesenchymal transition in cutaneous wound healing: Where we are and where we are heading. Dev Dyn 2018; 247(3): 473-80.
[http://dx.doi.org/10.1002/dvdy.24561] [PMID: 28795450]

[55] Muller M, Trocme C, Lardy B, Morel F, Halimi S, Benhamou PY. Matrix metalloproteinases and diabetic foot ulcers: the ratio of MMP-1 to TIMP-1 is a predictor of wound healing. Diabet Med 2008; 25(4): 419-26.
[http://dx.doi.org/10.1111/j.1464-5491.2008.02414.x] [PMID: 18387077]

[56] Caley MP, Martins VL, O'Toole EA. Metalloproteinases and wound healing. Adv Wound Care (New Rochelle) 2015; 4(4): 225-34.
[http://dx.doi.org/10.1089/wound.2014.0581] [PMID: 25945285]

[57] Grimstad Ø, Sandanger Ø, Ryan L, *et al.* Cellular sources and inducers of cytokines present in acute wound fluid. Wound Repair Regen 2011; 19(3): 337-47.
[http://dx.doi.org/10.1111/j.1524-475X.2011.00668.x] [PMID: 21371163]

[58] Pastar I, Stojadinovic O, Yin NC, *et al.* Epithelialization in wound healing: a comprehensive review. Adv Wound Care (New Rochelle) 2014; 3(7): 445-64.
[http://dx.doi.org/10.1089/wound.2013.0473]

[59] Mascré G, Dekoninck S, Drogat B, *et al.* Distinct contribution of stem and progenitor cells to epidermal maintenance. Nature 2012; 489(7415): 257-62.
[http://dx.doi.org/10.1038/nature11393] [PMID: 22940863]

[60] Hirsch T, Rothoeft T, Teig N, *et al.* Regeneration of the entire human epidermis using transgenic stem cells. Nature 2017; 551(7680): 327-32.
[http://dx.doi.org/10.1038/nature24487] [PMID: 29144448]

[61] Fisher G, Rittié L. Restoration of the basement membrane after wounding: a hallmark of young human skin altered with aging. J Cell Commun Signal 2018; 12(1): 401-11. [http://dx.doi.org/10.1007/s12079-017-0417-3] [PMID: 29086203]

[62] Schmidt BA, Horsley V. Intradermal adipocytes mediate fibroblast recruitment during skin wound healing. Development 2013; 140(7): 1517-27. [http://dx.doi.org/10.1242/dev.087593] [PMID: 23482487]

[63] Gauglitz GG, Korting HC, Pavicic T, Ruzicka T, Jeschke MG. Hypertrophic scarring and keloids: pathomechanisms and current and emerging treatment strategies. Mol Med 2011; 17(1-2): 113-25. [http://dx.doi.org/10.2119/molmed.2009.00153] [PMID: 20927486]

[64] Oliveira GV, Hawkins HK, Chinkes D, *et al.* Hypertrophic *versus* non hypertrophic scars compared by immunohistochemistry and laser confocal microscopy: type I and III collagens. Int Wound J 2009; 6(6): 445-52. [http://dx.doi.org/10.1111/j.1742-481X.2009.00638.x] [PMID: 20051096]

[65] Werner S, Grose R. Regulation of wound healing by growth factors and cytokines. Physiol Rev 2003; 83(3): 835-70. [http://dx.doi.org/10.1152/physrev.2003.83.3.835] [PMID: 12843410]

[66] Eming SA, Brachvogel B, Odorisio T, Koch M. Regulation of angiogenesis: wound healing as a model. Prog Histochem Cytochem 2007; 42(3): 115-70. [http://dx.doi.org/10.1016/j.proghi.2007.06.001] [PMID: 17980716]

[67] Eming SA, Hubbell JA. Extracellular matrix in angiogenesis: dynamic structures with translational potential. Exp Dermatol 2011; 20(7): 605-13. [http://dx.doi.org/10.1111/j.1600-0625.2011.01309.x] [PMID: 21692861]

[68] Tziotzios C, Profyris C, Sterling J. Cutaneous scarring: Pathophysiology, molecular mechanisms, and scar reduction therapeutics Part II. Strategies to reduce scar formation after dermatologic procedures. J Am Acad Dermatol 2012; 66(1): 13-24. [http://dx.doi.org/10.1016/j.jaad.2011.08.035] [PMID: 22177632]

[69] Xue M, Jackson CJ. Extracellular matrix reorganization during wound healing and its impact on abnormal scarring. Adv Wound Care (New Rochelle) 2015; 4(3): 119-36. [http://dx.doi.org/10.1089/wound.2013.0485] [PMID: 25785236]

[70] Van De Water L, Varney S, Tomasek JJ. Mechanoregulation of the myofibroblast in wound contraction, scarring, and fibrosis: opportunities for new therapeutic intervention. Adv Wound Care (New Rochelle) 2013; 2(4): 122-41. [http://dx.doi.org/10.1089/wound.2012.0393] [PMID: 24527336]

[71] Granato G, Ruocco MR, Iaccarino A, *et al.* Generation and analysis of spheroids from human primary skin myofibroblasts: an experimental system to study myofibroblasts deactivation. Cell Death Discov 2017; 3: 17038. [http://dx.doi.org/10.1038/cddiscovery.2017.38] [PMID: 28725488]

[72] Darby IA, Zakuan N, Billet F, Desmoulière A. The myofibroblast, a key cell in normal and pathological tissue repair. Cell Mol Life Sci 2016; 73(6): 1145-57. [http://dx.doi.org/10.1007/s00018-015-2110-0] [PMID: 26681260]

[73] Ojeh N, Pastar I, Tomic-Canic M, Stojadinovic O. Stem cells in skin regeneration, wound healing, and their clinical applications. Int J Mol Sci 2015; 16(10): 25476-501. [http://dx.doi.org/10.3390/ijms161025476] [PMID: 26512657]

[74] O'Brien J, Hayder H, Zayed Y, Peng C. Overview of microRNA biogenesis, mechanisms of actions, and circulation. Front Endocrinol (Lausanne) 2018; 9: 402. [http://dx.doi.org/10.3389/fendo.2018.00402] [PMID: 30123182]

[75] Ha M, Kim VN. Regulation of microRNA biogenesis. Nat Rev Mol Cell Biol 2014; 15(8): 509-24. [http://dx.doi.org/10.1038/nrm3838] [PMID: 25027649]

[76] Yoda M, Kawamata T, Paroo Z, *et al.* ATP-dependent human RISC assembly pathways. Nat Struct Mol Biol 2010; 17(1): 17-23. [http://dx.doi.org/10.1038/nsmb.1733] [PMID: 19966796]

[77] Mulholland EJ, Dunne N, McCarthy HO. MicroRNA as therapeutic targets for chronic wound healing. Mol Ther Nucleic Acids 2017; 8: 46-55. [http://dx.doi.org/10.1016/j.omtn.2017.06.003] [PMID: 28918046]

[78] Soliman AM, Das S, Abd Ghafar N, Teoh SL. Role of microRNA in proliferation phase of wound healing. Front Genet 2018; 9: 38. [http://dx.doi.org/10.3389/fgene.2018.00038] [PMID: 29491883]

[79] Wang T, Feng Y, Sun H, *et al.* miR-21 regulates skin wound healing by targeting multiple aspects of the healing process. Am J Pathol 2012; 181(6): 1911-20. [http://dx.doi.org/10.1016/j.ajpath.2012.08.022] [PMID: 23159215]

[80] Yang X, Wang J, Guo SL, *et al.* miR-21 promotes keratinocyte migration and re-epithelialization during wound healing. Int J Biol Sci 2011; 7(5): 685-90. [http://dx.doi.org/10.7150/ijbs.7.685] [PMID: 21647251]

[81] Wang J, Qiu Y, Shi NW, *et al.* microRNA-21 mediates the TGF-β1-induced migration of keratinocytes *via* targeting PTEN. Eur Rev Med Pharmacol Sci 2016; 20(18): 3748-59. [PMID: 27735045]

[82] Han Z, Chen Y, Zhang Y, *et al.* MiR-21/PTEN Axis Promotes Skin Wound Healing by Dendritic Cells Enhancement. J Cell Biochem 2017; 118(10): 3511-9. [http://dx.doi.org/10.1002/jcb.26026] [PMID: 28374893]

[83] Hu Y, Rao SS, Wang ZX, *et al.* Exosomes from human umbilical cord blood accelerate cutaneous wound healing through miR-21-3p-mediated promotion of angiogenesis and fibroblast function. Theranostics 2018; 8(1): 169-84. [http://dx.doi.org/10.7150/thno.21234] [PMID: 29290800]

[84] Madhyastha R, Madhyastha H, Nakajima Y, Omura S, Maruyama M. MicroRNA signature in diabetic wound healing: promotive role of miR-21 in fibroblast migration. Int Wound J 2012; 9(4): 355-61. [http://dx.doi.org/10.1111/j.1742-481X.2011.00890.x] [PMID: 22067035]

[85] Li D, Li XI, Wang A, *et al.* MicroRNA-31 promotes skin wound healing by enhancing keratinocyte proliferation and migration. J Invest Dermatol 2015; 135(6): 1676-85. [http://dx.doi.org/10.1038/jid.2015.48] [PMID: 25685928]

[86] Shi J, Ma X, Su Y, *et al.* MiR-31 Mediates inflammatory signaling to promote re-epithelialization during skin wound healing. J Invest Dermatol 2018; 138(10): 2253-63. [http://dx.doi.org/10.1016/j.jid.2018.03.1521] [PMID: 29605672]

[87] Peng H, Kaplan N, Hamanaka RB, *et al.* microRNA-31/factor-inhibiting hypoxia-inducible factor 1 nexus regulates keratinocyte differentiation. Proc Natl Acad Sci USA 2012; 109(35): 14030-4. [http://dx.doi.org/10.1073/pnas.1111292109] [PMID: 22891326]

[88] Wang X, Zhang Y, Jiang BH, *et al.* Study on the role of Hsa-miR-31-5p in hypertrophic scar formation and the mechanism. Exp Cell Res 2017; 361(2): 201-9. [http://dx.doi.org/10.1016/j.yexcr.2017.09.009] [PMID: 29056521]

[89] Viticchiè G, Lena AM, Cianfarani F, *et al.* MicroRNA-203 contributes to skin re-epithelialization. Cell Death Dis 2012; 3e435 [http://dx.doi.org/10.1038/cddis.2012.174] [PMID: 23190607]

[90] Xu Y, Ji Y, Lan X, Gao X, Chen HD, Geng L. miR-203 contributes to IL-17-induced VEGF secretion by targeting SOCS3 in keratinocytes. Mol Med Rep 2017; 16(6): 8989-96. [http://dx.doi.org/10.3892/mmr.2017.7759] [PMID: 29039484]

[91] Zhou Z, Shu B, Xu Y, *et al.* microRNA-203 modulates wound healing and scar formation *via*

suppressing Hes1 expression in epidermal stem cells. Cell Physiol Biochem 2018; 49(6): 2333-47. [http://dx.doi.org/10.1159/000493834] [PMID: 30261495]

[92] Pulkkinen K, Malm T, Turunen M, Koistinaho J, Ylä-Herttuala S. Hypoxia induces microRNA miR-210 *in vitro* and *in vivo* ephrin-A3 and neuronal pentraxin 1 are potentially regulated by miR-210. FEBS Lett 2008; 582(16): 2397-401. [http://dx.doi.org/10.1016/j.febslet.2008.05.048] [PMID: 18539147]

[93] Chan YC, Banerjee J, Choi SY, Sen CK. miR-210: the master hypoxamir. Microcirculation 2012; 19(3): 215-23. [http://dx.doi.org/10.1111/j.1549-8719.2011.00154.x] [PMID: 22171547]

[94] Fasanaro P, D'Alessandra Y, Di Stefano V, *et al.* MicroRNA-210 modulates endothelial cell response to hypoxia and inhibits the receptor tyrosine kinase ligand Ephrin-A3. J Biol Chem 2008; 283(23): 15878-83. [http://dx.doi.org/10.1074/jbc.M800731200] [PMID: 18417479]

[95] Biswas S, Roy S, Banerjee J, *et al.* Hypoxia inducible microRNA 210 attenuates keratinocyte proliferation and impairs closure in a murine model of ischemic wounds. Proc Natl Acad Sci USA 2010; 107(15): 6976-81. [http://dx.doi.org/10.1073/pnas.1001653107] [PMID: 20308562]

[96] Wen Z, Huang W, Feng Y, *et al.* MicroRNA-377 regulates mesenchymal stem cell-induced angiogenesis in ischemic hearts by targeting VEGF. PLoS One 2014; 9(9)e104666 [http://dx.doi.org/10.1371/journal.pone.0104666] [PMID: 25251394]

[97] Li B, Xu WW, Han L, *et al.* MicroRNA-377 suppresses initiation and progression of esophageal cancer by inhibiting CD133 and VEGF. Oncogene 2017; 36(28): 3986-4000. [http://dx.doi.org/10.1038/onc.2017.29] [PMID: 28288140]

[98] Rosca AM, Mitroi DN, Cismasiu V, *et al.* Collagen regulates the ability of endothelial progenitor cells to protect hypoxic myocardium through a mechanism involving miR-377/VE-PTP axis. J Cell Mol Med 2018; 22(10): 4700-8. [http://dx.doi.org/10.1111/jcmm.13712] [PMID: 30044046]

[99] Nayerossadat N, Maedeh T, Ali PA. Viral and nonviral delivery systems for gene delivery. Adv Biomed Res 2012; 1: 27. [http://dx.doi.org/10.4103/2277-9175.98152] [PMID: 23210086]

[100] Yang N. An overview of viral and nonviral delivery systems for microRNA. Int J Pharm Investig 2015; 5(4): 179-81. [http://dx.doi.org/10.4103/2230-973X.167646] [PMID: 26682187]

[101] Herrera-Carrillo E, Liu YP, Berkhout B. Improving miRNA delivery by optimizing miRNA expression cassettes in diverse virus vectors. Hum Gene Ther Methods 2017; 28(4): 177-90. [http://dx.doi.org/10.1089/hgtb.2017.036] [PMID: 28712309]

[102] Daya S, Berns KI. Gene therapy using adeno-associated virus vectors. Clin Microbiol Rev 2008; 21(4): 583-93. [http://dx.doi.org/10.1128/CMR.00008-08] [PMID: 18854481]

[103] Song J, Li X, Li Y, *et al.* Biodegradable and biocompatible cationic polymer delivering microRNA-221/222 promotes nerve regeneration after sciatic nerve crush. Int J Nanomedicine 2017; 12: 4195-208. [http://dx.doi.org/10.2147/IJN.S132190] [PMID: 28652727]

[104] Mondal G, Almawash S, Chaudhary AK, Mahato RI. EGFR-targeted cationic polymeric mixed micelles for codelivery of Gemcitabine and miR-205 for treating advanced pancreatic cancer. Mol Pharm 2017; 14(9): 3121-33. [http://dx.doi.org/10.1021/acs.molpharmaceut.7b00355] [PMID: 28719220]

[105] Maroof H, Islam F, Dong L, *et al.* Liposomal delivery of miR-34b-5p induced cancer cell death in

thyroid carcinoma cells 2018; 7(12): pii: E265.

[106] Vanić Ž, Hafner A, Bego M, Škalko-Basnet N. Characterization of various deformable liposomes with metronidazole. Drug Dev Ind Pharm 2013; 39(3): 481-8. [http://dx.doi.org/10.3109/03639045.2012.670247] [PMID: 22455377]

[107] Ascenso A, Raposo S, Batista C, *et al.* Development, characterization, and skin delivery studies of related ultradeformable vesicles: transfersomes, ethosomes, and transethosomes. Int J Nanomedicine 2015; 10: 5837-51. [http://dx.doi.org/10.2147/IJN.S86186] [PMID: 26425085]

[108] Romero EL, Morilla MJ. Highly deformable and highly fluid vesicles as potential drug delivery systems: theoretical and practical considerations. Int J Nanomedicine 2013; 8: 3171-86. [http://dx.doi.org/10.2147/IJN.S33048] [PMID: 23986634]

[109] Hattori Y, Date M, Arai S, Kawano K, Yonemochi E, Maitani Y. Transdermal Delivery of Small Interfering RNA with Elastic Cationic Liposomes in Mice. J Pharm (Cairo) 2013; 2013149695 [http://dx.doi.org/10.1155/2013/149695] [PMID: 26555966]

[110] Bracke S, Carretero M, Guerrero-Aspizua S, *et al.* Targeted silencing of DEFB4 in a bioengineered skin-humanized mouse model for psoriasis: development of siRNA SECosome-based novel therapies. Exp Dermatol 2014; 23(3): 199-201. [http://dx.doi.org/10.1111/exd.12321] [PMID: 24428418]

[111] Jose A, Labala S, Venuganti VV. Co-delivery of curcumin and STAT3 siRNA using deformable cationic liposomes to treat skin cancer. J Drug Target 2017; 25(4): 330-41. [http://dx.doi.org/10.1080/1061186X.2016.1258567] [PMID: 27819148]

[112] Ferguson SW, Wang J, Lee CJ, *et al.* The microRNA regulatory landscape of MSC-derived exosomes: a systems view. Sci Rep 2018; 8(1): 1419. [http://dx.doi.org/10.1038/s41598-018-19581-x] [PMID: 29362496]

[113] Cheng L, Zhang K, Wu S, Cui M, Xu T. Focus on mesenchymal stem cell-derived exosomes: opportunities and challenges in cell-free therapy. Stem Cells Int 2017; 20176305295 [http://dx.doi.org/10.1155/2017/6305295] [PMID: 29410682]

[114] Phan J, Kumar P, Hao D, Gao K, Farmer D, Wang A. Engineering mesenchymal stem cells to improve their exosome efficacy and yield for cell-free therapy. J Extracell Vesicles 2018; 7(1)1522236 [http://dx.doi.org/10.1080/20013078.2018.1522236] [PMID: 30275938]

[115] Ma X, Wang J, Li J, *et al.* Loading miR-210 in endothelial progenitor cells derived exosomes boosts their beneficial effects on hypoxia/reoxygeneation-injured human endothelial cells *via* protecting mitochondrial function. Cell Physiol Biochem 2018; 46(2): 664-75. [http://dx.doi.org/10.1159/000488635] [PMID: 29621777]

[116] Ross K. Towards topical microRNA-directed therapy for epidermal disorders. J Control Release 2018; 269: 136-47. [http://dx.doi.org/10.1016/j.jconrel.2017.11.013] [PMID: 29133119]

[117] Guo S, Dipietro LA. Factors affecting wound healing. J Dent Res 2010; 89(3): 219-29. [http://dx.doi.org/10.1177/0022034509359125] [PMID: 20139336]

[118] Kanji S, Das H. Advances of stem cell therapeutics in cutaneous wound healing and regeneration. Mediators Inflamm 2017; 20175217967 [http://dx.doi.org/10.1155/2017/5217967] [PMID: 29213192]

[119] Schepeler T, Page ME, Jensen KB. Heterogeneity and plasticity of epidermal stem cells. Development 2014; 141(13): 2559-67. [http://dx.doi.org/10.1242/dev.104588] [PMID: 24961797]

[120] Barrandon Y, Green H. Three clonal types of keratinocyte with different capacities for multiplication. Proc Natl Acad Sci USA 1987; 84(8): 2302-6. [http://dx.doi.org/10.1073/pnas.84.8.2302] [PMID: 2436229]

[121] Pellegrini G, Golisano O, Paterna P, *et al.* Location and clonal analysis of stem cells and their differentiated progeny in the human ocular surface. J Cell Biol 1999; 145(4): 769-82. [http://dx.doi.org/10.1083/jcb.145.4.769] [PMID: 10330405]

[122] Beaver CM, Ahmed A, Masters JR. Clonogenicity: holoclones and meroclones contain stem cells. PLoS One 2014; 9(2)e89834 [http://dx.doi.org/10.1371/journal.pone.0089834] [PMID: 24587067]

[123] De Rosa L, Secone Seconetti A, De Santis G, *et al.* Laminin 332-Dependent YAP Dysregulation Depletes Epidermal Stem Cells in Junctional Epidermolysis Bullosa. Cell Rep 2019; 27(7): 2036-2049.e6. [http://dx.doi.org/10.1016/j.celrep.2019.04.055] [PMID: 31091444]

[124] Jiménez F, Garde C, Poblet E, *et al.* A pilot clinical study of hair grafting in chronic leg ulcers. Wound Repair Regen 2012; 20(6): 806-14. [http://dx.doi.org/10.1111/j.1524-475X.2012.00846.x] [PMID: 23110506]

[125] Martínez ML, Escario E, Poblet E, *et al.* Hair follicle-containing punch grafts accelerate chronic ulcer healing: A randomized controlled trial. J Am Acad Dermatol 2016; 75(5): 1007-14. [http://dx.doi.org/10.1016/j.jaad.2016.02.1161] [PMID: 27745629]

[126] Thakur P, Sacchidanand S, Nataraj HV, Savitha AS. A study of hair follicular transplantation as a treatment option for vitiligo. J Cutan Aesthet Surg 2015; 8(4): 211-7. [http://dx.doi.org/10.4103/0974-2077.172192] [PMID: 26865785]

[127] Zhao H, Chen Y, Zhang C, Fu X. Autologous epidermal cell suspension: A promising treatment for chronic wounds. J Tissue Viability 2016; 25(1): 50-6. [http://dx.doi.org/10.1016/j.jtv.2015.11.003] [PMID: 26706649]

[128] Hartmann B, Ekkernkamp A, Johnen C, Gerlach JC, Belfekroun C, Küntscher MV. Sprayed cultured epithelial autografts for deep dermal burns of the face and neck. Ann Plast Surg 2007; 58(1): 70-3. [http://dx.doi.org/10.1097/01.sap.0000250647.39784.bb] [PMID: 17197946]

[129] Auxenfans C, Menet V, Catherine Z, *et al.* Cultured autologous keratinocytes in the treatment of large and deep burns: a retrospective study over 15 years. Burns 2015; 41(1): 71-9. [http://dx.doi.org/10.1016/j.burns.2014.05.019] [PMID: 24996248]

[130] Sood R, Roggy DE, Zieger MJ, Nazim M, Hartman BC, Gibbs JT. A comparative study of spray keratinocytes and autologous meshed split-thickness skin graft in the treatment of acute burn injuries. Wounds 2015; 27(2): 31-40. [PMID: 25785905]

[131] Yun EJ, Yon B, Joo MK, Jeong B. Cell therapy for skin wound using fibroblast encapsulated poly(ethylene glycol)-poly(L-alanine) thermogel. Biomacromolecules 2012; 13(4): 1106-11. [http://dx.doi.org/10.1021/bm2018596] [PMID: 22394182]

[132] Venugopal SS, Yan W, Frew JW, *et al.* A phase II randomized vehicle-controlled trial of intradermal allogeneic fibroblasts for recessive dystrophic epidermolysis bullosa. J Am Acad Dermatol 2013; 69(6): 898-908.e7. [http://dx.doi.org/10.1016/j.jaad.2013.08.014] [PMID: 24075228]

[133] You HJ, Han SK, Rhie JW. Randomised controlled clinical trial for autologous fibroblast-hyaluronic acid complex in treating diabetic foot ulcers. J Wound Care 2014; 23(11): 521-522, 524, 526-530. [http://dx.doi.org/10.12968/jowc.2014.23.11.521] [PMID: 25375400]

[134] Frykberg RG, Marston WA, Cardinal M. The incidence of lower-extremity amputation and bone resection in diabetic foot ulcer patients treated with a human fibroblast-derived dermal substitute. Adv Skin Wound Care 2015; 28(1): 17-20. [http://dx.doi.org/10.1097/01.ASW.0000456630.12766.e9] [PMID: 25407083]

[135] Friedenstein AJ, Chailakhyan RK, Latsinik NV, Panasyuk AF, Keiliss-Borok IV. Stromal cells responsible for transferring the microenvironment of the hemopoietic tissues. Cloning *in vitro* and

retransplantation in vivo. Transplantation 1974; 17(4): 331-40.
[http://dx.doi.org/10.1097/00007890-197404000-00001] [PMID: 4150881]

[136] Friedenstein AJ, Gorskaja JF, Kulagina NN. Fibroblast precursors in normal and irradiated mouse hematopoietic organs. Exp Hematol 1976; 4(5): 267-74.
[PMID: 976387]

[137] Dominici M, Le Blanc K, Mueller I, *et al.* Minimal criteria for defining multipotent mesenchymal stromal cells. The International Society for Cellular Therapy position statement. Cytotherapy 2006; 8(4): 315-7.
[http://dx.doi.org/10.1080/14653240600855905] [PMID: 16923606]

[138] Marfia G, Navone SE, Di Vito C, *et al.* Mesenchymal stem cells: potential for therapy and treatment of chronic non-healing skin wounds. Organogenesis 2015; 11(4): 183-206.
[http://dx.doi.org/10.1080/15476278.2015.1126018] [PMID: 26652928]

[139] Falanga V, Iwamoto S, Chartier M, *et al.* Autologous bone marrow-derived cultured mesenchymal stem cells delivered in a fibrin spray accelerate healing in murine and human cutaneous wounds. Tissue Eng 2007; 13(6): 1299-312.
[http://dx.doi.org/10.1089/ten.2006.0278] [PMID: 17518741]

[140] Yoshikawa T, Mitsuno H, Nonaka I, *et al.* Wound therapy by marrow mesenchymal cell transplantation. Plast Reconstr Surg 2008; 121(3): 860-77.
[http://dx.doi.org/10.1097/01.prs.0000299922.96006.24] [PMID: 18317135]

[141] Dash NR, Dash SN, Routray P, Mohapatra S, Mohapatra PC. Targeting nonhealing ulcers of lower extremity in human through autologous bone marrow-derived mesenchymal stem cells. Rejuvenation Res 2009; 12(5): 359-66.
[http://dx.doi.org/10.1089/rej.2009.0872] [PMID: 19929258]

[142] Bey E, Prat M, Duhamel P, *et al.* Emerging therapy for improving wound repair of severe radiation burns using local bone marrow-derived stem cell administrations. Wound Repair Regen 2010; 18(1): 50-8.
[http://dx.doi.org/10.1111/j.1524-475X.2009.00562.x] [PMID: 20082681]

[143] Zhou H, Guo M, Bian C, *et al.* Efficacy of bone marrow-derived mesenchymal stem cells in the treatment of sclerodermatous chronic graft-versus-host disease: clinical report. Biol Blood Marrow Transplant 2010; 16(3): 403-12.
[http://dx.doi.org/10.1016/j.bbmt.2009.11.006] [PMID: 19925878]

[144] Lu D, Chen B, Liang Z, *et al.* Res. Clin Pract 2011; 92(1): 26-36.

[145] Ravari H, Hamidi-Almadari D, Salimifar M, Bonakdaran S, Parizadeh MR, Koliakos G. Treatment of non-healing wounds with autologous bone marrow cells, platelets, fibrin glue and collagen matrix. Cytotherapy 2011; 13(6): 705-11.
[http://dx.doi.org/10.3109/14653249.2011.553594] [PMID: 21284564]

[146] Gupta PK, Chullikana A, Parakh R, *et al.* A double blind randomized placebo controlled phase I/II study assessing the safety and efficacy of allogeneic bone marrow derived mesenchymal stem cell in critical limb ischemia. J Transl Med 2013; 11: 143.
[http://dx.doi.org/10.1186/1479-5876-11-143] [PMID: 23758736]

[147] Das AK, Bin Abdullah BJ, Dhillon SS, Vijanari A, Anoop CH, Gupta PK. Intra-arterial allogeneic mesenchymal stem cells for critical limb ischemia are safe and efficacious: report of a phase I study. World J Surg 2013; 37(4): 915-22.
[http://dx.doi.org/10.1007/s00268-012-1892-6] [PMID: 23307180]

[148] Lee HC, An SG, Lee HW, *et al.* Safety and effect of adipose tissue-derived stem cell implantation in patients with critical limb ischemia: a pilot study. Circ J 2012; 76(7): 1750-60.
[http://dx.doi.org/10.1253/circj.CJ-11-1135] [PMID: 22498564]

[149] de la Portilla F, Alba F, García-Olmo D, Herrerías JM, González FX, Galindo A. Expanded allogeneic

adipose-derived stem cells (eASCs) for the treatment of complex perianal fistula in Crohn's disease: results from a multicenter phase I/IIa clinical trial. Int J Colorectal Dis 2013; 28(3): 313-23. [http://dx.doi.org/10.1007/s00384-012-1581-9] [PMID: 23053677]

[150] Păunescu V, Deak E, Herman D, *et al.* *In vitro* differentiation of human mesenchymal stem cells to epithelial lineage. J Cell Mol Med 2007; 11(3): 502-8. [http://dx.doi.org/10.1111/j.1582-4934.2007.00041.x] [PMID: 17635641]

[151] Sasaki M, Abe R, Fujita Y, Ando S, Inokuma D, Shimizu H. Mesenchymal stem cells are recruited into wounded skin and contribute to wound repair by transdifferentiation into multiple skin cell type. J Immunol 2008; 180(4): 2581-7. [http://dx.doi.org/10.4049/jimmunol.180.4.2581] [PMID: 18250469]

[152] Wu Y, Zhao RC, Tredget EE. Concise review: bone marrow-derived stem/progenitor cells in cutaneous repair and regeneration. Stem Cells 2010; 28(5): 905-15. [PMID: 20474078]

[153] Chen D, Hao H, Fu X, Han W. Insight into Reepithelialization: How Do Mesenchymal Stem Cells Perform? Stem Cells Int 2016; 20166120173 [http://dx.doi.org/10.1155/2016/6120173] [PMID: 26770209]

[154] Kamolz LP, Keck M, Kasper C. Wharton's jelly mesenchymal stem cells promote wound healing and tissue regeneration. Stem Cell Res Ther 2014; 5(3): 62. [http://dx.doi.org/10.1186/scrt451] [PMID: 25157597]

[155] Hofer HR, Tuan RS. Secreted trophic factors of mesenchymal stem cells support neurovascular and musculoskeletal therapies. Stem Cell Res Ther 2016; 7(1): 131. [http://dx.doi.org/10.1186/s13287-016-0394-0] [PMID: 27612948]

[156] Lin L, Du L. The role of secreted factors in stem cells-mediated immune regulation. Cell Immunol 2018; 326: 24-32. [http://dx.doi.org/10.1016/j.cellimm.2017.07.010] [PMID: 28778535]

[157] McBride JD, Rodriguez-Menocal L, Guzman W, Candanedo A, Garcia-Contreras M, Badiavas EV. Bone marrow mesenchymal stem cell-derived CD63+ exosomes transport Wnt3a exteriorly and enhance dermal fibroblast proliferation, migration, and angiogenesis in vitro. Stem Cells Dev 2017; 26(19): 1384-98. [http://dx.doi.org/10.1089/scd.2017.0087] [PMID: 28679315]

[158] Wang L, Hu L, Zhou X, *et al.* Exosomes secreted by human adipose mesenchymal stem cells promote scarless cutaneous repair by regulating extracellular matrix remodelling. Sci Rep 2017; 7(1): 13321. [http://dx.doi.org/10.1038/s41598-017-12919-x] [PMID: 29042658]

[159] Choi EW, Seo MK, Woo EY, Kim SH, Park EJ, Kim S. Exosomes from human adipose-derived stem cells promote proliferation and migration of skin fibroblasts. Exp Dermatol 2018; 27(10): 1170-2. [http://dx.doi.org/10.1111/exd.13451] [PMID: 28940813]

[160] Ng AP, Alexander WS. Haematopoietic stem cells: past, present and future. Cell Death Discov 2017; 3: 17002. [http://dx.doi.org/10.1038/cddiscovery.2017.2] [PMID: 28180000]

[161] Kirby GT, Mills SJ, Cowin AJ, Smith LE. Stem Cells for Cutaneous Wound Healing. BioMed Res Int 2015; 2015285869 [http://dx.doi.org/10.1155/2015/285869] [PMID: 26137471]

[162] Whiteley J, Chow T, Adissu H, Keating A, Rogers IM. Topical application of culture-expanded CD34+ umbilical cord blood cells from frozen units accelerates healing of diabetic skin wounds in mice. Stem Cells Transl Med 2018; 7(8): 591-601. [http://dx.doi.org/10.1002/sctm.17-0302] [PMID: 29752867]

[163] Cohen KS, Cheng S, Larson MG, *et al.* Circulating CD34(+) progenitor cell frequency is associated with clinical and genetic factors. Blood 2013; 121(8): e50-6.

[http://dx.doi.org/10.1182/blood-2012-05-424846] [PMID: 23287867]

[164] Takahashi K, Yamanaka S. Induction of pluripotent stem cells from mouse embryonic and adult fibroblast cultures by defined factors. Cell 2006; 126(4): 663-76. [http://dx.doi.org/10.1016/j.cell.2006.07.024] [PMID: 16904174]

[165] Guo XL, Chen JS. Research on induced pluripotent stem cells and the application in ocular tissues. Int J Ophthalmol 2015; 8(4): 818-25. [PMID: 26309885]

[166] Singh VK, Kalsan M, Kumar N, Saini A, Chandra R. Induced pluripotent stem cells: applications in regenerative medicine, disease modeling, and drug discovery. Front Cell Dev Biol 2015; 3: 2. [http://dx.doi.org/10.3389/fcell.2015.00002] [PMID: 25699255]

[167] Itoh M, Kiuru M, Cairo MS, Christiano AM. Generation of keratinocytes from normal and recessive dystrophic epidermolysis bullosa-induced pluripotent stem cells. Proc Natl Acad Sci USA 2011; 108(21): 8797-802. [http://dx.doi.org/10.1073/pnas.1100332108] [PMID: 21555586]

[168] Dinella J, Koster MI, Koch PJ. Use of induced pluripotent stem cells in dermatological research. J Invest Dermatol 2014; 134(8): 1-5. [http://dx.doi.org/10.1038/jid.2014.238] [PMID: 25029325]

[169] Kogut I, Roop DR, Bilousova G. Differentiation of human induced pluripotent stem cells into a keratinocyte lineage. Methods Mol Biol 2014; 1195: 1-12. [http://dx.doi.org/10.1007/7651_2014_79] [PMID: 24510784]

[170] Umegaki-Arao N, Pasmooij AM, Itoh M, *et al.* Induced pluripotent stem cells from human revertant keratinocytes for the treatment of epidermolysis bullosa. Sci Transl Med 2014; 6(264)264ra164 [http://dx.doi.org/10.1126/scitranslmed.3009342] [PMID: 25429057]

[171] Domingues S, Masson Y, Marteyn A, *et al.* Differentiation of nonhuman primate pluripotent stem cells into functional keratinocytes. Stem Cell Res Ther 2017; 8(1): 285. [http://dx.doi.org/10.1186/s13287-017-0741-9] [PMID: 29258610]

[172] Lei M, Schumacher LJ, Lai YC, *et al.* Self-organization process in newborn skin organoid formation inspires strategy to restore hair regeneration of adult cells. Proc Natl Acad Sci USA 2017; 114(34): E7101-10. [http://dx.doi.org/10.1073/pnas.1700475114] [PMID: 28798065]

[173] Lee J, Böscke R, Tang PC, Hartman BH, Heller S, Koehler KR. Hair Follicle Development in Mouse Pluripotent Stem Cell-Derived Skin Organoids. Cell Rep 2018; 22(1): 242-54. [http://dx.doi.org/10.1016/j.celrep.2017.12.007] [PMID: 29298425]

[174] Zhou H, Wang L, Zhang C, *et al.* Feasibility of repairing full-thickness skin defects by iPSC-derived epithelial stem cells seeded on a human acellular amniotic membrane. Stem Cell Res Ther 2019; 10(1): 155. [http://dx.doi.org/10.1186/s13287-019-1234-9] [PMID: 31151466]

[175] Sen CK, Gordillo GM, Roy S, *et al.* Human skin wounds: a major and snowballing threat to public health and the economy. Wound Repair Regen 2009; 17(6): 763-71. [http://dx.doi.org/10.1111/j.1524-475X.2009.00543.x] [PMID: 19903300]

[176] Han G, Ceilley R. Chronic Wound Healing: A Review of Current Management and Treatments. Adv Ther 2017; 34(3): 599-610. [http://dx.doi.org/10.1007/s12325-017-0478-y] [PMID: 28108895]

[177] Gould LJ. Topical Collagen-Based Biomaterials for Chronic Wounds: Rationale and Clinical Application. Adv Wound Care (New Rochelle) 2016; 5(1): 19-31. [http://dx.doi.org/10.1089/wound.2014.0595] [PMID: 26858912]

[178] Haddad AG, Giatsidis G, Orgill DP, Halvorson EG. Skin Substitutes and Bioscaffolds: Temporary and Permanent Coverage. Clin Plast Surg 2017; 44(3): 627-34.

[http://dx.doi.org/10.1016/j.cps.2017.02.019] [PMID: 28576252]

[179] Dickinson LE, Gerecht S. Engineered Biopolymeric Scaffolds for Chronic Wound Healing. Front Physiol 2016; 7: 341.
[http://dx.doi.org/10.3389/fphys.2016.00341] [PMID: 27547189]

[180] Curran MP, Plosker GL. Bilayered bioengineered skin substitute (Apligraf): a review of its use in the treatment of venous leg ulcers and diabetic foot ulcers. BioDrugs 2002; 16(6): 439-55.
[http://dx.doi.org/10.2165/00063030-200216060-00005] [PMID: 12463767]

[181] Stone RC, Stojadinovic O, Rosa AM, *et al.* A bioengineered living cell construct activates an acute wound healing response in venous leg ulcers. Sci Transl Med 2017; 9(371)eaaf8611
[http://dx.doi.org/10.1126/scitranslmed.aaf8611] [PMID: 28053158]

[182] Towler MA, Rush EW, Richardson MK, Williams CL. Randomized, Prospective, Blinded-Enrollment, Head-To-Head Venous Leg Ulcer Healing Trial Comparing Living, Bioengineered Skin Graft Substitute (Apligraf) with Living, Cryopreserved, Human Skin Allograft (TheraSkin). Clin Podiatr Med Surg 2018; 35(3): 357-65.
[http://dx.doi.org/10.1016/j.cpm.2018.02.006] [PMID: 29861018]

[183] Harding K, Sumner M, Cardinal M. A prospective, multicentre, randomised controlled study of human fibroblast-derived dermal substitute (Dermagraft) in patients with venous leg ulcers. Int Wound J 2013; 10(2): 132-7.
[http://dx.doi.org/10.1111/iwj.12053] [PMID: 23506344]

[184] clinicaltrials.gov/ct2/show/NC2018.accessed on 12.12.2018

[185] De Angelis B, Orlandi F, Fernandes Lopes Morais D'Autilio M, *et al.* Long-term follow-up comparison of two different bi-layer dermal substitutes in tissue regeneration: Clinical outcomes and histological findings. Int Wound J 2018; 15(5): 695-706.
[http://dx.doi.org/10.1111/iwj.12912] [PMID: 29590523]

[186] Swan J. Use of Cryopreserved, Particulate Human Amniotic Membrane and Umbilical Cord (AM/UC) Tissue: A Case Series Study for Application in the Healing of Chronic Wounds Surg Technol Int 2014; 25: 73-8.

[187] Ilic D, Vicovac L, Nikolic M, Lazic Ilic E. Human amniotic membrane grafts in therapy of chronic non-healing wounds. Br Med Bull 2016; 117(1): 59-67.
[http://dx.doi.org/10.1093/bmb/ldv053] [PMID: 26764358]

[188] Cole WE. DermACELL: Human Acellular Dermal Matrix Allograft A Case Report. J Am Podiatr Med Assoc 2016; 106(2): 133-7.
[http://dx.doi.org/10.7547/14-091] [PMID: 27031550]

[189] Zenn MR, Salzberg CA. A Direct Comparison of Alloderm-Ready to Use (RTU) and DermACELL in Immediate Breast Implant Reconstruction. Eplasty 2016; 16e23
[PMID: 27602176]

[190] Sheikholeslam M, Wright MEE, Jeschke MG, Amini-Nik S. Biomaterials for Skin Substitutes. Adv Healthc Mater 2018; 7(5)
[http://dx.doi.org/10.1002/adhm.201700897] [PMID: 29271580]

[191] Ma K, Laco F, Ramakrishna S, Liao S, Chan CK. Differentiation of bone marrow-derived mesenchymal stem cells into multi-layered epidermis-like cells in 3D organotypic coculture. Biomaterials 2009; 30(19): 3251-8.
[http://dx.doi.org/10.1016/j.biomaterials.2009.02.025] [PMID: 19285341]

[192] Samadikuchaksaraei A, Mehdipour A, Habibi Roudkenar M, *et al.* A Dermal Equivalent Engineered with TGF-β3 Expressing Bone Marrow Stromal Cells and Amniotic Membrane: Cosmetic Healing of Full-Thickness Skin Wounds in Rats. Artif Organs 2016; 40(12): E266-79.
[http://dx.doi.org/10.1111/aor.12807] [PMID: 27911032]

[193] Kolakshyapati P, Li X, Chen C, *et al.* Gene-activated matrix/bone marrow-derived mesenchymal stem

cells constructs regenerate sweat glands-like structure *in vivo*. Sci Rep 2017; 7(1): 17630. [http://dx.doi.org/10.1038/s41598-017-17967-x] [PMID: 29247230]

[194] Qi Y, Dong Z, Chu H, Zhao Q, *et al.* Denatured acellular dermal matrix seeded with bone marrow mesenchymal stem cells for wound healing in mice. Burns 2019. pii: S0305-4179(18)31116-1 [http://dx.doi.org/10.1016/j.burns.2019.04.017]

[195] Bayati V, Abbaspour MR, Dehbashi FN, Neisi N, Hashemitabar M. A dermal equivalent developed from adipose-derived stem cells and electrospun polycaprolactone matrix: an *in vitro* and *in vivo* study. Anat Sci Int 2017; 92(4): 509-20. [http://dx.doi.org/10.1007/s12565-016-0352-z] [PMID: 27329656]

[196] Leirós GJ, Kusinsky AG, Drago H, *et al.* Dermal papilla cells improve the wound healing process and generate hair bud-like structures in grafted skin substitutes using hair follicle stem cells. Stem Cells Transl Med 2014; 3(10): 1209-19. [http://dx.doi.org/10.5966/sctm.2013-0217] [PMID: 25161315]

[197] Ertl J, Pichlsberger M, Tuca AC, *et al.* Comparative study of regenerative effects of mesenchymal stem cells derived from placental amnion, chorion and umbilical cord on dermal wounds. Placenta 2018; 65: 37-46. [http://dx.doi.org/10.1016/j.placenta.2018.04.004] [PMID: 29908640]

[198] Yan WC, Davoodi P, Vijayavenkataraman S, *et al.* 3D bioprinting of skin tissue: From pre-processing to final product evaluation. Adv Drug Deliv Rev 2018; 132: 270-95. [http://dx.doi.org/10.1016/j.addr.2018.07.016] [PMID: 30055210]

[199] Augustine R. Skin bioprinting: a novel approach for creating artificial skin from synthetic and natural building blocks. Prog Biomater 2018; 7(2): 77-92. [http://dx.doi.org/10.1007/s40204-018-0087-0] [PMID: 29754201]

[200] Skardal A, Mack D, Kapetanovic E, *et al.* Bioprinted amniotic fluid-derived stem cells accelerate healing of large skin wounds. Stem Cells Transl Med 2012; 1(11): 792-802. [http://dx.doi.org/10.5966/sctm.2012-0088] [PMID: 23197691]

[201] Huang S, Yao B, Xie J, Fu X. 3D bioprinted extracellular matrix mimics facilitate directed differentiation of epithelial progenitors for sweat gland regeneration. Acta Biomater 2016; 32: 170-7. [http://dx.doi.org/10.1016/j.actbio.2015.12.039] [PMID: 26747979]

[202] Cubo N, Garcia M, Del Cañizo JF, Velasco D, Jorcano JL. 3D bioprinting of functional human skin: production and *in vivo* analysis. Biofabrication 2016; 9(1)015006 [http://dx.doi.org/10.1088/1758-5090/9/1/015006] [PMID: 27917823]

[203] Huang R, Wang J, Chen H, *et al.* The topography of fibrous scaffolds modulates the paracrine function of Ad-MSCs in the regeneration of skin tissues. Biomater Sci 2019; 7(10): 4248-59. [http://dx.doi.org/10.1039/C9BM00939F] [PMID: 31393466]

[204] Albanna M, Binder KW, Murphy SV, *et al.* *In Situ* Bioprinting of Autologous Skin Cells Accelerates Wound Healing of Extensive Excisional Full-Thickness Wounds. Sci Rep 2019; 9(1): 1856.

CHAPTER 5

MiRNA Mediated Stem Cell Therapy for Cardiac Arrhythmia

Priyadarshan Kathirvelu, Bhuvaneshwari Sampath and **Kavitha Sankaranarayanan***

Ion Channel Biology Laboratory, AU-KBC Research Center, Madras Institute of Technology campus of Anna University, Chennai, India

Abstract: Arrhythmias are electrical disturbances resulting in irregular heartbeats. Multiple ion channels orchestrate the coordinated propagation of electric stimuli within the heart. Dysfunction of the ion channels, which may result from genetic alterations in ion channel genes or their aberrant expression, can render electrical disturbances predisposing to cardiac arrhythmias. MicroRNA (miRNA) is a type of small RNA belonging to non-protein coding RNAs and is involved in RNA interference, thereby functioning as a regulator of gene expression. Recently, the role of miRNA in cardiac conduction diseases, arrhythmogenesis and their therapeutic potential has been implicated. Stem cell therapies for cardiac arrhythmia have attempted to modulate defective ionic currents by delivering engineered cells. The modulation or engineering of stem cells with appropriate miRNAs could improvise stem cell-based therapies for arrhythmia. This chapter would review the role of miRNAs in cardiac rhythmic disorders and its potential in diagnosis and treatment of cardiac arrhythmia.

Keywords: Arrhythmia, Channelopathy, Conduction Disorder, Epigenetic Regulation, Ion Channels, MicroRNA, Regenerative Medicine, Stem Cells, Stem Cell Therapy, SCD, Target Prediction.

INTRODUCTION

Cardiac arrhythmia is a worldwide serious problem responsible for high rate of cardiovascular mortality and morbidity. An arrhythmia is a condition where the heart beat becomes irregular. Cardiac arrhythmias are significantly associated with increased risks of cardiovascular complications and sudden death [1]. Arrhythmias affect people worldwide, atrial fibrillation alone is estimated to affect 33.5 million people [2]. Rhythmic disturbances of cardiac function are underlying symptoms of a wide range of diseases, disorders or side effects of

* **Corresponding author Dr. Kavitha Sankaranarayanan**: Ion Channel Biology Laboratory, AU-KBC Research Centre, MIT Campus of Anna University, Chromepet, Chennai 44, Tamil Nadu, India; Tel: + 91-044-22232711; E-mail: skavitham@yahoo.com

Atta-ur-Rahman and Shazia Anjum (Eds.)

drugs, which could result in either tachycardia or bradycardia. Arrhythmia are classified as atrial tachycardia, sinus bradycardia or ventricular arrhythmia depending on the origin of the rhythm disturbance. It also may be further classified on the basis of the rate and mechanism (triggered, ectopic, reentry).

Arrhythmias could manifest themselves as ones with mild to serious effects with some being fatal enough to result in sudden cardiac death. Ventricular arrhythmias are the most important cause of sudden cardiac death (SCD). Occurrence of SCD is estimated to be 4 to 5 million cases per year worldwide [3, 4], thereby highlighting arrhythmias and SCD as the most serious cardiovascular diseases worldwide. Arrhythmias do not manifest at any specific age and could be acquired or inherited. In addition to genetics, the occurrence of arrhythmia is also influenced by other factors like smoking, caffeine, drugs, ischemia, infarction, *etc.* Acquired arrhythmias are a manifestation of diverse abnormalities (electrical, structural, metabolic, neurohormonal, or molecular alterations) in multiple pathological conditions (heart failure, hypertension, diabetes mellitus, hyperthyroidism, aging, *etc.*) [5].

The rhythm of the heart beat is controlled by the electrical conduction system (CS) of the heart. The conduction system is responsible for the maintenance of normal blood pressure and ensures proper blood circulation. The electrical activity of the heart is mediated by ion channels and transporters. Ion channels, which are integral membrane protein complexes, facilitate a selective passage of ions across the cell membrane through various gating mechanisms including voltage, pH, ligands *etc.*, thereby being pivotal in the generation of electrical impulses in excitable cells. The coordinated activity of these ion channels is instrumental in the generation and propagation of the cardiac action potential as well as normal ECG pattern Fig. (**1**). The unique signature pattern of ion channels across different regions of the heart is responsible for the unique action potentials generated by them. The normal cardiac rhythm is regulated by the precise action potential in each type of cell in the various compartments of the heart. The electrocardiogram (ECG) is a result of the cumulative effect of the cellular action potential generated by the meticulously timed action of different ion channels. Dysregulation or functional impairment of ion channels, transporters, intracellular Ca^{2+}-handling proteins *etc.* can result in the generation of structural and metabolic substrates, which have the potential to culminate in electrical disturbances predisposing the individual to conduction abnormalities and arrhythmias [6].

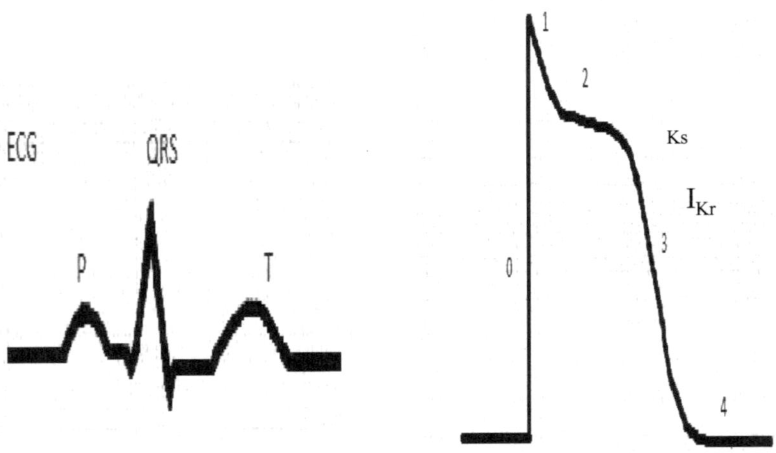

Fig. (1). The normal ECG pattern and the ionic currents involved in myocardial action potential.

CARDIAC CHANNELOPATHIES AND ARRHYTHMIA

Electrocardiographic abnormalities are the result of underlying changes in the cellular action potentials which arise due to aberrant cell-to-cell coupling, diseased conditions of the heart, congenital ion channel abnormalities, drug intervention, or electrolyte imbalance [7 - 9]. Primary electrical diseases of the heart refer to rare inherited cardiac arrhythmias in the absence of structural abnormalities of the heart that are associated with mutations in ion channel genes. Wilde and Bezzina (2005) have summarized the ion channel genes that are associated with these inherited arrhythmia conditions [10]. These findings were based on studies of various mutants and highlighting the important contribution of ion channels in maintenance of normal cardiac rhythm. The term cardiac channelopathies is used to refer to ion channel disorders resulting in anomalies of cardiac function, including but not limited to arrhythmias. Molecular studies revealed mutations in genes coding for specific ion channels being responsible for heritable arrhythmogenic disorders such as LQTS, SQTS, Brugada syndrome and idiopathic atrial fibrillation. The findings of genetic basis of the inherited arrhythmia syndromes promoted better understanding of the molecular genetics of cardiac arrhythmias and the role of different ion channels. The development of molecular diagnostic tests has also increased the chances of an early detection of the subjects at high risk of developing arrhythmias.

LQTS, an inherited disorder characterized by QT prolongation is due to altered ion channel gene expression resulting in increased depolarization and reduced repolarization ultimately prolonging the action potential. Mutations in potassium channels (KCNE1, KCNE2, KCNQ1, KCNH2), sodium channel (SCN5A) have been reported to be associated with LQTS. The pattern of inheritance could be autosomal recessive or dominant, where only 1 copy of the mutated gene is necessary to produce the clinical syndrome in the latter form. Mutation in ANK2 coding for Ankyrin B, a non-channel scaffolding protein associated with the function of many ion channels and transporters have been reported in LQTS4.

Brugada syndrome is also an inherited arrhythmia diagnosed by ST abnormalities in the ECG. This syndrome is linked with mutations in the sodium channel alpha subunit coded by SCN5A. Different mutations of SCN5A has been reported in this syndrome. CPVT is another heritable disorder recognised by arrhythmias with delayed after-depolarization. Ryanodine receptor (RYR2) and Calsequestrin (CSQ) are involved in calcium homeostasis of cardiomyocytes. Inherited abnormalities in the ryanodine receptor or associated proteins have been shown to result in a leaky sarcoplasmic reticulum and cytosolic calcium overload, thus providing a link between abnormal calcium handling and triggered arrhythmias [11]. This kind of disorder also serves to understand the pathophysiology of acquired arrhythmias.

Unlike inherited arrhythmias, many forms of arrhythmia are induced as a result of structural modification of hearts as an outcome of inflammation, changes in myocardium structure *etc.* Also these acquired forms are found to be heterogenic. Structural and electrical remodelling in response to myocardial injury, altered hemodynamic loads, and changes in neuro-hormonal signalling can lead to alterations in ion channel function, intracellular calcium handling, intercellular communication, and the composition of the intercellular matrix creates a substrate for both atrial and ventricular arrhythmias [12]. Reduced functional expression of L-type calcium channels [13] is thought to underlie the shortening of action potential duration associated with tachycardia remodelling of the atria. Dobrev *et al.* (2001) [14] has reported down regulation of transient outward current (*I*to) and alterations in the inward rectifier current (*I*K1) and acetylcholine-sensitive potassium channels (*I*KACh) in chronic human atrial fibrillation.

STEM CELL THERAPY FOR CARDIAC ARRHYTHMIAS

Channelopathies, or dysfunction of ion channels, which may result from genetic alterations in ion channel genes or aberrant expression of these genes, can render electrical disturbances predisposing to cardiac arrhythmias [15]. As discussed previously cardiac electrical activity depends on the coordinated propagation of excitatory stimuli through the heart and, as a consequence, the generation of action potentials in individual cardiomyocytes [16]. Implantable electronic devices [17] can serve as surrogate pacemakers to sustain heart rate, or as defibrillators to treat excessively rapid rhythms. Such devices are expensive, and implantation involves a number of acute and chronic risks (pulmonary collapse, bacterial infection, lead or generator failure) [18]. These drawbacks associated with electronic pacemakers necessitate the need for a biological alternative. The biological therapies seek to achieve functional re-engineering of cardiac tissue, so as to alter a specific electrical property of the tissue [15]. Biological pacemaking can be achieved by modulating ionic currents by gene transfer or by delivering engineered pacemaker cells into normally quiescent myocardium [15]. The term "biological pacemaker" refers to a biological substrate that can induce automaticity specifically in the heart, with a hope to replace electronic pacemakers. This approach could include modification of ionic currents by suppression or activation of ion channel expression/function, stem cell therapy, genetically engineered novel synthetic pacemaker channels, and adult somatic cell-fusion approach [19]. The first biological pacemaker generated *in-vitro* was by the overexpression of β2-adrenergic receptors in porcine atrium [20]. This will promote increased binding of catecholamine to β2-adrenergic receptors thereby increasing cAMP synthesis and pacemaking rate. Similarly, the first attempt to manipulate ion channel activity was performed by injecting a dominant negative construct of the repolarizing K current IK1 into guinea pig ventricle [21]. But this resulted in prolonged repolarization setting a stage for arrhythmogenesis. The other approach is to increase or to create a variant on the inward current If. HCN2 and HCN4 isoforms are largely explored in generating biological pacemakers. HCN2 in an adenoviral construct was found to express well in atrium [22], ventricle [23] and also generated pacemaker function which was sensitive to both catechol-amines and vagal stimulation [22]. Viral vectors carrying the gene of interest has been used in the gene therapy strategy. Since these vectors possess the risk of transmitting viral based infections, other strategy namely cell therapy was developed.

Cell therapy, a sub-type of regenerative medicine based on stem cells involves the process of introducing stem cells into tissue either directly or with gene therapy for required channel. Stem cells possess the capability to differentiate into various cell types under certain physiological or experimental conditions. Cell-based therapy for cardiac regeneration has been evaluated in the clinic in 3 distinct clinical scenarios: (1) recent acute MI, (2) chronic myocardial ischemia in no-option revascularization patients, and (3) chronic infarct-related heart failure [24].

ES cells can give rise to almost all cell lineages and are the most promising cells for regenerative medicine. ESCs have been phenotyped in terms of ionic current makeup, intracellular Ca2 handling, and connexin expression [25]. Embryonic Stem Cells (ESCs) have been shown to have at least fast sodium current, L-type Ca2 current, If, IK [25 - 27] and immature excitation-contraction (EC) coupling [28]. The implanted ESCs would form gap junctions not only with fellow ESCs but also with surviving myocytes in the damaged region. ESCs have shown slow upstroke action potentials and triggered activity [29 - 32]. This inherent pacemaking activity is thought to be due to high input resistance and high sodium current density [27].

Human ESCs injected into the ventricles of pigs in complete heart block, delivered acceptable pacemaker function for several months [29] and the cells were coupled with the porcine myocytes such that the signal transfer from hESCs to the heart was robust [33]. Use of ESCs are however limited by the facts that they are allogeneic, hence requiring immuno-suppressive therapy, also its potential to form teratomas which can be handled however if targeted differentiation only into cardiomyocytes is achieved. The use of hESC will be anyways limited by ethical problems. Other cells like MSC, iPSCs, CSCs are the hope givers in place of ESCs.

MSCs are present in almost all tissues. MSCs are found to be effective in the treatment of tissue injury and degenerative diseases. The implanted hMSCs would undergo site-specific differentiation into chondrocytes, adipocytes, myocytes and cardiomyocytes, bone marrow stromal cells and thymic stroma [34]. MSCs has a complement of ionic currents in both the undifferentiated and differentiated MSC [35, 36] resulting in cells with resting potentials in the -30 to -40 mV range. Other studies have described critical cardiac proteins such as troponin-I and connexin 43 (Cx43) [37]. However, MSCs do not have the complete ion channel components to generate a cardiac action potential. Hence they were engineered to overexpress HCN genes. But hMSCs are superior over hESCs in being immunoprivileged [38, 39] eliminating the need for immunosuppression. MSCs also has a tendency to migrate (home) to damaged tissue sites with inflammation [40, 41], The inflammation-directed MSC homing has been demonstrated to involve several important cell trafficking-related molecules. MSCs either *de novo* mobilized or exogenously administered have been found to migrate to tumors and adjacent tissue sites [42]. This property was utilized to use MSCs as vehicles for the delivery of nanoparticles to tumor sites enhancing their tumoricidal effects [43, 44]. hMSCs injected into the anterior left ventricular walls of dogs in complete heart block showed pacemaker firing at 50 to 60 bpm in these animals and catecholamine responsiveness [45].

MSCs produce a large amount of growth factors that perform multiple functions for tissue regeneration [46 - 48]. Pre-treatment with MSC growth factors or gene modification of MSCs can enhance the therapeutic efficacy for myocardial infarction and other wound-healing processes [49, 50].

Cardiac stem cells [51, 52] show promise for use in stem cell therapy. For example, c-Kit cardiac-derived cells isolated from a normal rat heart when delivered *via* the aortic root seem to invade the infarcted myocardium and regenerate muscle to improve left ventricular (LV) function [53]. The autologous skeletal myoblast (SkM)–derived cell show some excitability and expresses connexins and form gap junctions. But these cells will slowly lack both mechanical synchronization and electrical integration within tissue. Experimental work with SkMs has reported that grafted myoblasts differentiate into peculiar hyperexcitable cells [54]. SkMs and BMCs might have some beneficial indirect effects on the myocardium (paracrine mechanisms, potential to induce angiogenesis) [55], but they do not differentiate into cardiomyocytes.

MIRNA-BIOGENESIS AND FUNCTION

MicroRNAs (miRNAs) have recently emerged as one of the central players of gene expression regulation [56]. MicroRNAs are natural, endogenous, single-stranded molecules consisting of approximately 22 nucleotides that regulate target genes [57]. MicroRNAs do not encode for any proteins, these are transcribed to longer molecules called pri-miRs by RNA polymerase II. RNase III type enzyme Drosha cleaves the pri-miR and yields pre-miR which is stem looped structure of approximately 70 bases. The pre –miR is exported from nucleus to cytoplasm by exportin-5 where the Dicer is involved in processing the pre-miR so as to form double stranded structure. One strand of miRNA called the "passenger strand" is removed yielding a single-stranded miRNA, which along with Argonaute forms the RNA-induced silencing complex (RISC). The mechanism of action of miRNA is similar to siRNA from this step. The miRNA strand guides the RISC to their target mRNAs.

Target recognition and the mode of regulation depends on the specific interaction of miRNA with its target mRNA. The miRNA recognition elements are present in the 3'UTR of mRNA with which perfect base pairing of 2-8 nucleotides at 5'end of miRNA called as seed region should happen so as to elucidate gene regulation. The mode of regulation can be translational inhibition or mRNA degradation resulting in the suppression of the target protein levels. MicroRNA binding to the 5'UTR of transcripts has also been demonstrated to be able to activate or suppress target genes [58]. Some miRNAs can also bind to the open reading frame (ORF) of mRNA transcripts and repress translation [59]. MicroRNA may also bind at the promoter of target gene causing repression of gene translation.

Some miRNAs mediate their function by binding to RNA-binding proteins that regulate the expression of mRNA transcripts [60]. Salmena *et al.* (2011) [61] proposed the competing endogenous RNA (ceRNA) hypothesis according to which, miRNAs, long non-coding RNAs (lncRNAs) and target mRNAs are in a finely tuned interaction through miRNA response elements (MREs) and their competition for target binding based on their total concentration in the cytoplasm defines a higher level of regulation. There are approximately 500–1000 different mammalian miRNA genes. MicroRNAs typically exert inhibitory effects on several mRNAs, which often encode proteins that govern the same biological process or several components of a molecular pathway [57, 62]. A single miRNA may have multiple gene targets and single mRNA may harbour binding sites for different miRNA. The multiplicity of miRNA targets may also promote combinatorial regulation by miRNAs that individually target various mRNAs whose protein products contribute to one particular regulatory axis [5]. A biological response happens only after co-expression of several miRNAs that cooperatively target various components of a functional network or are all required to sufficiently repress a single target [63, 64].

MICRORNA REGULATION OF CARDIAC ION CHANNEL GENES

The biology of miRNA is highly orchestrated mode of gene regulation, potentially influencing almost all biological processes in mammals. Cardiac specific miRNAs have been shown to regulate a number of critical cellular functions in the developing and normal adult heart [65]. MicroRNA expression is dynamically regulated and altered miRNA expression can render expression deregulation of ion channel genes leading to channelopathies-arrhythmogenesis [66]. Evidence has emerged indicating the crucial role of miRNAs in controlling cardiac excitability by regulating expression of ion channel genes at the posttranscriptional level [67]. miRNAs are critically involved in the pathological processes of adult hearts, including heart failure [68, 69], cardiomyopathy [70], angiogenesis [71] and arrhythmogenesis [72 - 77]. Aberrant expression of these miRNAs has also been shown to have dramatic impacts in the development of cardiac hypertrophy, fibrosis, and angiogenesis [78]. Muscle-specific miRNA miR-133 was shown to regulate pacemaker channels, HCN2 and HCN4 and contributing to the re-expression of these channels in hypertrophy heart [74]. This miRNA had also been shown to repress HERG K+ channel gene KCNH2 contributing to the abnormal QT prolongation in an animal model of diabetes mellitus [75]. Both *miR-1* and *miR-133* may be involved in the spatial patterns of tissue distribution of ion channels [76]. MicroRNAs are being increasingly recognized as important regulators of cardiac conduction and the development of arrhythmias. The miR-133 has been shown to repress ERG translation, which encodes for the I_{kr} potassium channel in a diabetic rabbit model. Down-regulation of ERG translation repressed the repolarization in diabetic rabbit hearts resulting in prolongation of the QT interval [79]. Dysregulation in the fine balance of miRNA homeostasis has been shown to provoke electrical instability and cause sudden death [80]. Different miRNAs involved in the regulation of genes functional in cardiac conduction, cardiac repolarization and automaticity have been summarized [56]. Up-regulation of pacemaker channels HCN2/ HCN4 has been reported following reduction in miR-1/ miR-133. Yang *et al.* (2008) [56] have also described other gene targets of miR-1 and miR-133 and their role in arrhythmogenesis (Fig. **2**).

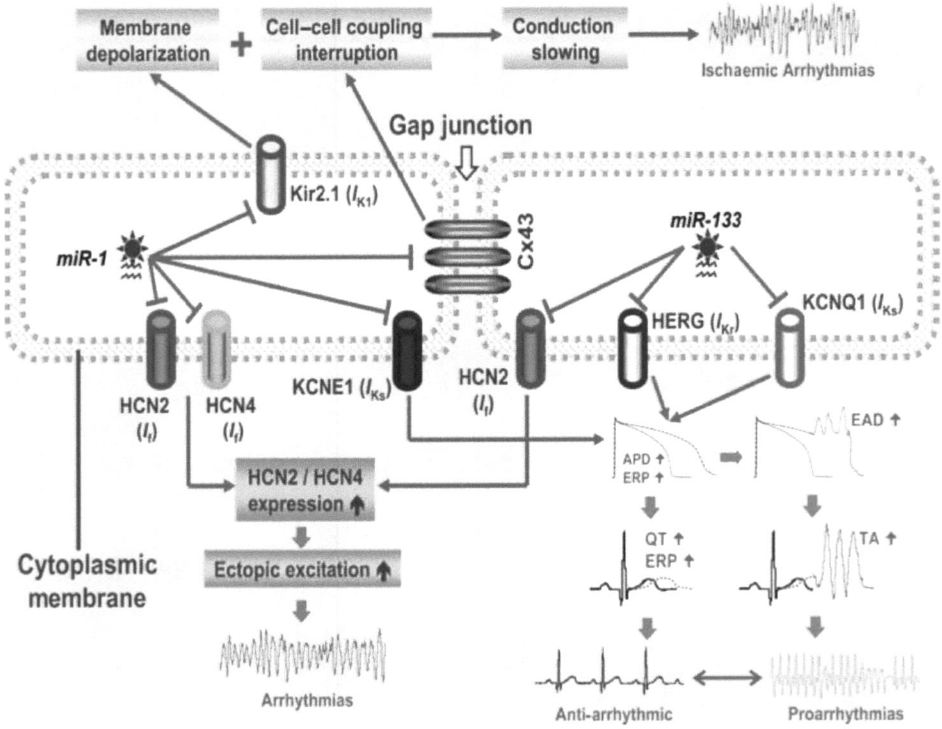

Fig. (2). Schematic illustration of the proposed role of the muscle-specific miRNAs miR-1 and miR-133 in arrhythmogenesis. Dashed arrows, post-transcriptional repression; solid arrows, indicating the cause–effect relationships; Cx43, connexin-43; IK1, inward rectifier Kþ current; IKr, rapid delayed rectifier Kþ current; IKs, slow delayed rectifier Kþ current; If, pacemaker non-selective cation current or funny current; Kir2.1, a pore-forming Kþ channel a-subunit for IK1; HERG, a poreformingKþ channel a-subunit for IKr; KCNQ1, a pore-forming Kþ channel a-subunit for IKs; KCNE1, an auxiliary b-subunit for IKs; HCN2 and HCN4, a-subunits for If;APD, action potential duration; ERP, effective refractory period; EAD, early afterdepolarization; TA, triggered activities [56].

THERAPEUTIC POTENTIAL OF MIRNA IN CARDIAC ARRHYTHMIAS

MicroRNAs play fundamental roles in different biological and pathological process. Lu *et al.* (2005) [81] describes microRNAs would exhibit tissue-specific and time-dependent expression patterns which makes them useful as both diagnostic and therapeutic targets. First anti-cancer miRNA-based drug, MRX-34 (a liposome-based miR-34 mimic) developed by Mirna Therapeutics came to the clinic in 2013 for the treatment of hepatocellular carcinoma (mirnarx.com; NCT0182997) [82]. MicroRNAs play critical roles in cardiovascular development and also their expression profile changes according to different pathological conditions. Cardiac resynchronization therapy (CRT) following heart failure improves the heart arrhythmia and it also affects the myocardial miRNA expression importantly miR-26 (affected by modified ionic channel function) [83]. The muscle-specific miRNA miR-1 has been shown to be involved in cardiac conduction disturbance in myocardial infarction. Elimination of miR-1 by an antisense inhibitor in infarcted rat hearts relieved the severity of the subsequent arrhythmogenesis. In addition, overexpression of miR-1 in the rat heart slowed conduction by repression of KCNJ2, encoding the potassium-channel subunit Kir2.1 and GAJ1 (Cx43) [56]. vanRooij *et al.* (2015) [84] used cholesterol modified anti-miR-29b oligonucleotides after myocardial infarction and observed upregulation of ECM proteins leading to cardiac fibrosis. Also, Bonauer *et al.* (2009) [85] demonstrated that intravenous injection of miR-92a antagomir, improved the function of damaged tissue in models of myocardial infarction. Inhibition of miR-92a results in neoangiogenic effects and functional recovery of ischemic tissues [85]. MiR-133, miR-30 and miR-328 have been linked to atrial fibrillation which mainly results from structural remodeling and fibrosis [86].

In patients with AMI and myocardial injury, higher levels of miR-208a [87, 88] and miR-499 [89] were detected in plasma, which showed higher specificity than conventional biomarkers such as cardiac troponinI (TnI) for diagnosis of the disease. LNA-modified anti-miR-208 oligonucleotides have successfully prevented pathologically associated cardiac remodeling during diastolic heart disease [90].

Upregulated levels of miR-328 targeting the L-type calcium channel was reported in the atria of dogs with atrial fibrillation (AF). Strikingly, inhibition of miR-328 levels with an antagomir reversed the conditions. Thus miR-328 could be used as a target for AF diagnosis and treatment [91].

Recently the miRNAs secreted from exosomes are shown to regulate different cellular functions and are emerging as novel diagnostic biomarkers and potential therapeutic agents. Cells secrete membrane vesicles like exosomes for communicating with other cells. Exosomes are secreted by many cell types including immune cells [92 - 96], tumor cells [97] and also found in almost all bodily fluids including breast milk [98, 99], cerebrospinal fluid [100], *etc.* The exosomes may contain proteins, lipids and nucleic acids (mRNA, miRNA, long non-coding RNA, and occasionally genomic DNA). Exosomes from the receptor cells can bind to the plasma membrane of the acceptor cell and internalized through either endocytosis or micro pinocytosis [101]. Exosomes can be taken up by a variety of neighbouring cells, as well as by cells at a distance [102, 103]. Exosomes serve as mediators for cellular communication in both normal physiological processes (lactation [104], intracellular communications) and pathophysiological process such as, immune response [98], neurodegenerative diseases [105], tumour induction and cardiovascular disease [106]. Khalyfa and Gozal (2014) [107] has suggested the use of exosomal miRNAs as biomarkers for diagnosis of children with obesity at increased risk of CVD. The transfer of miRNAs from cancer cells to endothelial cells has been shown promote tumour metastasis [108] and those from immune cells to Cancer cells were shown to have an anti-proliferative effect on the tumour cells [109]. Hence it is obvious that different cell types can secrete or receive miRNAs as a form of intercellular communication. Bang *et al.* (2014) [110] observed exosome-secretion from cardiomyocytes with a new miRNA profile being involved in reprograming the cardiac fibroblasts leading to cardiac hypertrophy following heart infarction. Thus, the miRNAs contained within exosomes are unique and those released under disease condition are of clinical interest. Studies have shown that the packaging of RNAs into exosomes is selective and dependent on 3-end uridylated isoforms [111], because the RNA profiles in exosomes do not fully reflect the RNA profiles observed in the parental cells [112, 113], but are consistent of distinct set of miRNAs [114 - 118]. The amounts of specific miRNAs contained within exosomes were also found to be less than one copy per exosome [119]. Also miRNA was reported to be very stable in human serum and plasma making them ideal as a novel non-invasive useful biomarker for cardiovascular diseases and cancer [120 - 122]. Matsumoto *et al.* (2013) [123] reported a significant association of serum levels of miR-192, particularly in exosomes, in patients who experienced development of heart failure within 1 year after MI, as compared to matched controls without subsequent cardiovascular events after discharge. Also miR-133 and miR-328 are increased in plasma in patients following MI thereby representing novel biomarkers of acute MI [124]. These evidences support the fact that majority of circulating miRNAs may be transferred *via* exosomal cargos.

COMPUTATIONAL PREDICTION OF MIRNA TARGETS OF ION CHANNELS

The involvement of miRNAs in the cardiac functions and also in dysfunctions promotes for therapeutic applications of miRNA in cardiac disease. Identification and validation of miRNA targets is required for the development of miRNA-based therapeutics. Since this is difficult to be accomplished only by experimental studies, several computational prediction methods for identifying miRNA targets has been developed. Some of these algorithms widely used include miRanda, TargetScan, Pictar, PITA and TargetBoost. Most algorithms search for targets in the 3'-UTR regions of mRNAs, where almost all miRNA-target interactions occur in vertebrates. Molecular studies of cardiac excitability have revealed the fact that altered expression of miRNA can change the expression pattern of cardiac ion channels. Channelopathies are not induced merely by mutations, whereas it could be possible by other regulations such as post transcriptional repression as a one mediated by miRNAs. Thus, identifying miRNAs as a novel mechanism for channelopathies paves the way for new treatment strategies, including engineered cell and gene therapy. Thus, acknowledging the importance of miRNAs, using computational methods we have predicted the miRNA targets of ion channel genes encoding sodium, calcium, potassium and transient receptor potential channels (Table **1**). The miRNA targets were predicted using the TargetScan algorithm version 6.2 [125], accessible at http://www.targetscan.org/ [126]. Also, we identified functionally related miRNA and mRNA. Since the objective of our study was to identify the possible miRNA targets of ion channel genes using bioinformatics tool, to make the results more informative, in addition to those predicted by TargetScan, we also have reported the positively and negatively correlated miRNAs for ion channel genes associated with certain channel disorders. The TargetScan uses precise criterion of seed match restricted to 6-8 nucleotides and also high degree of conservation search, hence there is possibility of missing miRNAs that may be potential target. To overcome this restriction, miRNA targets with positive or negative correlation with specific ion channel genes with p value cut off of 0.01 and 0.001 integrated with data from TargetScan, identified by mimiRNA [127] are reported in Fig. (**3**). Talayeh *et al.* (2010) also discusses the advantages of using combined multiple algorithms to produce more accurate target predictions [128].

Table 1. MicroRNAs targeting of Human Cardiac ion channel genes predicted with Target Scan 6.2.

Gene	Protein	Location	Predicted Target	Description
SCN1A	$Na_v1.1$	Central neurons, [peripheral neurons] and cardiac myocytes	miR-30a,miR-30b,miR-30c,miR-30d,miR-30e	*SCN1,* voltage gated type I, α-subunit (280kDa)
SCN3A	$Na_v1.3$	Central neurons, peripheral neurons and cardiac myocytes	miR-30a, miR-30b, miR-30c, miR-30d, miR-30e	Voltage gated type III, α -subunit (280 kDa)
SCN5A	$Na_v1.5$	Cardiac myocytes, skeletal muscle, central neurons	miR-26a, miR-26b, miR-4465, miR-1297	Voltage gated type V, α –subunit

(Table 1) cont.....

Gene	Protein	Location	Predicted Target	Description
SCN8A	$Na_v1.6$	Central neurons, dorsal root ganglia, peripheral neurons, heart, glia cells	miR-30a, miR-30b, miR-30c, iR-30d, miR-30e	Voltage gated type VIII, α –subunit
SCN7A	$Na_v2.1$	heart, uterus, skeletal muscle, astrocytes, dorsal root ganglion cells	miR-506, miR-124	Voltage gated type VII, α –subunit
SCN1B	$Na_v\beta1$	Central Neurons, Peripheral Neurons, skeletal muscle, heart, glia	miR-137	Hs.89634, voltage gated type I, β1-subunit (38 kDa)
SCN2B	$Na_v\beta2$	Central Neurons, peripheral neurons, heart, glia	miR-211, miR-204	Voltage gated type II, β2-subunit (33 kDa)
SCN4B	$Na_v\beta4$	heart, skeletal muscle, central and peripheral neurons	miR-424, miR-497, miR-15a, miR-15b, miR-16, miR-195	Voltage gated type IV, β –subunit
KCNA1	$K_{v1.1}$	Brain, heart, Skeletal muscle	miR-18a,miR-18b,miR-4735-3p	*Shaker* homolog 1
KCNA2	$K_{v1.2}$	Brain, Nerve, Heart	miR-298, miR-520d,miR-524, miR-4699-3p	Kv1.2, *Shaker* homolog 2
KCNA4	$K_{v1.4}$	Brain, nerve, heart, fetal skeletal muscle, pancreatic islet	miR-153	*Shaker* homolog 4
KCNA5	$K_{v1.5}$	Brain, heart, kidney, lung, skeletal muscle, pancreatic islet	miR-613, miR-206, miR-1	*Shaker* homolog 5
KCNB1	$K_{v2.1}$	Brain, heart, kidney, retina, skeletal muscle	miR-153	*Shab* homolog 1
KCNB2	$K_{v2.2}$	Brain, heart, retina	miR-137	*Shab* homolog 2
KCND2	$K_{v4.2}$	Brain, heart, aorta	miR-218	*Shal* homolog 2
KCND3	$K_{v4.3}$	Heart	miR-4673, miR-4645	*Shal* homolog 3
KCNF1	$K_{v5.1}$	Heart, Brain	miR-3188, miR-4459, miR-4722-5p	Potassium voltage-gated channel, subfamily F, member 1
KCNG3	$K_{v6.3}$	Heart	miR-1178, miR-188-5p, miR-548g, miR-548i, miR-607, miR-1264	γ subunit: Modifies Kv2.1 channel activity

(Table 1) cont.....

Gene	Protein	Location	Predicted Target	Description
KCNG4	$K_{v6.4}$	Brain, heart, kidney, retina, skeletal muscle	miR-486-3p, miR-127-5p	Potassium voltage-gated channel, subfamily G, member 4
KCNH1	$K_{v10.1}$	Heart	miR-22	Potassium voltage-gated channel, subfamily H (eag-related), member 1
KCNH2	$K_{v11.1}$	Heart	miR-365	Potassium voltage-gated channel, subfamily H (eag-related), member 2
KCNH3	$K_{v12.2}$	Heart	miR-125a-5p, miR-125b	Potassium voltage-gated channel, subfamily H (eag-related), member 3
KCNJ2	Kir2.1	Muscle, neural tissue, heart	miR-455-5p	IRK1, Kir2.1
KCNJ3	Kir3.1	Heart, cerebellum	miR-365	GIRK1, Kir3.1
KCNJ4	Kir2.3	Heart, skeletal muscle, brain	miR-574-5p	HIR, HRK1, HIRK2, Kir2.3
KCNJ5	Kir3.4	Heart, pancreas	miR-149,miR-635,miR-4731-5p	CIR, KATP1, GIRK4, Kir3.4
KCNJ8	Kir6.1	Brain, heart, skeletal, smooth muscle, others	miR-520a-3p, miR-520b, miR-520c-3p, miR-520d-3p, miR-520e, miR-373, miR-372, miR-302a, miR-302b, miR-302c, miR-302d, miR-302e	Kir6.1, uKATP, ubiquitous KATP α –subunit
KCNJ9	Kir3.3	Heart	miR-3064, miR-1231, miR-4711-3p, miR-4691-3p, miR-3974, miR-493,miR-661	Potassium inwardly-rectifying channel, subfamily J, member 9
KCNN1	$K_{Ca}2.1$	Brain, heart	miR-4456, miR-548, miR-3926, miR-3145, miR-4680, miR-3545, miR-2861	Potassium intermediate/small conductance calcium-activated channel, subfamily N, member 1
KCNN3	$K_{Ca}2.3$	Brain, heart, (human embryonic) skeletal muscle, liver	miR-208a, miR-208b	Potassium intermediate/small conductance calcium-activated channel, subfamily N, member 3

(Table 1) cont.....

Gene	Protein	Location	Predicted Target	Description
KCNQ1	$K_{v7.1}$	Heart, cochlea, kidney, lung, placenta, colon	miR-365	KCNA9, (KV)LQT1, KQT-like subfamily member 1
KCNE1	minK protein	Kidney, submandibular gland, uterus, heart, cochlea, retina	miR-193a-3p, miR-193b	MinK, ISK, voltage gated Isk homolog 1 (129 aa)
CACNA1C	$Ca_v1.2$	Skeletal muscle, bone (osteoblasts), ventricular myocytes, dendrites and dendritic spines of cortical	miR-490-3p, miR-19a, miR-19b	*CCHL1A1,* L-type A 1C-subunit
CACNA1D	$Ca_v1.3$	Skeletal muscle, bone (osteoblasts), ventricular myocytes, dendrites and dendritic spines of cortical	miR-135a, miR-135b	*CCHL1A2,* L-type α 1D-subunit
CACNA1F *CACNA1D*	$Ca_v1.4$ $Ca_v1.3$	Skeletal muscle, bone (osteoblasts), ventricular myocytes, dendrites and dendritic spines of cortical Skeletal muscle, bone (osteoblasts), ventricular myocytes, dendrites and dendritic spines of cortical	miR-370miR-135a, miR-135b	α 1F-Subunit *CCHL1A2,* L-type α 1D-subunit
CACNA1G *CACNA1E*	$Ca_v3.1$ $Ca_v2.3$	neurons, cells that have pacemaker activity, bone (osteocytes)Cerebellar granule cells, other neurons	miR-137miR-135a, miR-135b	T-type α1G-subunitR-type α 1E-subunit
CACNA1H *CACNA1F*	$Ca_v3.2$ $Ca_v1.4$	neurons, cells that have pacemaker activity, bone (osteocytes)Skeletal muscle, bone (osteoblasts), ventricular myocytes, dendrites and dendritic spines of cortical	miR-490-3pmiR-370	T-type α_{1H} subunitα 1F-Subunit

(Table 1) cont.....

Gene	Protein	Location	Predicted Target	Description
CACNA1I *CACNA1G*	$Ca_v3.3$ $Ca_v3.1$	neurons, cells that have pacemaker activity, bone (osteocytes)	hsa let-7b, hsa let-7e, miR-98, hsa let-7a, hsa let-7c, hsa let-7d, hsa let-7i, hsa let-7f, miR-137	T-type α_{1I} subunitT-type α1G-subunit
CACNA1S *CACNA1H*	$Ca_v1.1$ $Ca_v3.2$	Skeletal muscle, bone (osteoblasts), ventricular myocytes, dendrites and dendritic spines of cortical neurons, cells that have pacemaker activity	miR-490-3p	L-type α 1S-subunit (5% 212, 95% 190 kDa)T-type α_{1H} subunit
CACNB1 *CACNA1I*	β_1 $Ca_v3.3$	Skeletal muscle, brain, heart, spleen neurons, cells that have pacemaker activity, bone (osteocytes)	miR-214hsa let-7b, hsa let-7e, miR-98, hsa let-7a, hsa let-7c, hsa let-7d, hsa let-7i, hsa let-7f,	β1-Subunit (524 aa, 54 kDa)T-type α_{1I} subunit
CACNB2 *CACNA1S*	β_2 $Ca_v1.1$	Brain, Heart, Lung Skeletal muscle, bone (osteoblasts), ventricular myocytes, dendrites and dendritic spines of cortical	miR-31	MYSB, β2-subunitL-type α 1S-subunit (5% 212, 95% 190 kDa)
CACNB3 *CACNB1*	β_3 β_1	Brain, Heart, Lung, Skeletal smooth muscle, Ovary, brain,	miR-125a-5p, miR-125b, miR-214	β3-subunit (482 aa) β1-Subunit (524 aa, 54 kDa)
CACNB2	β_2	Brain, Heart, Lung	miR-129-5p, miR-31	β4-subunit, *lethargic* miceMYSB, β2-subunit
CACNA2D1 *CACNB3*	β_3	Skeletal muscle, Heart, Brain, Ileum	miR-107, miR-103miR-125a-5p, miR-125b	*CACNA2, CACNA2D1,* α2/d-subunit (175 kDa)β3-subunit (482 aa)
CACNA2D1		Skeletal muscle, Heart, Brain, Ileum	miR-107, miR-103	*CACNA2, CACNA2D1,* α2/d-subunit (175 kDa)
TRPC1	TRPC1	Heart, brain, lung, liver, spleen, kidney, testes	miR-135a, miR-135b	Transient receptor potential cation channel, subfamily C, member 1

(Table 1) cont.....

Gene	Protein	Location	Predicted Target	Description
TRPC2	TRPC2	Testes, sperm, heart, kidney, brain	Not Predicted	Transient receptor potential cation channel, subfamily C, member 2,
TRPC3	TRPC3	Pituitary gland > brain, heart, lung, dorsal root ganglia	miR-218	Transient receptor potential cation channel, subfamily C, member 3
TRPC4	TRPC4	Heart, brain, pancreas, placenta, kidney	miR-1283	Transient receptor potential cation channel, subfamily C, member 4
TRPC5	TRPC5	Brain, testes, kidney, uterus, Heart	miR-204, miR-211	Transient receptor potential cation channel, subfamily C, member 5
TRPC6	TRPC6	Heart, lung, kidney, muscle, intestine, stomach, pancreas, prostate, bone, brain	miR-26a, miR-26b	Transient receptor potential cation channel, subfamily C, member 6
TRPM2	TRPM2	Brain, spinal cord, bone marrow, spleen, heart, lung, eye, neutrophils, leukocytes, pancreatic β-cells.	miR-4755-3p, miR-3184, miR-423-5p	Transient receptor potential cation channel, subfamily M, member 2
TRPM4	TRPM4	Small intestine, prostate, colon, kidneys, testes, heart, lymphocytes, spleen, lung, brain, pituitary, skeletal muscle, stomach,	miR-199a-5p, miR-199b-5p	Transient receptor potential cation channel, subfamily M, member 4
TRPM6	TRPM6	Kidney, small intestine, colon, brain, heart	hsa let-7b, hsa let-7e, miR-98, hsa let-7a, hsa let-7c, hsa let-7d, hsa let-7i, hsa let-7f, hsa let-7g	Transient receptor potential cation channel, subfamily M, member6
TRPM7	TRPM7	Ubiquitously expressed, especially heart, pituitary, bone, adipose tissue	miR-135a, miR-135b	Transient receptor potential cation channel, subfamily M, member7
TRPP1	TRPP1	Widely Expressed. Kidney, lung, brain, heart, liver, testis, muscle	miR-17, miR-20a, miR-20b, miR-93, miR-106a, miR-106b, miR-519d	Transient receptor potential cation channel, subfamily P, member 1

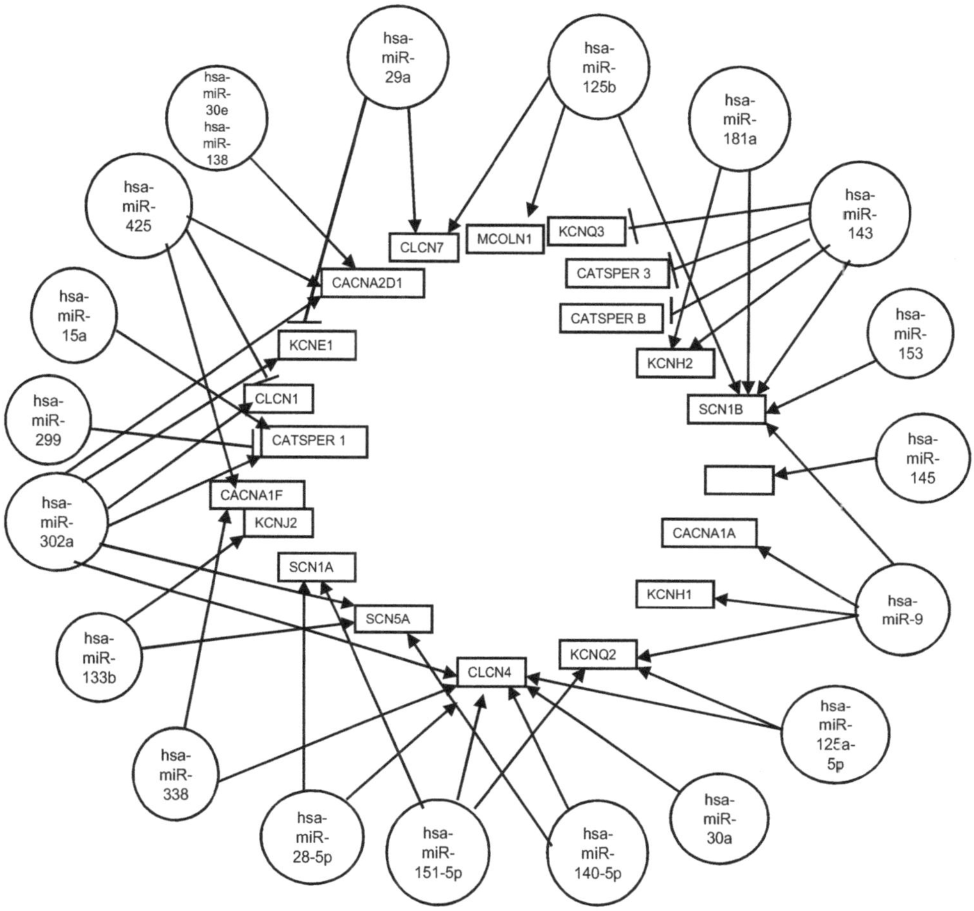

Fig. (3). Positively and negatively correlated miRNA targets of ion channel genes. The functional relativeness of miRNAs and the associated ion channel genes of clinical significance is identified using mimiRNA [127]. The predicted human miRNAs are given in the circles and ion channel gene targets are shown in the rectangular boxes. The Arrow head and —| indicates miRNAs exhibiting positive and negative correlations respectively with ion channel genes.

MIRNA MEDIATED STEM CELL THERAPY- DELIVERY TOOLS FOR MIRNAS

In the field of cardiac regenerative medicine especially in myocardial ischemia/reperfusion injuries, mesenchymal stem cell-derived exosomes have been investigated [129]. Though induced pluripotent stem cells are capable of enhancing survival and proliferation in ischemic tissues following myocardial infarction, its therapeutic application is limited by its tumorigenic potential. But the exosomes secreted by iPS cells showed reduced cardiomyocyte apoptosis and protected against myocardial ischemia/reperfusion (MIR) injury by delivering cardio protective miRNAs, including Nanog-regulated miR-21 and HIF-1--regulated miR-210. Thus, iPS-EXo has been proposed to serve as an off-the-shelf therapy to rescue ischemic cardiomyocytes in conditions such as MIR [130]. Many reports have demonstrated that pre-treatment with growth factors or gene modification of MSCs can enhance the therapeutic efficacy for myocardial infarction and other wound healing processes [49, 131]. A further understanding of the molecular pathways involved in growth factor production will be helpful to develop better strategies for MSC-based therapies [132]. The gene modification could be achieved by delivering the desired miRNAs in to MSCs thereby making it to express genes required to treat cardiovascular diseases. Using a guinea pig model, Shan *et al.* (2012) showed that increasing MiR-1 and MiR-133A, induced QT prolongation, while antisense silencing of MiR-1 and MiR-133A abolished Arsenic tri oxide induced QT prolongation [133].

The therapeutic potential of miRNA as a drug in clinical application is hampered by a lack of appropriate delivery systems. Recently the exosomes, natural carriers of miRNA are exploited in RNA drug delivery system (Fig. **4**). Exosomal transfer of signals has been shown to confer cardiovascular protection after a disease. For example, in heart ischemia, secretory exosomes with miRNA cargos from cardiomyocytes have been shown to activate and induce the bone marrow-derived stem cells. These latter cells release these cold subset of exosomes that lead to myocardial regeneration or protection [134]. MicroRNAs released in circulation are of highly stable forms, not prone to RNases degradation. These miRNAs are involved in epigenetic reprogramming of the recipient cells. The transfer of exosome-derived unique miRNAs to recipient cells is an alternative mechanism enabling gene-based communication between cells [135]. Exosome mediated RNA transfer is a type of cell signalling impacting the biological processes in the recipient cells [136]. Alexander *et al.* (2015) [137] predicted functional miRNA transfer *via* exosomes to be most relevant in defined micro environments such as stem cell niches, where exosome concentrations might be at their highest. Exosomes are also natural carriers of exogenous siRNA to human cells *in vitro* [138]. Alvarez-Erviti *et al.* (2011) recently used exosomes with modified membranes containing a neuron-specific peptide to deliver small interfering RNA (siRNA) to mouse brain tissue [139]. Thus, this ability of exosomes to naturally carry and deliver RNA between cells could be used as a natural way to transfer miRNAs, thereby portraying the therapeutic potential of exosomes as delivery vehicles for specific miRNAs or their inhibitors [95, 98, 105, 106, 140, 141]. Exosomes isolated from patients own cells promote autologous therapy and the miRNA amount in exosomes also can be manipulated. In addition, the exosome mediated delivery is also advantageous in promoting site targeted delivery. Ohno *et al.* (2013) [142] have shown the efficiency of exosomes in delivering microRNA (miRNA) to epidermal growth factor receptor (EGFR) expressing breast cancer cells. Exosomal membranes contain certain proteins which have binding affinity for specific receptors on the surface of recipient cells, hence they can selectively target cell types of interest and manipulate their miRNA components, thereby providing promising tools for the future of personalized medicine [82].

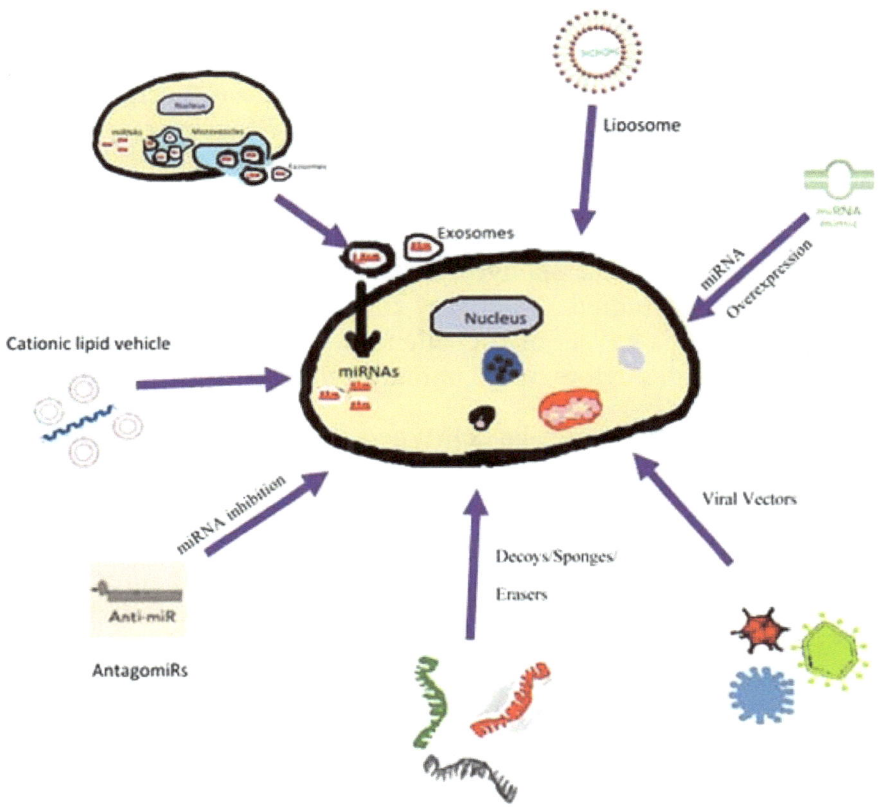

Fig. (4). Delivery vehicles for miRNA.

Synthetic nucleotides with similar and complementary bases for a particular miRNA called miRNA mimics and inhibitors respectively have been developed. In principle, miRNA mimics are used to increase the expression of a required miRNAs in disease settings. Alternatively, miRNA inhibitors are meant for blocking the activity of miRNAs responsible for disease. Single-stranded oligonucleotides termed, antagomirs are effective in inactivating miRNAs *in vivo* [143]. These oligonucleotides contain locked nucleotides for increased binding specificity and cholesterol-conjugation for enhanced circulation time and to promote cellular uptake. RNA molecules with complementary bases to the mature target miRNA. Antagomirs are shown to be very stable *in vivo* potential to silence target miRNA in different organs for long periods. LNA-modified anti-miR- 208 oligonucleotides have successfully prevented pathologically associated cardiac remodelling during diastolic heart disease [90].

In addition to chemically modified antisense oligonucleotides, where modulation of miRNAs is achieved either by over expression or suppression of a miRNA, modulation also can be established by preventing the miRNA from functioning referred as soaking up of miRNA. This technique has been named miRNA erasers, sponges, or decoys [144]. Expression vectors expressing miRNA target sites (sponges or decoy) are used to scavenge a miRNA and prevent it from regulating its natural targets [5]. Addition of a reporter such as GFP or luciferase in vector harbouring multiple miRNA binding sites can indicate effective scavenging of miRNA by the sponges. Lipid-based vehicles, viral systems, and cationic polymers are studied as delivery tools for miRNA-based therapeutics. Each of these strategies has its own challenges and still needs improvements to address problems such as cytotoxicity, immunogenicity, and low efficiency [145].

Fiedler *et al.* (2011) [146] showed a cell-type specific tendency or mechanism for cholesterol-based antagomir uptake by cardiomyocytes and cardiac endothelial cells. Cholesterol-based antagomirs for silencing miR-21 were also proved to inhibit cardiac fibrosis and dysfunction [147]. A common issue in LNA modified oligonucleotides is it's *in vivo* instability and the low homing efficiency of resulting in modest changes in the expression levels of the target protein [80]. Hence viral vectors were studied in order to develop an efficient delivery tool for miRNAs. Adeno-associated viruses, AAV9, a specific serotype, has tropism for myocardiocytes and enriches in the heart [148]. This tropism promotes an efficient tool for targeted delivery of miRNA-based therapeutics for cardiac gene transfer. Santulli *et al.* (2014b) used an adeno-viral vector containing miR-126-3p target sites for its down regulation, which inhibits proliferative vascular smooth muscle cells and prevents restenosis in animal models [149]. Care *et al.* [2007] also delivered miRNA decoys using an adenoviral system in cardiac tissue to suppress specific miRNA function [150]. The decoy technology also can be used to validate computational target predictions and to assay loss-of-function phenotypes.

CONCLUSION

MiRNAs have been investigated for their role in regulating cardiac ion channel genes. The important role of ion channels in pathophysiology of arrhythmogenesis have been well reported. Dysfunction of different ion channels lead to different types of cardiac arrhythmia. The miRNAs targeting cardiac ion channel genes are computationally predicted and some are experimentally validated. Thus, by combining the miRNA manipulation along with existing stem cell therapy, therapeutic potential of the later can be well improved for cardiac arrhythmia. Stem cells modulated with specific miRNAs targeting altered ion channel genes would be a new approach, thereby making therapies available individually for different types of arrhythmia.

CONSENT FOR PUBLICATION

Not applicable.

CONFLICT OF INTEREST

The author confirms that she has no conflict of interest to declare for this publication.

ACKNOWLEDGEMENTS

Declared none.

REFERENCES

[1] Murakoshi N, Aonuma K. Epidemiology of Arrhythmias and Sudden Cardiac Death in Asia. Circulation 2013; 77(10): 2419-31.
[http://dx.doi.org/10.1253/circj.CJ-13-1129]

[2] Couceiro R, Henriques J, Paiva RP, Antunes M, Carvalho P. Physiologically motivated detection of Atrial Fibrillation In: 39th Annual International Conference of the IEEE Engineering in Medicine and Biology Society (EMBC); Seogwipo. 2017; pp. 1278-81.
[http://dx.doi.org/10.1109/EMBC.2017.8037065]

[3] Winkle RA. The effectiveness and cost effectiveness of public-access defibrillation. Clin Cardiol 2010; 33: 396-9.
[http://dx.doi.org/10.1002/clc.20790]

[4] Junttila MJ, Castellanos A, Huikuri HV, Myerburg RJ. Risk markers of sudden cardiac death in standard 12-lead electrocardiograms. Ann Med 2012; 44: 717-32.
[http://dx.doi.org/10.3109/07853890.2011.594807]

[5] Kim GH. MicroRNA Regulation of Cardiac Conduction and Arrhythmias. Transl Res 2013; 161(5): 381-92.
[http://dx.doi.org/10.1016/j.trsl.2012.12.004]

[6] Martin CA, Matthews GD, Huang CL. Sudden cardiac death and inherited channelopathy: the basic electrophysiology of the myocyte and myocardium in ion channel disease. Heart 2012; 98(7): 536-43.
[http://dx.doi.org/10.1136/heartjnl-2011-300953]

[7] Antzelevitch C, Yan G, Shimuzu W, Burashnikov A. Electrical heterogeneity, the ECG, and cardiac arrhythmias.Cardiac Electrophysiology: From Cell to Bedside. Philidelphia, PA: Saunders 2000; pp. 222-38.

[8] Gima K, Rudy Y. Ionic current basis of electrocardiographic waveforms: a model study. Circ Res 2002; 90: 889-96.
[http://dx.doi.org/10.1161/01.RES.0000016960.61087.86]

[9] Roden DM, Balser JR, George AL, Anderson ME. Cardiac ion channels. Annu Rev Physiol 2002; 64: 431-75.
[http://dx.doi.org/10.1146/annurev.physiol.64.083101.145105]

[10] Wilde AAM, Bezzina CR. Genetics of Cardiac Arrhythmias. Heart 2005; 91: 1352-8.
[http://dx.doi.org/10.1136/hrt.2004.046334]

[11] Wehrens XH, Lehnart SE, Huang F, *et al.* FKBP12.6 deficiency and defective calcium release channel (ryanodine receptor) function linked to exercise-induced sudden cardiac death. Cell 2003; 113: 829-40.
[http://dx.doi.org/10.1016/S0092-8674(03)00434-3]

[12] Shah M, Akar FG, Tomaselli GF. Molecular Basis of Arrhythmias. Circulation 2005; 112: 2517-29.
[http://dx.doi.org/10.1161/CIRCULATIONAHA.104.494476]

[13] Van Wagoner DR, Pond AL, Lamorgese M, Rossie SS, McCarthy PM, Nerbonne JM. Atrial L-type Ca2_ currents and human atrial fibrillation. Circ Res 1999; 85: 428-36.
[http://dx.doi.org/10.1161/01.RES.85.5.428]

[14] Dobrev D, Graf E, Wettwer E, *et al.* Molecular basis of downregulation of G-protein-coupled inward rectifying K(+) current (I(KACh)) in chronic human atrial fibrillation: decrease in GIRK4 mRNA correlates with reduced I(KACh) and muscarinic receptor-mediated shortening of action potentials. Circulation 2001; 104: 2551-7.
[http://dx.doi.org/10.1161/hc4601.099466]

[15] Marban E. Cardiac channelopathies. Nature 2002; 415: 213-8.
[http://dx.doi.org/10.1038/415213a]

[16] Amin AS, Hanno LT, Arthur AMW. Cardiac ion channels in health and disease. Heart Rhythm 2010; 7: 117-26.
[http://dx.doi.org/10.1016/j.hrthm.2009.08.005]

[17] Kusumoto FM, Goldschlager N. Cardiac pacing. N Engl J Med 1996; 334: 89-97.
[http://dx.doi.org/10.1056/NEJM199601113340206]

[18] Bernstein AD, Parsonnet V. Survey of cardiac pacing and implanted defibrillator practice patterns in the United States in 1997. Pacing Clin Electrophysiol 2001; 24: 842-55.
[http://dx.doi.org/10.1046/j.1460-9592.2001.00842.x]

[19] Marban E, Cho HC. Biological pacemakers as a therapy for cardiac arrhythmias. Curr Opin Cardiol 2008; 23: 46-54.
[http://dx.doi.org/10.1097/HCO.0b013e3282f30416]

[20] Edelberg JM, Huang DT, Josephson ME, Rosenberg RD. Molecular enhancement of porcine cardiac chronotropy. Heart 2001; 86: 559-62.
[http://dx.doi.org/10.1136/heart.86.5.559]

[21] Miake J, Marbán E, Nuss HB. Gene therapy: biological pacemaker created by gene transfer. Nature 2002; 419: 132-3.
[http://dx.doi.org/10.1038/419132b]

[22] Qu J, Plotnikov AN, Danilo P, *et al.* Expression and function of a biological pacemaker in canine heart. Circulation 2003; 107: 1106-9.
[http://dx.doi.org/10.1161/01.CIR.0000059939.97249.2C]

[23] Plotnikov AN, Sosunov EA, Qu J, *et al.* A biological pacemaker implanted in the canine left bundle branch provides ventricular escape rhythms having physiologically acceptable rates. Circulation 2004; 109: 506-12.
[http://dx.doi.org/10.1161/01.CIR.0000114527.10764.CC]

[24] Macia and Boyden. Stem Cell Therapy Is Proarrhythmic. Circulation 2009; 119: 1814-23.
[http://dx.doi.org/10.1161/CIRCULATIONAHA.108.779900]

[25] Laflamme MA, Zbinden S, Epstein SE, Murry CE. Cell-based therapy for myocardial ischemia and

infarction: pathophysiological mechanisms. Annu Rev Pathol 2007; 2: 307-39. [http://dx.doi.org/10.1146/annurev.pathol.2.010506.092038]

[26] Sartiani L, Bettiol E, Stillitano F, Mugelli A, Cerbai E, Jaconi ME. Developmental changes in cardiomyocytes differentiated from human embryonic stem cells: a molecular and electrophysiological approach. Stem Cells 2007; 25: 1136-44. [http://dx.doi.org/10.1634/stemcells.2006-0466]

[27] Satin J, Kehat I, Caspi O, *et al.* Mechanism of spontaneous excitability in human embryonic stem cell derived cardiomyocytes. J Physiol 2004; 559: 479-96. [http://dx.doi.org/10.1113/jphysiol.2004.068213]

[28] Dolnikov K, Shilkrut M, Zeevi-Levin N, *et al.* Functional properties of human embryonic stem cell–derived cardiomyocytes: intracellular ca2_ handling and the role of sarcoplasmic reticulum in the contraction. Stem Cells 2006; 24: 236-45. [http://dx.doi.org/10.1634/stemcells.2005-0036]

[29] Kehat I, Khimovich L, Caspi O, *et al.* Electromechanical integration of cardiomyocytes derived from human embryonic stem cells. Nat Biotechnol 2004; 22: 1282-9. [http://dx.doi.org/10.1038/nbt1014]

[30] Zhang YM, Hartzell C, Narlow M, Dudley SC Jr. Stem cell– derived cardiomyocytes demonstrate arrhythmic potential. Circulation 2002; 106: 1294-9. [http://dx.doi.org/10.1161/01.CIR.0000027585.05868.67]

[31] Singla DK, Lyons GE, Kamp TJ. Transplanted embryonic stem cells following mouse myocardial infarction inhibit apoptosis and cardiac remodeling. Am J Physiol Heart Circ Physiol 2007; 293: H1308-14. [http://dx.doi.org/10.1152/ajpheart.01277.2006]

[32] Igelmund P, Fleischmann BK, Fischer IR, *et al.* Action potential propagation failures in long-term recordings from embryonic stem cell–derived cardiomyocytes in tissue culture. Pflugers Arch 1999; 437: 669-79. [http://dx.doi.org/10.1007/s004240050831]

[33] Rosen MR, Brink PR, Cohen IS, Robinson RB. Cardiac Pacing. Circ Arrhythm Electrophysiol 2008; 1: 54-61. [http://dx.doi.org/10.1161/CIRCEP.108.764621]

[34] Liechty KW, MacKenzie TC, Shaaban AF, *et al.* Human mesenchymal stem cells engraft and demonstrate site-specific differentiation after in utero transplantation in sheep. Nat Med 2000; 6: 1282-6. [http://dx.doi.org/10.1038/81395]

[35] Heubach JF, Graf EM, Leutheuser J, *et al.* Electrophysiological properties of human mesenchymal stem cells. J Physiol 2004; 554: 659-72. [http://dx.doi.org/10.1113/jphysiol.2003.055806]

[36] Li GR, Deng XL, Sun H, Chung SSM, Tse HF, Lau CP. Ion channels in mesenchymal stem cells from rat bone marrow. Stem Cells 2006; 24: 1519-28. [http://dx.doi.org/10.1634/stemcells.2005-0307]

[37] Amado LC, Saliaris AP, Schuleri KH, *et al.* Cardiac repair with intramyocardial injection of allogeneic mesenchymal stem cells after myocardial infarction. Proc Natl Acad Sci USA 2005; 102: 11474-9. [http://dx.doi.org/10.1073/pnas.0504388102]

[38] Groh ME, Maitra B, Szekely E, Koc ON. Human mesenchymal stem cells require monocyte-mediated activation to suppress alloreactive T cells. Exp Hematol 2005; 33: 928-34. [http://dx.doi.org/10.1016/j.exphem.2005.05.002]

[39] Di Nicola M, Carlo-Stella C, Magni M, *et al.* Human bone marrow stromal cells suppress T-lymphocyte proliferation induced by cellular or nonspecific mitogenic stimuli. Blood 2002; 99: 3838-

43.
[http://dx.doi.org/10.1182/blood.V99.10.3838]

[40] Horwitz EM, Gordon PL, Koo WK, *et al.* Isolated allogeneic bone marrow-derived mesenchymal cells engraft and stimulate growth in children with osteogenesis imperfecta: Implications for cell therapy of bone. Proc Natl Acad Sci USA 2002; 99: 8932-7.
[http://dx.doi.org/10.1073/pnas.132252399]

[41] Mahmood A, Lu D, Lu M, Chopp M. Treatment of traumatic brain injury in adult rats with intravenous administration of human bone marrow stromal cells. Neurosurgery 2003; 53: 697-702.
[http://dx.doi.org/10.1227/01.NEU.0000079333.61863.AA]

[42] Spaeth E, Klopp A, Dembinski J, Andreeff M, Marini F. Inflammation and tumor microenvironments: defining the migratory itinerary of mesenchymal stem cells. Gene Ther 2008; 15: 730-8.
[http://dx.doi.org/10.1038/gt.2008.39]

[43] Roger M, Clavreul A, Venier-Julienne MC, *et al.* Mesenchymal stem cells as cellular vehicles for delivery of nanoparticles to brain tumors. Biomaterials 2010; 31: 8393-401.
[http://dx.doi.org/10.1016/j.biomaterials.2010.07.048]

[44] Cheng H, Kastrup CJ, Ramanathan R, *et al.* Nanoparticulate cellular patches for cell-mediated tumoritropic delivery. ACS Nano 2010; 4: 625-31.
[http://dx.doi.org/10.1021/nn901319y]

[45] Plotnikov AP, Shlapakova I, Szabolcs MJ, *et al.* Xenografted adult human mesenchymal stem cells provide a platform for sustained biological pacemaker function in canine heart. Circulation 2007; 116: 706-13.
[http://dx.doi.org/10.1161/CIRCULATIONAHA.107.703231]

[46] Crisostomo PR, Wang Y, Markel TA, Wang M, Lahm T, Meldrum DR. Human mesenchymal stem cells stimulated by TNF-alpha, LPS, or hypoxia produce growth factors by an NF kappa B- but not JNKdependent mechanism. Am J Physiol Cell Physiol 2008; 294: C675-82.
[http://dx.doi.org/10.1152/ajpcell.00437.2007]

[47] Caplan AI, Dennis JE. Mesenchymal stem cells as trophic mediators. J Cell Biochem 2006; 98: 1076-84.
[http://dx.doi.org/10.1002/jcb.20886]

[48] Jing Y, Han Z, Zhang S, Liu Y, Wei L. Epithelial-mesenchymal transition in tumor microenvironment. Cell Biosci 2011; 1: 29.
[http://dx.doi.org/10.1186/2045-3701-1-29]

[49] Hahn JY, Cho HJ, Kang HJ, *et al.* Pretreatment of mesenchymal stem cells with a combination of growth factors enhances gap junction formation, cytoprotective effect on cardiomyocytes, and therapeutic efficacy for myocardial infarction. J Am Coll Cardiol 2008; 51: 933-43.
[http://dx.doi.org/10.1016/j.jacc.2007.11.040]

[50] Schnabel LV, Lynch ME, van der Meulen MC, Yeager AE, Kornatowski MA, Nixon AJ. Mesenchymal stem cells and insulin-like growth factor-I gene-enhanced mesenchymal stem cells improve structural aspects of healing in equine flexor digitorum superficialis tendons. J Orthop Res 2009; 27: 1392-8.
[http://dx.doi.org/10.1002/jor.20887]

[51] Beltrami AP, Barlucchi L, Torella D, *et al.* Adult cardiac stem cells are multipotent and support myocardial regeneration. Cell 2003; 114: 763-76.
[http://dx.doi.org/10.1016/S0092-8674(03)00687-1]

[52] Smith RR, Barile L, Messina E. Marb_n E. Stem cells in the heart: what's the buzz all about?–part 1: preclinical considerations. Heart Rhythm 2008; 5: 749-57.
[http://dx.doi.org/10.1016/j.hrthm.2008.02.010]

[53] Dawn B, Stein AB, Urbanek K, *et al.* Cardiac stem cells delivered intravascularly traverse the vessel

barrier, regenerate infarcted myocardium, and improve cardiac function. Proc Natl Acad Sci USA 2005; 102: 3766-71.
[http://dx.doi.org/10.1073/pnas.0405957102]

[54] Leobon B, Garcin I, Menasché P, Vilquin JT, Audinat E, Charpak S. Myoblasts transplanted into rat infarcted myocardium are functionally isolated from their host. Proc Natl Acad Sci USA 2003; 100: 7808-11.
[http://dx.doi.org/10.1073/pnas.1232447100]

[55] Pagani FD, DerSimonian H, Zawadzka A, *et al.* Autologous skeletal myoblasts transplanted to ischemia-damaged myocardium in humans: histological analysis of cell survival and differentiation. J Am Coll Cardiol 2003; 41: 879-88.
[http://dx.doi.org/10.1016/S0735-1097(03)00081-0]

[56] Yang B, Lu Y, Wang Z. Control of cardiac excitability by microRNAs. Cardiovasc Res 2008; 79: 571-80.
[http://dx.doi.org/10.1093/cvr/cvn181]

[57] Ambros V. The functions of animal microRNAs. Nature 2004; 431: 350-5.
[http://dx.doi.org/10.1038/nature02871]

[58] Lee I, Ajay S, Yook J, Kim H, Hong S, Kim N, *et al.* New class of microRNA targets containing simultaneous 5′-UTR and 3′-UTR interaction sites. Genome Res 19: 1175-83.
[http://dx.doi.org/10.1101/gr.089367.108]

[59] Tay Y, Zhang J, Thomson A, Lim B, Rigoutsos I. MicroRNAs to Nanog, Oct4 and Sox2 coding regions modulate embryonic stem cell differentiation. Nature 2008; 455: 1124-8.
[http://dx.doi.org/10.1038/nature07299]

[60] Eiring A, Harb J, Neviani P, *et al.* miR- 328 functions as an RNA decoy to modulate hnRNPE2 regulation of mRNA translation in leukemic blasts. Cell 2010; 140: 652-65.
[http://dx.doi.org/10.1016/j.cell.2010.01.007]

[61] Salmena L, Poliseno L, Tay Y, Kats L. Pan dolfi P. AceRNA hypothesis: the rosetta stone of a hidden rna language. Cell 146: 353-8.
[http://dx.doi.org/10.1016/j.cell.2011.07.014]

[62] Lagos-Quintana M, Rauhut R, Lendeckel W, Tuschl T. Identification of novel genes coding for small expressed RNAs. Science 2001; 294(5543): 853-8.
[http://dx.doi.org/10.1126/science.1064921]

[63] Schlesinger J, Schueler M, Grunert M, *et al.* The cardiac transcription network modulated by Gata4, Mef2a, Nkx2.5, Srf, histone modifications, and microRNAs. PLoS Genet 2011; 7(2): e1001313.
[http://dx.doi.org/10.1371/journal.pgen.1001313]

[64] Chen CY, Chen ST, Fuh CS, Juan HF, Huang HC. Coregulation of transcription factors and microRNAs in human transcriptional regulatory network. BMC Bioinformatics 2011; 12 (Suppl. 1): S41.
[http://dx.doi.org/10.1186/1471-2105-12-S1-S41]

[65] Boettger T, Braun T. A new level of complexity: the role of microRNAs in cardiovascular development. Circ Res 2012; 110(7): 1000-13.
[http://dx.doi.org/10.1161/CIRCRESAHA.111.247742]

[66] Luo X, Zhang H, Xiao J, Wang Z. Regulation of Human Cardiac Ion Channel Genes by MicroRNAs: Theoretical Perspective and Pathophysiological Implications. Cell Physiol Biochem 2010; 25: 571-86.
[http://dx.doi.org/10.1159/000315076]

[67] Robertson JA. Human embryonic stem cell research: ethical and legal issues. Nat Rev Genet 2001; 2: 74-8.
[http://dx.doi.org/10.1038/35047594]

[68] Valencia-Sanchez MA, Liu J, Hannon GJ, Parker R. Control of translation and mRNA degradation by

miRNAs and siRNAs. Genes Dev 2006; 20: 515-24.
[http://dx.doi.org/10.1101/gad.1399806]

[69] Forstemann K, Horwich MD, Wee L, Tomari Y, Zamore PD. Drosophila microRNAs are sorted into functionally distinct argonaute complexes after production by Dicer-1. Cell 2007; 130: 287-97.
[http://dx.doi.org/10.1016/j.cell.2007.05.056]

[70] Vasudevan S, Tong Y, Steitz JA. Switching from repression to activation: microRNAs can up-regulate translation. Science 2007; 318: 1931-4.
[http://dx.doi.org/10.1126/science.1149460]

[71] Farh KK-H, Grimson A, Jan C, Lewis BP, Johnston WK, Lim LP, *et al.* The widespread impact of mammalian microRNAs on mRNA repression and evolution. Science 2005; 310: 1817-21.
[http://dx.doi.org/10.1126/science.1121158]

[72] Wightman B, Ha I, Ruvkun G. Posttranscriptional regulation of the heterochronic gene Lin-14 by Lin-4 mediates temporal pattern-formation in C. elegans. Cell 1993; 75: 855-62.
[http://dx.doi.org/10.1016/0092-8674(93)90530-4]

[73] Darnell DK, Kaur S, Stanislaw S, Konieczka JH, Yatskievych TA, Antin PB. MicroRNA expression during chick embryo development. Dev Dyn 2006; 235: 3156-65.
[http://dx.doi.org/10.1002/dvdy.20956]

[74] Giraldez AJ, Cinalli RM, Glasner ME, Enright AJ, Thomson JM, Baskerville S, *et al.* MicroRNAs regulate brain morphogenesis in zebrafish. Science 2005; 308: 833-8.
[http://dx.doi.org/10.1126/science.1109020]

[75] Lakshmipathy U, Love B, Goff LA, Jornsten R, Graichen R, Hart RP, *et al.* MicroRNA expression pattern of undifferentiated and differentiated human embryonic stem cells. Stem Cells Dev 2007; 16: 1003-16.
[http://dx.doi.org/10.1089/scd.2007.0026]

[76] Thum T, Galuppo P, Wolf C, Fiedler J, Kneitz S. van Laake LW*et al.* Micro-RNAs in the human heart: a clue to fetal gene reprogramming in heart failure. Circulation 2007; 116: 258-67.
[http://dx.doi.org/10.1161/CIRCULATIONAHA.107.687947]

[77] Ivey KN, Muth A, Arnold J, King FW, Yeh RF, Fish JE, *et al.* MicroRNA regulation of cell lineages in mouse and human embryonic stem cells. Cell Stem Cell 2008; 2: 219-29.
[http://dx.doi.org/10.1016/j.stem.2008.01.016]

[78] Dorn GW. MicroRNAs in Cardiac Disease. Transl Res 2011; 157(4): 226-35.
[http://dx.doi.org/10.1016/j.trsl.2010.12.013]

[79] Xiao J, Yang B, Lin H, Lu Y, Luo X, Wang Z. Novel approaches for gene-specific interference *via* manipulating actions of microRNAs: Examination on the pacemaker channel genes HCN2 and HCN4. J Cell Physiol 2007; 212(2): 285-92.
[http://dx.doi.org/10.1002/jcp.21062]

[80] Anderson ME, Mohler P. MicroRNA may have macro effect on sudden death. J Nat Med 2007; 13: 410.
[http://dx.doi.org/10.1038/nm0407-410]

[81] MicroRNA expression profiles classify human cancers. Nature 2005; 435: 834-8.
[http://dx.doi.org/10.1038/nature03702]

[82] Nouraee N, Mowla SJ. miRNA therapeutics in cardiovascular diseases: promises and problems. Front Genet 2015; 6: 232-8.
[http://dx.doi.org/10.3389/fgene.2015.00232]

[83] Sardu C, Marfella R, Santulli G, Paolisso G. Functional role of miRNA in cardiac resynchronization therapy. Pharmacogenomics 2015; 15: 1159-68.
[http://dx.doi.org/10.2217/pgs.14.76]

[84] Van Rooij E, Sutherland LB, Thatcher JE, *et al.* Dysregulation of microRNAs after myocardial infarction reveals a role of miR-29 in cardiac fibrosis. Proc Natl Acad Sci USA 2008; 105(35): 13027-32.
[http://dx.doi.org/10.1073/pnas.0805038105]

[85] Bonauer A, Carmona G, Iwasaki M, *et al.* MicroRNA-92a controls angiogenesis and functional recovery of Ischemic tissues in mice. Science 2009; 324: 1710-3.
[http://dx.doi.org/10.1126/science.1174381]

[86] Santulli G, Iaccarino G, DeLuca N, Trimarco B, Condorelli G. Atrial fibrillation and microRNAs. Front Physiol 2014; 5: 15. a
[http://dx.doi.org/10.3389/fphys.2014.00015]

[87] Takahashi R, Hiura Y,Hirokawa G, FukushimaY, Iwai,N.(2009). Plasma miR-208 as a biomarker of myocardial injury. Clin Chem 2009; 55: 1944-9.
[http://dx.doi.org/10.1373/clinchem.2009.125310]

[88] Wang G, Zhu J, Zhang J, *et al.* (2010).Circulating microRNA: a novel potential biomarker for early diagnosis of acute myocardial infarction in humans. Eur Heart J 2010; 31: 659-66.
[http://dx.doi.org/10.1093/eurheartj/ehq013]

[89] Adachi T, Nakanishi M, Otsuka Y, *et al.* Plasma microRNA 499 as a biomarker of acute myocardial infarction. Clin Chem 2010; 56: 1183-5.
[http://dx.doi.org/10.1373/clinchem.2010.144121]

[90] van Rooij E, Sutherland L, Qi X, Richardson J, Hill J, Olson E. Control of stress-dependent cardiac growth and gene expression by a microRNA. Science 2007; 316: 575-9.
[http://dx.doi.org/10.1126/science.1139089]

[91] Ono K. microRNAs and Cardiovascular Remodeling. Adv Exp Med Biol 2015; 888: 197-213.
[http://dx.doi.org/10.1007/978-3-319-22671-2_10]

[92] Raposo G, Nijman HW, Stoorvogel W, *et al.* B lymphocytes secrete antigen-presenting vesicles. J Exp Med 1996; 183: 1161-72.
[http://dx.doi.org/10.1084/jem.183.3.1161]

[93] Peters PJ, Geuze HJ, Van der Donk HA, *et al.* Molecules relevant for T cell-target cell interaction are present in cytolytic granules of human T lymphocytes. Eur J Immunol 1989; 19: 1469-75.
[http://dx.doi.org/10.1002/eji.1830190819]

[94] Zitvogel L, Regnault A, Lozier A, *et al.* Eradication of established murine tumors using a novel cell-free vaccine: dendritic cell-derived exosomes. Nat Med 1998; 4: 594-600.
[http://dx.doi.org/10.1038/nm0598-594]

[95] Raposo G, Tenza D, Mecheri S, Peronet R, Bonnerot C, Desaymard C. Accumulation of major histocompatibility complex class II molecules in mast cell secretory granules and their release upon degranulation. Mol Biol Cell 1997; 8: 2631-45.
[http://dx.doi.org/10.1091/mbc.8.12.2631]

[96] Heijnen HF, Schiel AE, Fijnheer R, Geuze HJ, Sixma JJ. Activated platelets release two types of membrane vesicles: microvesicles by surface shedding and exosomes derived from exocytosis of multivesicular bodies and alpha-granules. Blood 1999; 94: 3791-9.

[97] Wolfers J, Lozier A, Raposo G, *et al.* Tumor-derived exosomes are a source of shared tumor rejection antigens for CTL cross-priming. Nat Med 2001; 7: 297-303.
[http://dx.doi.org/10.1038/85438]

[98] Admyre C, Johansson SM, Qazi KR, *et al.* Exosomes with immune modulatory features are present in human breast milk. J Immunol 2007; 179: 1969-78.
[http://dx.doi.org/10.4049/jimmunol.179.3.1969]

[99] Lasser C, Alikhani VS, Ekstrom K, *et al.* Human saliva, plasma and breast milk exosomes contain

RNA: uptake by macrophages. J Transl Med 2011; 9: 9.
[http://dx.doi.org/10.1186/1479-5876-9-9]

[100] Saman S, Kim W, Raya M, *et al.* Exosome-associated tau is secreted in Tauopathy models and is selectively phosphorylated in cerebrospinal fluid in early Alzheimer disease. J Biol Chem 2012; 287: 3842-9.
[http://dx.doi.org/10.1074/jbc.M111.277061]

[101] Tian T, Wang Y, Wang H, Zhu Z, Xiao Z. Visualizing of the cellular uptake and intracellular trafficking of exosomes by live-cell microscopy. J Cell Biochem 2010; 111(2): 488-96.
[http://dx.doi.org/10.1002/jcb.22733]

[102] Th'ery C. Exosomes: secreted vesicles and intercellular communications. F1000 Biol Rep 2011; 3: 15.

[103] Quesenberry PJ, Goldberg LR, Aliotta JM, *et al.* Cellular phenotype and extracellular vesicles: basic and clinical considerations. Stem Cells Dev 2014; 23(13): 1429-36.
[http://dx.doi.org/10.1089/scd.2013.0594]

[104] Harding C, Heuser J, Stahl P. Receptor-mediated endocytosis of transferrin and recycling of the transferrin receptor in rat reticulocytes. J Cell Biol 1983; 97: 329-39.
[http://dx.doi.org/10.1083/jcb.97.2.329]

[105] Vella LJ, Sharples RA, Nisbet RM, Cappai R, Hill AF. The role of exosomes in the processing of proteins associated with neurodegenerative diseases. Eur Biophys J 2008; 37: 323-32.
[http://dx.doi.org/10.1007/s00249-007-0246-z]

[106] Cosme J, Liu PP, Gramolini AO. The cardiovascular exosome: current perspectives and potential. Proteomics 2013; 13: 1654-9.
[http://dx.doi.org/10.1002/pmic.201200441]

[107] Khalyfa A, Gozal D. Exosomal miRNAs as potential biomarkers of cardiovascular risk in children. J Transl Med 2014; 12: 162.
[http://dx.doi.org/10.1186/1479-5876-12-162]

[108] Zhou W, *et al.* Cancer-secreted miR-105 destroys vascular endothelial barriers to promote metastasis. Cancer Cell 2014; 25: 501-15.
[http://dx.doi.org/10.1016/j.ccr.2014.03.007]

[109] Aucher A, Rudnicka D, Davis DM. MicroRNAs transfer from human macrophages to hepato-carcinoma cells and inhibit proliferation. J Immunol 2013; 2013(191): 6250-60.
[http://dx.doi.org/10.4049/jimmunol.1301728]

[110] Bang C, Batkai S, Dangwal S, *et al.* Cardiac fibroblast–derived microRNA passenger strand-enriched exosomes mediate cardiomyocytes hypertrophy. J Clin Invest 2014; 124: 2136-46.
[http://dx.doi.org/10.1172/JCI70577]

[111] Koppers-Lalic D, *et al.* Nontemplated nucleotide additions distinguish the small RNA composition in cells from exosomes. Cell Reports 2014; 8: 1649-58.
[http://dx.doi.org/10.1016/j.celrep.2014.08.027]

[112] Valadi H, Ekstrom K, Bossios A, Sjostrand M, Lee JJ, Lotvall JO. Exosome mediated transfer of mRNAs and microRNAs is a novel mechanism of genetic exchange between cells. Nat Cell Biol 2007; 9: 654-9.
[http://dx.doi.org/10.1038/ncb1596]

[113] Skog J, Wurdinger T, van Rijn S, *et al.* Glioblastoma microvesicles transport RNA and proteins that promote tumour growth and provide diagnostic biomarkers. Nat Cell Biol 2008; 10: 1470-6.
[http://dx.doi.org/10.1038/ncb1800]

[114] Villarroya-Beltri C, *et al.* Sumoylated hnRNPA2B1 controls the sorting of miRNAs into exosomes through binding to specific motifs. Nat Commun 2013; 4: 2980.
[http://dx.doi.org/10.1038/ncomms3980]

[115] Squadrito ML, *et al.* Endogenous RNAs modulate microRNA sorting to exosomes and transfer to acceptor cells. Cell Reports 2014; 8: 1432-46. [http://dx.doi.org/10.1016/j.celrep.2014.07.035]

[116] Villarroya-Beltri C, Baixauli F, Gutie'rrez-Va'zquez C, Sa'nchez-Madrid F, Mittelbrunn M. Sorting it out: regulation of exosome loading. Semin Cancer Biol 2014; 28: 3-13. [http://dx.doi.org/10.1016/j.semcancer.2014.04.009]

[117] Gibbings DJ, Ciaudo C, Erhardt M, Voinnet O. Multivesicular bodies associate with components of miRNA effector complexes and modulate miRNA activity. Nat Cell Biol 2009; 11: 1143-9. [http://dx.doi.org/10.1038/ncb1929]

[118] Guduric-Fuchs J, *et al.* Selective extracellular vesicle-mediated export of an overlapping set of microRNAs from multiple cell types. BMC Genomics 2012; 13: 357. [http://dx.doi.org/10.1186/1471-2164-13-357]

[119] Chevillet JR, *et al.* Quantitative and stoichiometric analysis of the microRNA content of exosomes. Proc Natl Acad Sci USA 2014; 111: 14888-93. [http://dx.doi.org/10.1073/pnas.1408301111]

[120] Scholer N, Langer C, Dohner H, Buske C, Kuchenbauer F. Serum microRNAs as a novel class of biomarkers: a comprehensive review of the literature. Exp Hematol 2010; 38: 1126-30. [http://dx.doi.org/10.1016/j.exphem.2010.10.004]

[121] Scholer N, Langer C, Kuchenbauer F. Circulating microRNAs as biomarkers -True Blood? Genome Med 2011; 3: 72. [http://dx.doi.org/10.1186/gm288]

[122] Van Aelst LN, Heymans S. MicroRNAs as biomarkers for ischemic heart disease. J Cardiovasc Transl Res 2013; 6: 458-70. [http://dx.doi.org/10.1007/s12265-013-9466-z]

[123] Matsumoto S, Sakata Y, Suna S, *et al.* Circulating p53- Responsive MicroRNAs are predictive indicators of heart failure after acute myocardial infarction. Circ Res 2013; 113(3): 322-6. [http://dx.doi.org/10.1161/CIRCRESAHA.113.301209]

[124] Wang R, Li N, Zhang Y, Ran Y, Pu J. Circulating microRNAs are promising novel biomarkers of acute myocardial infarction. Intern Med 2011; 50(17): 1789-95. [http://dx.doi.org/10.2169/internalmedicine.50.5129]

[125] Lewis BP, Shih LH, Jones-Rhoades MW, *et al.* Prediction of mammalian microRNA targets. Cell 2003; 115: 787-98. [http://dx.doi.org/10.1016/S0092-8674(03)01018-3]

[126] Whitehead Institute for Biomedical Research TargetScan Human: Prediction of microRNA targets http://www.targetscan.org/ (Accessed April 11, 2011)

[127] Ritchie W, Flamant S, John E, Rasko J. mimiRNA: a microRNA expression profiler and classification resource designed to identify functional correlations between microRNAs and their targets. Bioinformatics 2010; 26(2): 223-7. [http://dx.doi.org/10.1093/bioinformatics/btp649]

[128] Ghezelayagh T. Analysis of the Computational Methods Available for MicroRNA Target Prediction. Biochemistry: Computation. Mol Biol 2010; 218: 1-17.

[129] Lai RC, Arslan F, Lee MM, *et al.* Exosome secreted by MSC reduces myocardial ischemia/reperfusion injury. Stem Cell Res (Amst) 2010; 3: 214-22. [http://dx.doi.org/10.1016/j.scr.2009.12.003]

[130] Wang Y, Zhang L, Li Y, *et al.* Exosomes/microvesicles from induced pluripotent stem cells deliver cardioprotective miRNAs and prevent cardiomyocyte apoptosis in the ischemic myocardium. Int J Cardiol 2015; 192: 61-9.

[http://dx.doi.org/10.1016/j.ijcard.2015.05.020]

[131] Schnabel LV, Lynch ME, van der Meulen MC, *et al.* Mesenchymal stem cells and insulin-like growth factor-I gene-enhanced mesenchymal stem cells improve structural aspects of healing in equine flexor digitorum superficialis tendons. J Orthop Res 2009; 27: 1392-8. [http://dx.doi.org/10.1002/jor.20887]

[132] Wei X, Yang X, Han ZP, Qu FF, Shao L, Shi YF. Mesenchymal stem cells: a new trend for cell therapy. Acta Pharmacol Sin 2013; 34: 747-54. [http://dx.doi.org/10.1038/aps.2013.50]

[133] Shan H, Zhang Y, Cai B, *et al.* Upregulation of microRNA-1 and microRNA-133 contributes to arsenic-induced cardiac electrical remodeling. Int J Cardiol 2012.

[134] Sahoo S, Losordo DW. Exosomes and cardiac repair after myocardial infarction. Circ Res 2014; 114(2): 333-44. [http://dx.doi.org/10.1161/CIRCRESAHA.114.300639]

[135] Kosaka N, Iguchi H, Yoshioka Y, Takeshita F, Matsuki Y, Ochiya T. Secretory mechanisms and intercellular transfer of microRNAs in living cells. J Biol Chem 2010; 285: 17442-52. [http://dx.doi.org/10.1074/jbc.M110.107821]

[136] Mittelbrunn M, Gutierrez-Vazquez C, Villarroya-Beltri C, *et al.* Unidirectional transfer of microRNA-loaded exosomes from T cells to antigen presenting cells. Nat Commun 2011; 2: 282. [http://dx.doi.org/10.1038/ncomms1285]

[137] Alexander M, Hu R, Runtsch MC, *et al.* Exosome-delivered microRNAs modulate the inflammatory response to endotoxin. Nat Commun 2015; 6: 7321. [http://dx.doi.org/10.1038/ncomms8321]

[138] Shtam TA, Kovalev RA, Varfolomeeva EY, Makarov EM, Kil YV, Filatov MV. Exosomes are natural carriers of exogenous siRNA to human cells *in vitro*. Cell Commun Signal 2013; 11: 88. [http://dx.doi.org/10.1186/1478-811X-11-88]

[139] Alvarez-Erviti L, Seow Y, Yin H, Betts C, Lakhal S, Wood MJ. Delivery of siRNA to the mouse brain by systemic injection of targeted exosomes. Nat Biotechnol 2011; 29: 341-5. [http://dx.doi.org/10.1038/nbt.1807]

[140] Masyuk AI, Masyuk TV, Larusso NF. Exosomes in the pathogenesis, diagnostics and therapeutics of liver diseases. J Hepatol 2013; 59: 621-5. [http://dx.doi.org/10.1016/j.jhep.2013.03.028]

[141] Thery C. Exosomes: secreted vesicles and intercellular communications. F1000 Biol Rep 2011; 3: 15. [http://dx.doi.org/10.3410/B3-15]

[142] Ohno S, Takanashi M, Sudo K, *et al.* Systemically Injected Exosomes Targeted to EGFR Deliver Antitumor MicroRNA to Breast Cancer Cells. Mol Ther 2013; 21(1): 185-91. [http://dx.doi.org/10.1038/mt.2012.180]

[143] Elmen J, Lindow M, Schutz S, Lawrence M, Petri A, Obad S, *et al.* LNA-mediated microRNA silencing in non-human primates. Nature 2008; 452(7189): 896-9. [http://dx.doi.org/10.1038/nature06783]

[144] Lanford RE, Hildebrandt-Eriksen ES, Petri A, Persson R, Lindow M, Munk ME, *et al.* Therapeutic silencing of microRNA-122 in primates with chronic hepatitis C virus infection. Science 2010; 327(5962): 198-201. [http://dx.doi.org/10.1126/science.1178178]

[145] van Rooij E, Olson EN. MicroRNA therapeutics for cardiovascular disease: opportunities and obstacles. Nat Rev Drug Discov 2012; 11(11): 860-72. [http://dx.doi.org/10.1038/nrd3864]

[146] Fiedler J, Jazbutyte V, Kirchmaier B, *et al.* MicroRNA-24 regulates vascularity after myocardial

infarction. Circulation 2011; 124: 720-30.
[http://dx.doi.org/10.1161/CIRCULATIONAHA.111.039008]

[147] Kumarswamy R, Volkmann I, Jazbutyte V, Dangwal S, Park D, Thum T. Transforming growth factor-β-induced endothelial-to-mesenchymal transition is partly mediated by microRNA-21. Arterioscler Thromb Vasc Biol 2012; 32: 361-9.
[http://dx.doi.org/10.1161/ATVBAHA.111.234286]

[148] Bish L, Morine K, Sleeper M, Sanmiguel J, Wu D, Gao G, *et al.* Adeno-associated virus (AAV) serotype9 provides global cardiac gene transfer superior to AAV1, AAV6, AAV7, and AAV8 in the mouse and rat. Hum Gene Ther 2008; 19: 1359-68.
[http://dx.doi.org/10.1089/hum.2008.123]

[149] Santulli G, Wronska A, Uryu K, *et al.* A selective microRNA-based strategy inhibits restenosis while preserving endothelial function. J Clin Invest 2014; 124: 4102-14. b
[http://dx.doi.org/10.1172/JCI76069]

[150] Care A, Catalucci D, Felicetti F, *et al.* MicroRNA-133 controls cardiac hypertrophy. Nat Med 2007; 13(5): 613-8.
[http://dx.doi.org/10.1038/nm1582]

CHAPTER 6

Malaria Treatment and Stem Cells

Norma Rivera[1,*], Marcela Rojas-Lemus[2], Nancy G. Flores-Jiménez[1] and Teresa I. Fortoul[2]

[1] *Departamento de Microbiología y Parasitología, Facultad de Medicina, UNAM, CDMX 04510, Mexico*

[2] *Departamento de Biología Celular y Tisular, Facultad de Medicina, UNAM, CDMX 04510, Mexico*

Abstract: In some countries, Malaria is still a challenge. The highest rates of mortality are reported in sub-Saharan Africa, where children under five years of age, pregnant women and immunocompromised patients are the most vulnerable groups. People living in these endemic areas still do not receive proper antimalarial therapy. Insecticides resistance, antimalarial drug resistance and commercialization of counterfeit and substandard antimalarials, are key factors contributing to complexity in malaria control; trying to find new alternatives to treat and control malaria, some members of the scientific community, have recently started to work in the field of stem cell therapy in experimental malaria models. The purpose of the present chapter is to make a general review concerning various aspects of the use of stem cell therapy and how these findings could improve clinical aspects during malaria pathogenesis and could be used in the field of antimalarial drug design. An overview of the effect of the parasite on the stem cells production in the host, as in hematopoiesis and in neurogenesis, is also described.

Keywords: Adult Neurogenesis, Bone Marrow Stromal Cells, Cognitive Dysfunction, Hepatocytes, Hematopoiesis, Hippocampus, Malaria, Malaria Pathogenesis, Malaria Drug Design, Mesenchymal Cells, Memory, Murine Malaria, Neurogenesis, *Plasmodium falciparum*, *Plasmodium vivax*, Stem Cell Therapy.

INTRODUCTION

Malaria is a disease of great global impact due to the number of deaths it produces annually. The World Health Organization (WHO) in 2015 reported 214 million cases world wide and 438,000 deaths. The most serious situation is in the countries of sub-Saharan Africa, where children under the age of five are parti-

* **Corresponding author Norma Rivera:** Departamento de Microbiología y Parasitología, Facultad de Medicina, UNAM, Mexico; Tel: 52 55 56232382; E-mail: normariv@unam.mx

Atta-ur-Rahman and Shazia Anjum (Eds.)

cularly exposed and it is one of the main causes of infant mortality in that region [1].

Due to the limitations in the adoption and implementation of chemoprophylaxis worldwide, in 2014, 306,000 children under the age of five died of malaria and the percentage of pregnant women receiving preventive therapy was less than 20% and only half of the population in endemic areas slept under mosquito nets treated with insecticides [2].

Human malaria is attributed to four species of *Plasmodium*: *P. falciparum*, *P. vivax*, *P. malariae* and *P. ovale*, which have a particular geographic distribution. *P. falciparum* is distributed in all malaria areas of the world but predominates in sub-Saharan Africa, *P. malariae* has a similar distribution to that of *P. falciparum* but is much less frequent, *P. vivax* predominates in Central America and India, and *P. ovale* is found in Africa and is very rare outside of it [2]. Malaria is transmitted to the human by an Anopheles female mosquito that injects the parasites into the host blood stream while feeding.

Plasmodium vivax predominates in countries with the lowest incidence of malaria, accounting for more than 70% of cases in countries with less than 5000 cases reported each year. According to the WHO, 2015, the malaria caused by *P. vivax* is concentrated in all Latin America, Oceania and Southeast Asia. In all endemic regions of vivax malaria, severe cases and deaths have been reported. Due to the difficulty in controlling *P. vivax*, its incidence has decreased more slowly than that of *P. falciparum* in places where the two species coexist. *P. vivax* can then persist as the main cause of malaria and constitute the main challenge for its elimination [3].

Malaria life cycle can be observed in Fig. (**1**).

More than 50 years ago, it was reported that under laboratory conditions, some ape *Plasmodium* parasites were transmitted to humans [4]; it is now known that *Plasmodium knowlesi* is emerging as a very important zoonotic pathogen, which is being exported from endemic areas to other parts of the world through travelers; the infection can be potentially spread through blood transfusions, bone marrow transplantation and congenital infections [5].

Knowledge about the pathogenesis of the parasite that produces malaria has been obtained mostly from non-human animal models such as murine models [6]. The pathogenesis of malaria is complex and is the result of the destruction of the infected erythrocytes and the host's immune-response to the infection. Only the blood stage of the parasite is involved in the pathogenesis of the disease [7].

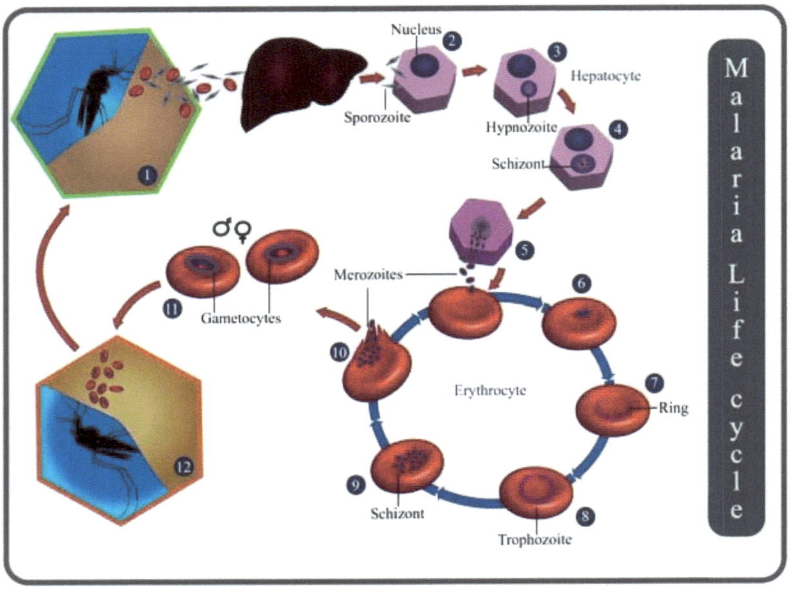

Fig. (1). Malaria life cycle. 1. Female Anopheles mosquito injects sporozoites into the host blood stream while feeding. The parasites replicate in the hepatocytes developing schizonts that liberates thousands of merozoites (2-5). *Plasmodium vivax* remains as hypnozoite inside the liver cells (3). Free merozoites invade the red blood cells and start the intraerythrocytic cycle (6-10). In the last schizogony, gametocytes (11) are formed and ingested by a female *Anopheles* which will develop new infective sporozoites (12).

Mice blood smears with different stages of *Plasmodium yoelii yoelii* can be observed in Fig. (**2**).

In malaria, different pathogenic mechanisms can be observed: block of microcirculation at different organs by sequestration of infected erythrocytes to the endothelium, exacerbated production of proinflammatory cytokines and oxidative stress. The products derived from the parasite during the intra-erythrocytic cycle induce the production of proinflammatory mediators such as TNF, IL-1, IL-6 and IFN-γ. These mediators induce the production of other cytokines and enzymes that perpetuate the inflammatory cascade. Serum levels of proinflammatory cytokines are increased in patients with complicated malaria or cerebral malaria [8, 9].

In malaria, oxidative stress occurs with increased generation of reactive oxygen species, such as superoxide anion, hydrogen peroxide, hydroxyl radical and peroxynitrite, which are produced by the host's inflammatory cells to control parasitemia, but are also produced directly by the intra-erythrocytic parasite, when in the digestive vacuole it transforms the hemoglobin into methemoglobin. The production of reactive oxygen species in malaria has been demonstrated by lipoperoxidation in the erythrocytes infected with *Plasmodium* during the acute phase of infection. Oxidative stress in malaria is also present by the decrease of enzymes and antioxidant substances such as albumin, plasma tocopherol and levels of enzymes such as catalase, glutathione peroxidase and superoxide dismutase. Severe malaria has been associated with complications such as cerebral malaria, because reactive oxygen species are involved in damaging the vascular endothelium of hosts with high parasitemia. In addition to the mechanical obstruction of the vascular microflow, an important inflammatory reaction coexists, presenting high levels of cytokines, lymphotoxins and nitric oxide (NO), which increase the interaction between vascular surface, erythrocytes, platelets and leukocytes. This triggers an inflammatory response at the local and systemic level and generates host cell apoptosis as well as endothelial damage, such as circulatory stasis and inflammation, hypoxia, tissue damage and organ dysfunction [10].

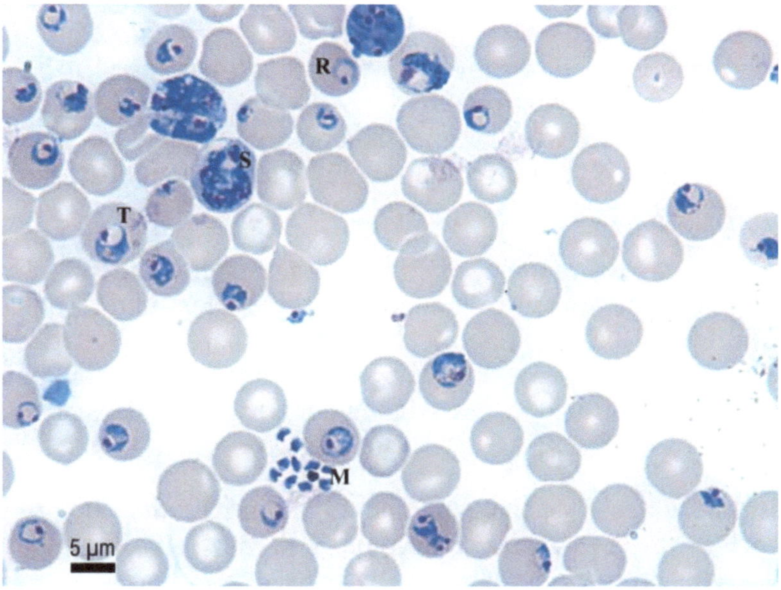

Fig. (2). Malaria blood smear showing different stages of *Plasmodium yoelii yoelii*. R rings, T trophozoite, S schizont and M merozoite.

Regardless of significant improvement in malaria elimination during the last decade, emerging *Plasmodium* resistance to the anti-malaria drugs [11] and limited efficacy of vaccine candidates [12, 13] are still major difficulties to control the disease. Most of malaria deaths are caused by *Plasmodium falciparum* infections, and the clinical features differ according to strain virulence, parasitemia, patients′ age and host immune response.

The Use of Stem Cell Therapy in Malaria

The use of stem cells to treat different human diseases mainly those involving immune compromise and tissue injuries, has shown satisfactory results. Since 1991, the use of multipotent hemopoietic cells on malaria models was evaluated, showing an excellent efficacy by enhancing the hosts′ immune system and survival time in mice infected with murine malaria [14].

Hematopoiesis is disrupted during a malaria infection as destruction of infected and non-infected red blood cells developing anemia in the host; this pathologic mechanism makes the bone marrow to meet the requirement of myeloid cells. In recent studies, researchers from the National Institute for Medical Research, UK, identified from malaria infected mice, atypical stem cells that may provide protection against malaria, as it was proved by the cure of mice infected with a lethal strain of *Plasmodium chabaudi*, after transplanting these linage cells in the mice [15].

Naïve mice infected with this *Plasmodium berghei* developed resistance to the infection after the transplantation of mesenchymal stem cells from secondary lymphoid organs; mesenchymal stem cells increased the production of cytokines such as IL-12, decreased the one of IL-10, and exerted a significant reduction of regulatory T cells [16].

In a *Plasmodium berghei* model, bone marrow mesenchymal cells were injected intravenously in murine models to reduce malaria pigment in different tissues and parasitemia, mortality, and damage to the brain in cerebral malaria; thus, these changes were observed for a short period. In this model, the bone marrow mesenchymal stem cells, also increased the count of glial cells, such as astrocytes and oligodendrocytes [17, 18].

In 1998, Shear *et al* observed that some hemoglobin variants, such as HbF, had a protective effect to *Plasmodium falciparum*, by retarding the parasite′s development. The protection seems to be caused by altering the digestion of hemoglobin in the parasites [19, 20]. It is now known that hemoglobin C, S, E and F, confer protection against *Plasmodium* infection. Respectively, these hemoglobin types act by different action mechanisms: by reducing the sequestration of infected red blood cells on the endothelium and the rosetting impairment, by improving the specific immunity and the phagocytosis of ring stages, by reducing the invasive capacity of the parasite, and, by delaying the parasite′s development because of disrupting the hemoglobin digestion. Therefore, some researchers have suggested that erythrocytes containing certain types of modified hemoglobin could be obtained by stem cell engineering, to improve the immune response in malaria [21].

In the field of experimental chemotherapy and development of vaccines against malaria, the liver stage of the parasite is a good target because it is in this early stage of infection, when the biologic cycle of *Plasmodium* can be interrupted. Nevertheless, the studies concerning these aspects seem to be difficult because of the failures while trying to reproduce parasite liver stages under *in vitro* conditions. Recently, a microscale human liver platform was designed to establish liver stages of both *P. falciparum* and *P. vivax*. The platform was made of cryopreserved micropatterned human primary hepatocytes and bounded by stromal cells. By using this design, the complete schizogony reproduction until the release of merozoite of *P. falciparum* was obtained, being these merozoites infective to red blood cells. This platform was validated as a potential tool to evaluate antimalarial drugs and vaccines [22, 23].

To study *P. falciparum* in a murine model, humanized-mice were designed by engrafting them with human hematopoietic stem cells in order to reconstitute human immune cells. At the beginning, in the murine model, the reconstitution of B and T cells was from robust to reasonable, respectively; while one of red blood cells was undetectable until an intravenously injection of plasmid DNA encoding human erythropoietin and interleukin-3. After this improvement, the reconstitution of red blood cells was enhanced, and the parasite was able to invade and reproduce in these host cells. Nevertheless, a sever dropping on the mice parasitemia was observed after the third cycle of replication [24].

Because of the evidence that they promote adaptive immunity under certain circumstances, mesenchymal stem cells are known to be used as gene therapy vectors in the treatment of different diseases; mainly in cancer and in immune diseases. Their use has also been reported in anti-microbial and anti-cancer vaccines. Recently, it was described that modified mesenchymal stem cells can express and secrete viral antigens that improve the production of antibodies *in vivo*; thus, enabling these cells as a novel vaccine platform that may be able to generate a broad array of epitopes that mimic infectious diseases [25].

The Effect of Malaria Infection on Neurogenesis

Neurogenesis is the generation of neurons from neural stem cells in the adult brain and it occurs in specific brain regions, such as the subventricular zone of the lateral ventricles and in the sub-granular zone of the dentate gyrus (Fig. **3**). The cells that are produced in the subventricular zone, travel through the rostral migratory stream to the olfactory bulb; new neurons generated in the sub-granular zone, migrate into the dentate gyrus adjacent cell layer. It is suggested that new neurons participate in plasticity and learning skills. Concerning to neurogenesis, the adult hippocampus is the one that attracts the most attention, because it is involved in processes related to cognitive, memory and affective skills. In the dentate gyrus, granule cells that are the principal excitatory neurons in this area, are generated [26]. The process of adult neurogenesis is regulated by several factors such as glucocorticoids, growth factors, inflammation, stress, drugs, components of the innate immune system and pathological events [27] as malaria.

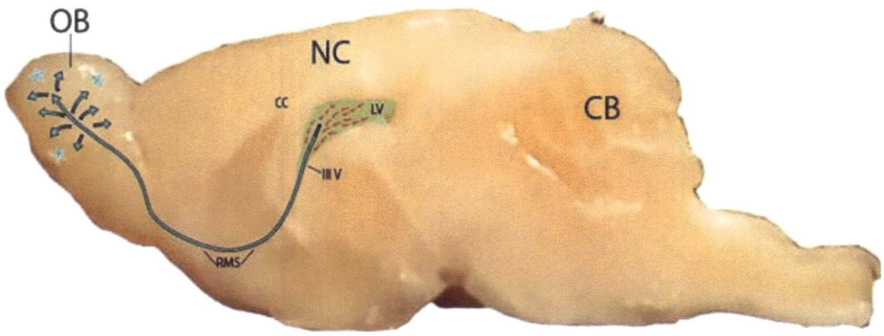

Fig. (3). Subventricular zone neurogenesis pathway. Sagittal image of an adult mouse brain. Progenitor neural cells are formed in the lateral ventricle (LV) and migrate through the olfactory bulb (OB) by the rostral migratory stream (RMS), where they develop and differentiate into new neurons (blue stars). NC neocortex, CB cerebellum, cc corpus callosum, III V third ventricle.

One percent of world-wide symptomatic malaria infections may become complicated and developed into severe malaria. Severe malaria in adults are quite rare cases, but in pregnant women and in newborn children are more frequent. According to the WHO, maternal and neonatal annual deaths, accounts for over 10,000 and 200,000 respectively. Severe or complicated malaria is defined by clinical or laboratory evidence of vital organ dysfunction, and the major complications include cerebral malaria (CM), pulmonary edema, acute renal failure and severe anemia. CM is the most severe neurological complication of infection with *Plasmodium falciparum*, and children in sub-Sahara Africa are the most affected; around 500,000 cases of cerebral malaria were reported in 2010 [28]. In CM, an unarousable coma without other direct causes is observed in adults and children with peripheral parasitemia, as well as clinical manifestations of cerebral malaria, like decreased consciousness and damage to the retina [29]. In Table **1** differences between adult and children cerebral malaria are described.

Table 1. Differences between adult and children cerebral malaria [40]

Adults	Children
Could manifest with multiorgan failure	Presented in a more isolated way
Higher mortality, about 30%	Less mortality, about 18%

(Table 1) cont.....

Adults	Children
There is a decrease in the incidence of convulsions with increasing age, these are presented in about 31% of the patients	Presented by 12% of the patients
30% of the patients will develop coma with a resolution of 1 to 2 days	40% of the patients will develop coma with a resolution with 2 to 4 days
Neurocognitive deficits of 25%	Neurocognitive deficits of less than 5%

In Africa, CM is considered a significant cause of childhood neurodisability, since patients who overcome CM, might develop behavioral and cognitive impairments [28], reduced fine motor skills and poorer academic development [30]. Cognitive disabilities are caused by malaria falciparum but they can be observed in less severe malaria infections. Other complications involving the central nervous system include seizures, hallucinations and even psychosis [31]. The exact cellular and molecular mechanisms by which CM promotes cognitive impairment, remain unknown, though, exacerbated production of proinflammatory cytokines as well as nitric oxide are involved [32] (Table **2**).

Table 2. Pathogenesis of some specific impairments in children after cerebral malaria

Impairment	Epidemiology	Immunopathogenesis	References
Cognitive sequelae	Described in 25% of children with rates of 14-24%. Risk factors included hypoglycemia, seizures, depth and duration of coma and hyporeflexia	Elevated cerebrospinal fluid of TNF adversely affects long-term cognitive outcome	[28, 40 - 43]
Speech and language impairment	In the tropics, CM is considered the leading cause of this impairment in about 11.8% of the children that recuperate from CM	The immunopathogenesis is poorly understood	[44]
Epilepsy	Is observe months or years after the infection in about 10% of the children	It is poorly understood, although it could be a consequence of focal or global hypoxic or ischemic injury in the border-zone areas of cerebral circulation.	[45 - 47]
Behavior and neuro-psychiatric disorders	Some clinical features observed after 1-4 months after infection are: inattention, impulsiveness, hyperactivity and destructive behaviors.	Is poorly understood.	[48, 49]

Murine models are commonly used to evaluate the neurological damage and the cognitive impairment caused by acute and complicated malaria. *Plasmodium berghei* ANKA is one of the more commonly used strains because of their ability to adhere to the deep cerebral endothelium, while triggering an exacerbated production of proinflammatory cytokines, such as TNF [33, 34], and in experimental conditions, it damages the hippocampal host cells and impairs adult hippocampal neurogenesis, promoting memory impairment as a consequence [35]. *Plasmodium chabaudi adami*, a mild malaria murine strain, induces damage to the hippocampus of adult mice, despite its lack of ability to adhere to the endothelium and to break the blood brain barrier. The damage is caused during just one episode of mild malaria by inducing inflammation and by altering the brain cytokine levels. Behavioral changes are also observed, accompanied by temporary suppression of adult neurogenesis and hippocampal microglial activation and redistribution [36]. Behavioral changes consist of a deficiency in social skills and anxiety episodes [37]. An affection to hippocampus neurogenesis could explain, at least in part, CM cognitive impairments. It is suggested that the neural circuit in the hippocampal-cortical area, could be involved in the learning and memory processes, thus, when this system is affected, a cognitive dysfunction may appear [35].

Hoffmann *et al* 2016, reported that erythropoietin increases the number of nestin+ cells in the dentate gyrus of mice infected with *Plasmodium berghei* during an episode of cerebral malaria [38]. Core *et al*., 2010, observed that erythropoietin activates neural stem cells in mice infected with *Plasmodium berghei* ANKA [39]. This leads to the conclusion that adult neurogenesis is altered in mice models of experimental cerebral malaria.

CONCLUDING REMARKS

Until now, very little is known about the use of stem cell therapy to treat malaria. Nevertheless, it has been proved that the use of stem cells may improve some clinical features and immunodeficiency in murine malaria models. Given promising evidences to open new windows in severe malaria treatment, more research to explore possibility of stem cell therapy in malaria is recommended.

CONSENT FOR PUBLICATION

Not applicable.

CONFLICT OF INTEREST

The authors confirm that they have no conflict of interest to declare for this publication.

ACKNOWLEDGEMENTS

Declared none.

REFERENCES

[1] Charlette SL, Nongkynrih B. Can intermittent preventive treatment for malaria reduce child mortality? Natl Med J India 2013; 26(1): 37-48.
[PMID: 24066993]

[2] WHO: World Malaria Report 2016 [Internet] , 2016 [25 Nov 2017]; Available in: www.who.int/malaria

[3] Baird JK. Malaria zoonoses. Travel Med Infect Dis 2009; 7(5): 269-77.
[http://dx.doi.org/10.1016/j.tmaid.2009.06.004] [PMID: 19747661]

[4] Contacos PG. Primate malarias: man and monkeys. J Wildl Dis 1970; 6(4): 323-8.
[http://dx.doi.org/10.7589/0090-3558-6.4.323] [PMID: 16512134]

[5] Carrasco-Ramírez E, Malagón-Gutiérrez F. La malaria como una zoonosis. Med Lab (Ed Ital) 2013; 5(2): 4-10.

[6] Engwerda C, Belnoue E, Grüner AC, Rénia L. Experimental models of cerebral malaria. Curr Top Microbiol Immunol 2005; 297: 103-43.
[http://dx.doi.org/10.1007/3-540-29967-X_4] [PMID: 16265904]

[7] Miller LH, Ackerman HC, Su XZ, Wellems TE. Malaria biology and disease pathogenesis: insights for new treatments. Nat Med 2013; 19(2): 156-67.
[http://dx.doi.org/10.1038/nm.3073] [PMID: 23389616]

[8] Vázquez A, Tobón A. Mecanismos de patogenia en la malaria por Plasmodium falciparum. Biomed 2012; 32: 106-20.
[http://dx.doi.org/10.7705/biomedica.v32i0.447]

[9] Dunst J, Kamena F, Matuschewski K. Cytokines and Chemokines in Cerebral Malaria Pathogenesis. Front Cell Infect Microbiol 2017; 7: 324.
[http://dx.doi.org/10.3389/fcimb.2017.00324] [PMID: 28775960]

[10] Hossein N. Nitric Oxide Involvement during Malaria Infection; Immunological Concepts, Mechanisms and Complexities: a Novel Review. J Trop Med Parasitol 2004; 27: 93-101.

[11] Cui L, Mharakurwa S, Ndiaye D, Rathod PK, Rosenthal PJ. Antimalarial drug resistance: literature review and activities and findings of the ICEMR network. Am J Trop Med Hyg 2015; 93(3) (Suppl.): 57-68.
[http://dx.doi.org/10.4269/ajtmh.15-0007] [PMID: 26259943]

[12] White MT, Verity R, Griffin JT, *et al.* Immunogenicity of the RTS,S/AS01 malaria vaccine and implications for duration of vaccine efficacy: secondary analysis of data from a phase 3 randomised controlled trial. Lancet Infect Dis 2015; 15(12): 1450-8.
[http://dx.doi.org/10.1016/S1473-3099(15)00239-X] [PMID: 26342424]

[13] Olotu A, Fegan G, Wambua J, *et al.* Seven-Year efficacy of RTS,S/AS01 malaria vaccine among young African children. N Engl J Med 2016; 374(26): 2519-29.
[http://dx.doi.org/10.1056/NEJMoa1515257] [PMID: 27355532]

[14] Asami M, Owhashi M, Abe T, Nawa Y. Susceptibility of multipotent haemopoietic stem cell deficient W/Wv mice to Plasmodium berghei-infection. Immunol Cell Biol 1991; 69(Pt 5): 355-60.
[http://dx.doi.org/10.1038/icb.1991.51] [PMID: 1787005]

[15] Belyaev NN, Brown DE, Diaz AI, *et al.* Induction of an IL7-R(+)c-Kit(hi) myelolymphoid progenitor critically dependent on IFN-gamma signaling during acute malaria. Nat Immunol 2010; 11(6): 477-85.
[http://dx.doi.org/10.1038/ni.1869] [PMID: 20431620]

[16] Thakur RS, Tousif S, Awasthi V, *et al.* Mesenchymal stem cells play an important role in host protective immune responses against malaria by modulating regulatory T cells. Eur J Immunol 2013; 43(8): 2070-7.
[http://dx.doi.org/10.1002/eji.201242882] [PMID: 23670483]

[17] Sohni A, Verfaillie CM. Mesenchymal stem cells migration homing and tracking. Stem Cells Int 2013; 2013: 130763.
[http://dx.doi.org/10.1155/2013/130763] [PMID: 24194766]

[18] Wang W, Qian H, Cao J. Stem cell therapy: a novel treatment option for cerebral malaria? Stem Cell Res Ther 2015; 6(1): 141.
[http://dx.doi.org/10.1186/s13287-015-0138-6] [PMID: 26253514]

[19] Zhang Y, Mi JY, Rui YJ, Xu YL, Wang W. Stem cell therapy for the treatment of parasitic infections: is it far away? Parasitol Res 2014; 113(2): 607-12.
[http://dx.doi.org/10.1007/s00436-013-3689-4] [PMID: 24276645]

[20] Shear HL, Grinberg L, Gilman J, *et al.* Transgenic mice expressing human fetal globin are protected from malaria by a novel mechanism. Blood 1998; 92(7): 2520-6.
[PMID: 9746793]

[21] Saei AA, Ahmadian S. Stem cell engineering might be protective against severe malaria. Biosci Hypotheses 2009; 2(1): 48-9.
[http://dx.doi.org/10.1016/j.bihy.2008.10.008]

[22] March S, Ng S, Velmurugan S, *et al.* A microscale human liver platform that supports the hepatic stages of Plasmodium falciparum and vivax. Cell Host Microbe 2013; 14(1): 104-15.
[http://dx.doi.org/10.1016/j.chom.2013.06.005] [PMID: 23870318]

[23] Ng S, Schwartz RE, March S, *et al.* Human iPSC-derived hepatocyte-like cells support Plasmodium liver-stage infection in vitro. Stem Cell Reports 2015; 4(3): 348-59.
[http://dx.doi.org/10.1016/j.stemcr.2015.01.002] [PMID: 25660406]

[24] Amaladoss A, Chen Q, Liu M, *et al. De Novo* Generated Human Red Blood Cells in Humanized Mice Support Plasmodium falciparum Infection. PLoS One 2015; 10(6): e0129825.
[http://dx.doi.org/10.1371/journal.pone.0129825] [PMID: 26098918]

[25] Tomchuck SL, Norton EB, Garry RF, *et al.* Mesenchymal stem cells as a novel vaccine platform. Front Cell Infect Microbiol 2012; 2: 140.
[http://dx.doi.org/10.3389/fcimb.2012.00140] [PMID: 23162801]

[26] Kempermann G, Song H, Gage FH. Neurogenesis in the Adult Hippocampus. Cold Spring Harb Perspect Biol 2015; 7(9): a018812.
[http://dx.doi.org/10.1101/cshperspect.a018812] [PMID: 26330519]

[27] Jin K, Minami M, Lan JQ, *et al.* Neurogenesis in dentate subgranular zone and rostral subventricular zone after focal cerebral ischemia in the rat. Proc Natl Acad Sci USA 2001; 98(8): 4710-5.
[http://dx.doi.org/10.1073/pnas.081011098] [PMID: 11296300]

[28] Idro R, Marsh K, John CC, Newton CR. Cerebral malaria: mechanisms of brain injury and strategies for improved neurocognitive outcome. Pediatr Res 2010; 68(4): 267-74.
[http://dx.doi.org/10.1203/PDR.0b013e3181eee738] [PMID: 20606600]

[29] Hawkes M, Elphinstone RE, Conroy AL, Kain KC. Contrasting pediatric and adult cerebral malaria: the role of the endothelial barrier. Virulence 2013; 4(6): 543-55.
[http://dx.doi.org/10.4161/viru.25949] [PMID: 23924893]

[30] Al Serouri AW, Grantham-McGregor SM, Greenwood B, Costello A. Impact of asymptomatic malaria parasitaemia on cognitive function and school achievement of schoolchildren in the Yemen Republic. Parasitology 2000; 121(Pt 4): 337-45.
[http://dx.doi.org/10.1017/S0031182099006502] [PMID: 11072896]

[31] Kihara M, Carter JA, Newton CR. The effect of Plasmodium falciparum on cognition: a systematic review. Trop Med Int Health 2006; 11(4): 386-97. [http://dx.doi.org/10.1111/j.1365-3156.2006.01579.x] [PMID: 16553922]

[32] Comim CM, Reis PA, Frutuoso VS, *et al.* Effects of experimental cerebral malaria in memory, brain-derived neurotrophic factor and acetylcholinesterase activity [correction for activity] in the hippocampus of survivor mice. Neurosci Lett 2012; 523(2): 104-7. [http://dx.doi.org/10.1016/j.neulet.2012.06.051] [PMID: 22750161]

[33] BasirR.RahimanS.F.HasballahK.ChongW.TalibH.YamM.Plasmodium berghei ANKA Infection in ICR Mice as a Model of Cerebral Malaria. Iran J Parasitol 2012; 7(4): 62-74.

[34] de Souza JB, Hafalla JC, Riley EM, Couper KN. Cerebral malaria: why experimental murine models are required to understand the pathogenesis of disease. Parasitology 2010; 137(5): 755-72. [http://dx.doi.org/10.1017/S0031182009991715] [PMID: 20028608]

[35] de Miranda AS, Brant F, Campos AC, *et al.* Evidence for the contribution of adult neurogenesis and hippocampal cell death in experimental cerebral malaria cognitive outcome. Neuroscience 2015; 284(284): 920-33. [http://dx.doi.org/10.1016/j.neuroscience.2014.10.062] [PMID: 25451296]

[36] Wilson KD, Stutz SJ, Ochoa LF, *et al.* Behavioural and neurological symptoms accompanied by cellular neuroinflammation in IL-10-deficient mice infected with Plasmodium chabaudi. Malar J 2016; 15(1): 428. [http://dx.doi.org/10.1186/s12936-016-1477-1] [PMID: 27557867]

[37] Guha SK, Tillu R, Sood A, *et al.* Single episode of mild murine malaria induces neuroinflammation, alters microglial profile, impairs adult neurogenesis, and causes deficits in social and anxiety-like behavior. Brain Behav Immun 2014; 42: 123-37. [http://dx.doi.org/10.1016/j.bbi.2014.06.009] [PMID: 24953429]

[38] Hoffmann A, Pfeil J, Alfonso J, *et al.* Experimental Cerebral Malaria Spreads along the Rostral Migratory Stream. PLoS Pathog 2016; 12(3): e1005470. [http://dx.doi.org/10.1371/journal.ppat.1005470] [PMID: 26964100]

[39] Core A, Hempel C, Kurtzhals JA, Penkowa M. Plasmodium berghei ANKA: erythropoietin activates neural stem cells in an experimental cerebral malaria model. Exp Parasitol 2011; 127(2): 500-5. [http://dx.doi.org/10.1016/j.exppara.2010.09.010] [PMID: 21044627]

[40] Carter JA, Mung'ala-Odera V, Neville BG, *et al.* Persistent neurocognitive impairments associated with severe falciparum malaria in Kenyan children. J Neurol Neurosurg Psychiatry 2005; 76(4): 476-81. [http://dx.doi.org/10.1136/jnnp.2004.043893] [PMID: 15774431]

[41] John CC, Bangirana P, Byarugaba J, *et al.* Cerebral malaria in children is associated with long-term cognitive impairment. Pediatrics 2008; 122(1): e92-9. [http://dx.doi.org/10.1542/peds.2007-3709] [PMID: 18541616]

[42] John CC, Panoskaltsis-Mortari A, Opoka RO, *et al.* Cerebrospinal fluid cytokine levels and cognitive impairment in cerebral malaria. Am J Trop Med Hyg 2008; 78(2): 198-205. [http://dx.doi.org/10.4269/ajtmh.2008.78.198] [PMID: 18256412]

[43] Idro R, Carter JA, Fegan G, Neville BG, Newton CR. Risk factors for persisting neurological and cognitive impairments following cerebral malaria. Arch Dis Child 2006; 91(2): 142-8. [http://dx.doi.org/10.1136/adc.2005.077784] [PMID: 16326798]

[44] Carter JA, Lees JA, Gona JK, *et al.* Severe falciparum malaria and acquired childhood language disorder. Dev Med Child Neurol 2006; 48(1): 51-7. [http://dx.doi.org/10.1017/S0012162206000107] [PMID: 16359594]

[45] Newton CR, Peshu N, Kendall B, *et al.* Brain swelling and ischaemia in Kenyans with cerebral malaria. Arch Dis Child 1994; 70(4): 281-7.

[http://dx.doi.org/10.1136/adc.70.4.281] [PMID: 8185359]

[46] Crawley J, Smith S, Kirkham F, Muthinji P, Waruiru C, Marsh K. Seizures and status epilepticus in childhood cerebral malaria. QJM 1996; 89(8): 591-7. [http://dx.doi.org/10.1093/qjmed/89.8.591] [PMID: 8935480]

[47] Potchen MJ, Birbeck GL, Demarco JK, *et al.* Neuroimaging findings in children with retinopathy-confirmed cerebral malaria. Eur J Radiol 2010; 74(1): 262-8. [http://dx.doi.org/10.1016/j.ejrad.2009.02.010] [PMID: 19345538]

[48] Idro R, Ndiritu M, Ogutu B, *et al.* Burden, features, and outcome of neurological involvement in acute falciparum malaria in Kenyan children. JAMA 2007; 297(20): 2232-40. [http://dx.doi.org/10.1001/jama.297.20.2232] [PMID: 17519413]

[49] Falchook GS, Malone CM, Upton S, Shandera WX. Postmalaria neurological syndrome after treatment of Plasmodium falciparum malaria in the United States. Clin Infect Dis 2003; 37(2): e22-4. [http://dx.doi.org/10.1086/375269] [PMID: 12856230]

SUBJECT INDEX

A

Atta-ur-Rahman and Shazia Anjum (Eds.)

F

N

O

P

R

S

T

V

W

Z

www.ingramcontent.com/pod-product-compliance
Lightning Source LLC
LaVergne TN
LVHW070949210926
848482LV00002B/148

9781681087634